Advances in Marine and Brackishwater Aquaculture

Santhanam Perumal
Thirunavukkarasu A.R.
Perumal Pachiappan
Editors

Advances in Marine and Brackishwater Aquaculture

Springer

Editors
Santhanam Perumal
Department of Marine Science
Bharathidasan University
Tiruchirappalli, Tamil Nadu, India

Perumal Pachiappan
Department of Biotechnology
Periyar University
Salem, Tamil Nadu, India

Thirunavukkarasu A.R.
Fish Culture Division
Central Institute of Brackishwater Aquaculture
Chennai, Tamil Nadu, India

ISBN 978-81-322-2270-5 ISBN 978-81-322-2271-2 (eBook)
DOI 10.1007/978-81-322-2271-2

Library of Congress Control Number: 2015935239

Springer New Delhi Heidelberg New York Dordrecht London

Printed on acid-free paper

Springer (India) Pvt. Ltd. is part of Springer Science+Business Media (www.springer.com)

Dedicated to the coastal fisher-folks and fish farmers

Foreword

Fisheries play an important role in augmenting nutritional security and employment to millions of rural folk worldwide. The global fish production is in the order of 145 million tonnes more or less equally contributed by capture and culture fisheries showing annual growth rate of 4.7 % and it is one of the fastest growing food sectors. Production through capture fisheries in many areas is showing a declining trend or is stagnating and reaching the maximum sustainable yield (MSY). This has forced to look for aquaculture as a potential alternate for fish production. The demand for fish and fishery products is increasing at a faster phase, and to meet this demand, technologies for producing more fish is the need of the hour. In this context, capture fisheries has limitations since constant catch efforts and increased exploitation have reduced resources, which will be further depleted. It will become difficult to increase or restore and even sustain fishery. Hence, globally all-out efforts are made for aquaculture development through various technologies for growing diversified organisms adopting different practices in various culture systems.

Indian aquaculture is in the growing trend, but the phase of development is slow. Though a phenomenal growth was expected considering the vast potential in line with the developments witnessed in 1980s and 1990s, the development was hampered due to unexpected social and environmental issues coupled with the outbreak of uncontrollable diseases. The sustainability of aquaculture itself has become difficult. One of the reasons attributed was over dependence on a single group of organisms, like shrimp, for farming. This has prompted for diversification to other species farming and systems and practices.

Various research and development institutions are making efforts for developing technologies, transferring the same for field applications. However, the process is very slow and requires concerted efforts of all concerned with cooperation and commitment. This only will pave way for further development of technologies, their dissemination and application. In this context, the book *Advances in Marine and Brackishwater Aquaculture* is one of the meaningful efforts. An attempt has been made to bring out a comprehensive information on marine and brackishwater aquaculture of different organisms compiling the articles of experts in this field.

I am sure that this book would be highly useful to students, research workers, planners, farmers and entrepreneurs involved in aquaculture. I express my warm appreciation to all the authors and organizers for their great contributions in bringing out this book.

Marine Products Export Development Authority (MPEDA)
Govt. of India, Kochi, India

Aquaculture Foundation of India (AFI)
Chennai, India

M. Sakthivel

Preface

The marine and brackishwater aquaculture is defined as the rearing of marine or brackishwater organisms under controlled environmental and feed conditions. Aquaculture is a rapidly growing food production industry, accounting for over one-third of global fisheries production and is a lucrative industry in the domestic and international markets. The interest in marine and brackishwater aquaculture has dramatically increased with the recent failures in the capture fisheries. Fishery production supplied the world with about 154 million tonnes of fish in 2010 (FAO 2012). Of this, over 115 million tonnes was used as human food, providing an estimated apparent per capita supply of about 17 kg (live weight equivalent). Recently, Aquaculture production accounted for 46 % of total food fish supply, a slightly lower proportion than reported in 2008, because of major downward trend in aquaculture and capture fishery production. The per capita supply has remained fairly static in recent years as growth in supply from aquaculture has offset a small decline in capture fishery production and a rising population. In 2008, per capita food fish supply was estimated at 13.7 kg excluding China. In 2007, fish accounted for 15.7 % of the global population's intake of animal protein and 6.1 % of all protein consumed. Globally, fish provides more than 1.5 billion people with almost 20 % of their average per capita intake of animal protein and 3.0 billion people with at least 15 % of such protein. In 2007, the average annual per capita apparent fish supply in developing countries was 15.1 kg, and 14.4 kg in low-income food-deficit countries.

World population has been increasing rapidly during the last few decades with increased exploitation of protein resources. Aquaculture industry remains a growing, exciting and vital production sector to meet a chief protein food requirement. Aquaculture accounted for 46 % of total food fish supply. The FAO (FAO 2012) has reported that about 35 million peoples are fishing around the world with 20 million boats. About 170 million jobs depend directly or indirectly on the fisheries sector, bringing the total web of people financially linked to 520 million. As of 2008, the value of the world aquaculture harvest, excluding aquatic plants, was estimated at US$ 98.4 billion. Average annual per capita supply of food fish from aquaculture for human consumption has increased by ten times, from 0.7 kg in 1970 to 7.8 kg in 2008, at an average rate of 6.6 % per year (FAO 2010). The average per capita fish consumption is expected to increase to a level of 11 kg per annum for fish eating population.

The global fish production is 154 million tones more or less equally contributed by capture and culture methods (FAO 2012), with major contribution (80 %) from inland aquaculture besides marine capture fishery. However, the wild catches are getting diminishing due to over exploitation, climate change and pollution. The UNEP-Green Economy preview report stated that if the world remained on its path of overfishing, by 2050 the ocean fish stock could become uneconomic to exploit or extinct. Under these circumstances, aquaculture is considered to be a promising sector to fulfill our protein requirement. Marine and brackishwater aquaculture sector is very important because it contributes to food and nutritional security, employment, supports for livelihoods, and which raises the socioeconomic status of poor fishing communities. Hence, aquaculture production should be increased to achieve high yield to eradicate the hunger, malnutrition and poverty of our teeming millions. Therefore, the present book which is the compilation of research works in diverse fields would be immensely useful to the students, researchers and fish farmers.

Tiruchirappalli, TN, India — Santhanam Perumal
Chennai, TN, India — Thirunavukkarasu A.R.
Salem, TN, India — Perumal Pachiappan

References

FAO (2010) The state of world fisheries and aquaculture. FAO, Rome, 197 pp

FAO (2012) The state of world fisheries and aquaculture. FAO, Rome, 209 pp

Acknowledgements

The editors thank all the authors for their contributions to this book. The financial supports extended by the Department of Biotechnology (Ministry of Science and Technology), Department of Science and Technology, Ministry of Earth Sciences, National Agricultural Bank for Rural Development (Govt. of India) and the authorities of Bharathidasan University, to organize the workshop on Advances in Aquaculture Technology (at Bharathidasan University), are gratefully acknowledged with thanks. Mr. S. Ananth, Research Scholar, Dept. of Marine Science, Bharathidasan University, who has helped in computation work, deserves our special appreciation.

Santhanam Perumal
Thirunavukkarasu A.R.
Perumal Pachiappan

Contents

About the Editors

Dr. Thirunavukkarasu A.R. is a Retired Principal Scientist and Head, Fish Culture Division, Central Institute of Brackishwater Aquaculture (CIBA), Indian Council of Agricultural Research, Chennai, Tamil Nadu, India. He obtained his M.Sc. and Ph.D. Degrees in Marine Biology from the C.A.S. in Marine Biology, Annamalai University. He started his career as a Research Assistant in the Tamil Nadu State Fisheries Department and then joined as a scientist in the Central Marine Fisheries Research Institute (Govt. of India), where he served till 1991 and subsequently moved to the CIBA. His specialization include Marine Fisheries Management, Inland Fisheries Development and Brackishwater Aquaculture. He has about three decade research experience with over 200 publications including manuals, edited books besides symposium proceedings. He has successfully completed over 20 major research projects worth to the tune of Rs. 8-crore, funded by Indo-French scheme, NFDB, NAIP, DBT, NATP, Mega Seed Bank, etc. To the pinnacle of his achievements, during mid-nineties, he has developed a new technique for the induced breeding and seed production in the commercially valuable finfish, Asian Seabass-*Lates calcarifer* that has revolutionized the Indian Aquaculture industry with the year round sustained supply of seeds. The Seabass seed production technology was transferred to the Rajiv Gandhi Centre for Aquaculture (RGCA), Marine Products Export Development Authority (MPEDA – Govt. of India) and he has been actively involved in the consultancy services to the private agencies like M/S Seabass Hatchery and to the aqua-culturists at large. Subsequently, his research team made a technological break-through for viable land-based brood stock production of Cobia *Rachycentron canadum* in pond and RCC tank holding systems. Further, he has successfully standardized the techniques and developed protocols for the controlled breeding of Grey mullet (*Mugil cephalus*) through pond reared stock, for the development of land-based broodstock of brackishwater ornamental fishes Scat (*Scatophagus argus*) and Moon fish (*Monodactylus argenteus*), controlled breeding and hormone-based sex reversal in Grouper (*Epinephelus tauvina*), maintenance of captive broodstock of Milkfish (*Chanos chanos*), etc. Under the NATP Scheme, he has standardized culture technique for the controlled breeding and seed production of shrimps in inland saline water environs. He is the Chairman of Steering Committee for evaluating the status of Inland Fish seed

Production, Adviser to the Tamil Nadu government-Task Force on Fisheries Development besides an expert member in many universities and national level scientific committees. He was conferred with Gold Medal Award and Citation by the Aquaculture Foundation of India, Chennai and Hiralal Chowdhry Award by Fishing Chimes. He has visited several countries like Norway, Vietnam, France and Thailand.

Dr. Perumal Pachiappan is a Professor and Head, Department of Biotechnology, Periyar University, Salem, Tamil Nadu, India. Prior to his appointment in Periyar University, he had served as a Faculty-Research Associate, Lecturer, Reader and Professor in the C.A.S. in Marine Biology, Annamalai University, where he received his Ph.D. Degree in 1989. He is also qualified in UGC-National Educational Test, 1985. He has over 30 years of research experience with about 150 publications to his credit. His specialization includes molecular taxonomy and live feed aspects of plankton, fish larviculture, and antibiotic principles/drugs from the marine bioresources. He has about 25 years of PG-teaching experience, has successfully guided over 12 Ph.Ds, about 40 P.G. Dissertations and currently guiding many Ph.Ds, M.Phils and M.Sc. students. He has successfully completed many major research projects funded by ICAR, MoEs, DRDO (Govt. of India) and is currently operating a major research project funded by the UGC. He is a member of Academic Council, Senate, Board of Research Studies (of Periyar University), Chairman of PG-Board of Studies in Biotechnology and Bioinformatics (of Periyar University) besides member of Board of Studies in various colleges and universities. He is the Chairman of the Periyar University-Institutional Animals Ethical Committee and Co-ordinator of the Periyar University-Patent Facilitation Centre and Industrial MOU, an expert-member for Projects Review committee of European Commission (EC), Scientist-Member of Govt. Mohan Kumaramangalam Medical College – Institutional Animal Ethics committee and Scientific Advisory committees besides serving as reviewer/editor for many national and international journals. He is a widely travelled man and has visited countries like China, Singapore, Korea, Thailand, France, Spain and Scotland.

Dr. Santhanam Perumal is an Assistant Professor in the Department of Marine Science, School of Marine Sciences, Bharathidasan University, Tiruchirappalli, Tamil Nadu, India. He obtained his M.Sc., M. Phil. and Ph.D. Degrees in Marine Biology, from the C.A.S. in Marine Biology, Annamalai University. He has over 18 years of research and 8 years of teaching experience with over 90 publications in national and internationally reputed journals, books and proceedings. He has been specializing in the areas of Marine Planktonology and Aquaculture with reference to biodiversity, taxonomy, biology and culture of live feeds (marine microalgae and marine copepods) for fish larviculture. He has successfully guided several PG (35), M. Phil. (12) and Ph.D. (6) students and currently guiding many research scholars. He has successfully completed many major research projects funded by UGC, DST and DBT (Govt. of India) and is currently operating 2 major research projects funded by the DST and DBT (Govt. of India). He is a member in several scientific bodies and has been serving as editorial board member for over 4 journals and reviewer for over 20 reputed national and international journals. He has received DST-Young Scientist Award in 2007 and has been expert-member for various committees.

Isolation and Culture of Microalgae

Perumal Pachiappan, B. Balaji Prasath, Santhanam Perumal, S. Ananth, A. Shenbaga Devi, S. Dinesh Kumar, and S. Jeyanthi

Introduction

Marine microalgae or phytoplankton are the floating microscopic unicellular plants of the seawater which are generally free living, pelagic with the size range of 2–20 μm. The important components of microalgae are the diatoms, dinoflagellates, green algae, blue-green algae, and coccolithophores. Most microalgae have got immense value as they are rich sources of essential fatty acids, pigments, amino acids, and vitamins. They play a critical role in the coastal aquaculture of fish, molluscs, shrimps, and oysters, especially to meet the nutritional requirement of the larvae as well as for bioencapsulation. It is an established fact that the success of any hatchery operation mainly depends on the availability of the basic food, the phytoplankton. The maintenance and supply of the required species at appropriate time form a major problem being encountered by the algal culturists. The procedure for the phytoplankton culture involves aspects such as the isolation of the required species, preparation of the suitable culture media, and maintenance of the culture in the laboratory scale, as well as large scale under controlled conditions of light, temperature, and aeration, and their constant supply to the aqua farmers in different phases of growth. A culture may be defined as an artificial environment in which the microalgae grow. The culture of phytoplankton is an important aspect of planktonology, and the mass culture of phytoplankton is achieved under laboratory-controlled conditions and under field/outdoor conditions. Under laboratory conditions, sterilized or thoroughly cleaned containers are filled with filtered/sterilized seawater (28–34‰) and enriched with the addition of fertilizers, i.e., Guillard and Ryther's F medium, Walne's medium, or TMRL medium. The culture containers are inoculated with pure strains of the desired phytoplankton previously cultured in the laboratory. They are provided with heavy aeration and light using aerators and fluorescent bulbs respectively in a controlled laboratory with temperature of 25 ± 2 °C. The exponential growth phase is generally observed in 36 h to 3 days after inoculation. Cell density of 1.5–4.5 million cells per ml could be recorded. As a sufficient quantity of phytoplankton inoculums usually is present in the coarsely filtered seawater when the nutrients are added, a phytoplankton bloom develops in a course of few days under substantial sunlight. However, it happens sometimes that diatom bloom is inhibited by lack of sunlight or due to the nature of seawater in the tank. In such cases, the addition of new seawater and/or addition of ferric chloride in small amounts may stimulate instant resumption of the diatom in culture.

P. Pachiappan (✉)
Department of Biotechnology, Periyar University, Salem 636011, Tamil Nadu, India
e-mail: perumaldr@gmail.com

B.B. Prasath • S. Perumal • S. Ananth
A. Shenbaga Devi • S.D. Kumar • S. Jeyanthi
Department of Marine Science, Bharathidasan University, Tiruchirappalli 620024, Tamil Nadu, India

S. Perumal et al. (eds.), *Advances in Marine and Brackishwater Aquaculture*,
DOI 10.1007/978-81-322-2271-2_1, © Springer India 2015

History of Microalgae Culture

Microalgae culture is one of the methods of modern biotechnologies. The first unialgal culture was achieved by Beijerinck in 1890 with *Chlorella vulgaris*, and the use of such cultures for studying plant physiology was developed by Warburg in the early 1900s (Warburg 1919). Mass culture of microalgae really began to be a focus of research after 1948 at Stanford (USA), Essen (Germany), and Tokyo, and the classic book edited by Burlew (1953) summarizes many of these early studies. Interest in applied algal culture continued, especially with studies on the use of algae as photosynthetic gas exchangers for space travel and as microbial protein sources. Commercial large-scale culture of microalgae commenced in the early 1960s in Japan with the culture of *Chlorella* followed in the early 1970s with the establishment of a *Spirulina* harvesting and culturing facility in Lake Texcoco, Mexico, by Sosa Texcoco S. A. In 1977, Dainippon Ink and Chemicals Inc. established a commercial *Spirulina* plant in Thailand, and by 1980 there were 46 large-scale factories in Asia producing more than 1,000 kg of microalgae (mainly *Chlorella*) per month (Kawaguchi 1980), and in 1996 about 2,000 t of *Chlorella* were traded in Japan alone. Other *Spirulina* plants were established in the USA (e.g., Microbio in California and Cyanotech in Hawaii). Commercial production of *Dunaliella salina*, as a source of β-carotene, became the third major microalgae industry when production facilities were established by Western Biotechnology Ltd and Betatene Ltd (now Cognis Nutrition & Health) in Australia in 1986. These were soon followed by other commercial plants in Israel and the USA. As well as these algae, the large-scale production of cyanobacteria (blue-green algae) commenced in India at about the same time. More recently, several plants producing *Haematococcus pluvialis* as a source of astaxanthin have been established in the USA and India. Thus, in a short period of about 30 years, the industry of microalgal biotechnology has grown and diversified significantly.

Role of Microalgae in Aquaculture

Microalga is an important source of nutrition and is used widely in the aquaculture of other organisms, either directly or as an added source of basic nutrients. Aquaculture farms rearing larvae of molluscs, echinoderms, crustaceans, and fish use microalgae as a source of nutrition. Low-bacterial and high-microalgal biomass is a crucial food source for shellfish aquaculture.

Microalgae can form the start of a chain of further aquaculture processes. For example, a microalga is an important food source in the aquaculture of brine shrimp. Brine shrimp produces dormant eggs, called cysts, which can be stored for long periods and then hatched on demand to provide a convenient form of live feed for the aquaculture of larval fish and crustaceans.

Other applications of microalgae within aquaculture include increasing the aesthetic appeal of finfish bred in captivity. One such example can be noted in the aquaculture of salmon, where a microalga is used to make salmon flesh pinker. This is achieved by the addition of natural pigments containing carotenoids such as astaxanthin produced from the microalga *Haematococcus* to the diet of farmed fish.

Methods of Isolation of Microalgae Single Species

The isolation of the required species can be done by one of the following methods:

Washing method or centrifugation: Repeated washing or centrifuging the water samples results in the isolation of larger organisms.

By exploiting the phototactic movement: By this method, the phytoflagellates will move to one direction and with a micropipette can be isolated.

By agar plating method: For preparing the agar medium, 1.5 % agar is added to 1 L of suitable medium or even natural seawater, and this agar solution is sterilized in an autoclave for 15 min under 150 lbs pressure and 120 °C temperature. Then this medium is poured in sterilized Petri dishes and left for 24 h. In case of culture tubes, the medium is poured in 1/3 part in tubes and properly plugged with cotton before autoclaving.

Micromanipulation: The algal cell is to be isolated in drop of enrichment sample. While observing, the cell is sucked up into micropipette. The cell is transferred to a drop of sterile medium on agar plate. This process is repeated to "wash" the cell. The more times a cell is washed, the less likely is bacterial contamination. However, the risk of cell damage increases with the number of times a cell is handled. The optimum number of washes will depend on the type of algae. Then transfer the cell to dilute medium in a tissue culture plate, Petri dish, or culture tube. Culture vessel is placed under low light at appropriate constant temperature. Growth is checked under the microscope, or we have to wait until macroscopic growth can be detected (3–4 weeks after transfer). A colonial unialgal culture results from this method.

Serial dilution: Tubes have to be labeled as 10^{-1}–10^{-10} to indicate dilution factor. Aseptically 1 ml of enrichment sample is to be added to the test tube (10^{-1}) and mixed gently. 1 ml of this dilution is taken and added to the next tube (10^{-2}) and then mixed gently. This procedure is repeated for the remaining tubes (10^{-3}–10^{-10}). Test tubes are incubated under controlled temperature and light conditions: The cultures are examined microscopically after 2–4 weeks by withdrawing a small sample aseptically from each dilution tube. A unialgal culture may grow in one of the higher-dilution tubes, e.g., 10^{-6}–10^{-10}. If the tubes contain two or three different species, then micromanipulation can be used to obtain unialgal cultures.

Growth Dynamics

The growth of an axenic culture of microalgae is characterized by five phases. Growth usually refers to changes in the culture rather than changes in an individual organism. Growth denotes the increase in number beyond that present in the original inoculums. Five distinct phases of growth are described (Fig. 1):

The lag phase – After the addition of inoculums to a culture medium, the population remains temporarily unchanged. The cells at this point increase in size beyond their normal dimensions. Physiologically, they are very active and are synthesizing new protoplasm. The organisms are actually metabolizing, but there is a lag in cell division.

The logarithmic or exponential phase – The cells begin to divide steadily at a constant rate. Given optimum culture conditions, growth rate is maximal at this stage.

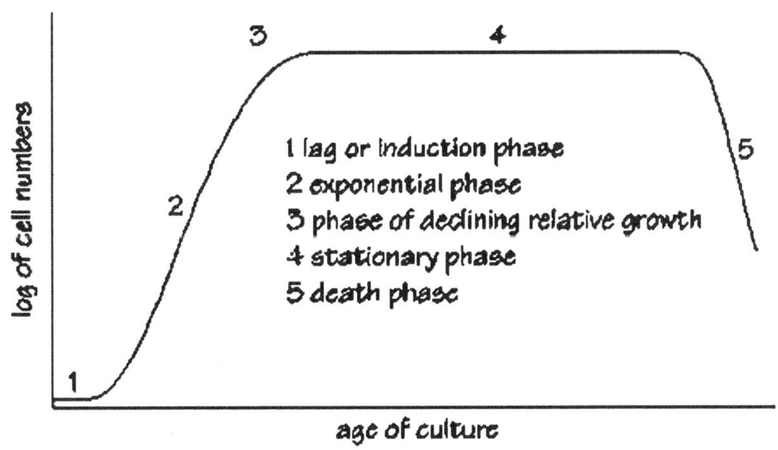

Fig. 1 Schematic diagram of algal growth stages

Phase of declining growth rate – Cell division slows down when nutrients, light, pH, carbon dioxide, or other physical and chemical factors begin to limit growth.

The stationary phase – At this point, the logarithmic phase of growth begins to taper off after several hours (or days) in a gradual fashion. The population more or less remains constant for a time, perhaps as a result of complete cessation of division or the balancing of reproduction rate by an equivalent death rate.

The phase of decline or death – The rate at which some cells die is faster than the rate of reproduction of new cells. The number of viable cells decreases geometrically.

Materials Required for Culture of Marine Microalgae

Culture Apparatus

Vessels made up of "Pyrex" or "Corning" glass are usually used for phytoplankton culturing. But from test tube to concrete tanks may be used, depending on the quantity of the culture required. For small-scale experiments, Erlenmeyer flasks equipped with inlet and outlet tubes for aeration are used. Glass tubes or a flask plugged with cotton provides enough aeration. The vessels should be cleaned well and sterilized in a hot air oven.

Light

As with all plants, microalgae photosynthesize, i.e., they assimilate inorganic carbon for conversion into organic matter. Light is the source of energy which drives this reaction, and in this regard intensity, spectral quality, and photoperiod need to be considered. Light intensity plays an important role, but the requirements vary greatly with the culture depth and the density of the algal culture: at higher depths and cell concentrations, the light intensity must be increased to penetrate through the culture (e.g., 1,000 lux is suitable for Erlenmeyer flasks; 5,000–10,000 is required for larger volumes). Light may be natural or supplied by fluorescent tubes. Too high light intensity (e.g., direct sunlight, small container close to artificial light) may result in photoinhibition. Also, overheating due to both natural and artificial illumination should be avoided. The duration of artificial illumination should be minimum 18 h of light per day, although cultivated phytoplankton develop normally under constant illumination. In controlled rooms, white daylight fluorescent lamps may be used. Outdoor cultures rely on sunlight for illumination.

Temperature

Temperature usually affects an organism's metabolic rate. Low temperatures are usually maintained in controlled rooms (18–23 °C). Transfer of algal starters or inoculate previously cultured in controlled rooms when scaled up for mass production should be done early morning to avoid stress brought about by sudden temperature increase. The optimal temperature for phytoplankton cultures is generally between 20 and 24 °C, although this may vary with the composition of the culture medium, the species, and the strain cultured. Most commonly cultured species of microalgae tolerate temperatures between 16 and 27 °C. Temperatures lower than 16 °C will slow down growth, whereas those higher than 35 °C are lethal for a number of species. If necessary, algal cultures can be cooled by a flow of cold water over the surface of the culture vessel or by controlling the air temperature with refrigerated air-conditioning units.

pH

The pH range for most cultured algal species is between 7 and 9, with the optimum range being 8.2–8.7. Complete culture collapse due to the disruption of many cellular processes can result from a failure to maintain an acceptable pH. The latter is accomplished by aerating the culture. In the case of high-density algal culture, the addition of carbon dioxide allows to correct for increased pH, which may reach limiting values of up to pH 9 during algal growth.

Aeration/Mixing

Mixing is necessary to prevent sedimentation of the algae, to ensure that all the cells of the population are equally exposed to the light and nutrients, to avoid thermal stratification (e.g., in outdoor cultures), and to improve gas exchange between the culture medium and the air. The latter is of primary importance as the air contains the carbon source for photosynthesis in the form of carbon dioxide. For very dense cultures, the CO_2 originating from the air (containing 0.03 % CO_2) bubbled through the culture is limiting the algal growth, and pure carbon dioxide may be supplemented to the air supply (e.g., at a rate of 1 % of the volume of air). CO_2 addition furthermore buffers the water against pH changes as a result of the CO_2/HCO_3^- balance. Depending on the scale of the culture system, mixing is achieved by stirring daily by hand (test tubes, Erlenmeyer), aerating (bags, tanks), or using paddle wheels and jet pumps (ponds). However, it should be noted that not all algal species can tolerate vigorous mixing.

Carbon Dioxide

Providing the algae with extra carbon, in the form of the gas carbon dioxide (CO_2), would facilitate much faster growth. CO_2 is supplied from compressed gas cylinders, and only very little is needed (about half of one percent) in the air supplied to the culture. The CO_2 has to be passed through a flowmeter to ensure that the amount used will keep the pH of the culture between 7.8 and 8.0. The pH can be checked with indicator papers, which change color with a change in pH, or with a pH meter. Both the air and the CO_2 should be filtered through an in-line filter unit of 0.3–0.5 μm before entering the culture, as this helps to prevent other, possibly contaminating, organisms from getting into the cultures.

Salinity

Marine phytoplankton are extremely tolerant to changes in salinity. Most species grow best at a salinity that is slightly lower than that of their native habitat, which is obtained by diluting seawater with tap water. Salinities of 20–24 $g.l^{-1}$ have been found to be optimal.

Nutrient Medium

In laboratory cultures, however, natural waters themselves are unsatisfactory for sustained algal growth mainly because some essential nutrients are usually present only in trace amounts. The concentration of these elements largely depends on dynamic equilibrium which is disturbed as soon as water is collected. Miquel (1890–93) observed that the waters of lakes, ponds, and the sea could not support in the laboratory continued and luxuriant growth of algae. Natural waters has to be enriched by the addition of some mineral salts that compounded in the famous solutions A and B. This marked the beginning of the use of enriched culture media where specific conditions are imposed to encourage growth of particular organisms. After the work of Allen and Nelson (1910) on Miquel's solutions, the newly formulated Allen-Nelson "Miquel seawater" medium became the standard for enriching seawater. Synthetic or artificial media were later developed for studies related to algal physiology and nutrition. Provasoli et al. (1957) have observed that artificial media show the most constant results for algal culture in contrast to enriched seawater media which may show varying results depending upon the time and place of collection of the seawater base. Some observations (Chu 1942), however, showed that although the heaviest cultures obtained have been in defined media, frequently growth fails in it and more often stops at a much earlier stage than in media containing natural materials. In the years that followed, the general trend was toward the improvement of the currently existing synthetic media and the introduction of new ones. Concentrations of cells in phytoplankton cultures are generally higher than those found in nature. Algal cultures must therefore be enriched with nutrients to make up for the deficiencies in the seawater. Macronutrients include nitrate, phosphate (in an approximate ratio of 6:1), and silicate. Silicate is specifically used for the growth of diatoms which utilize this compound for production of an external shell. Micronutrients consist of various trace metals and the vitamins thiamin (B_1), cyanocobalamin (B_{12}), and sometimes biotin. Several enrichment media that have been used extensively and are suitable for the growth of most algae are the Walne's medium and the Guillard's F/2 medium. Various specific recipes for algal culture media are described by Vonshak (1986). These recent advances in algal culture somehow present probable answers to the many problems in aquaculture.

Selection of the Culture Medium

On securing the desired organism, microalgal sample to be transfer into a series of Petri dishes, each containing different enriched media. They have to be exposed to sunlight or artificial light. This preparatory culture is used to select the suitable medium for the particular species. During this time, the organism multiplies in one of the media and provides enough material for further process of culturing. Pure cultures are sometime obtained only after several attempts. The preparatory cultures may be maintained till pure cultures are obtained.

Culture Media

The following are some of the culture media found suitable to most planktonic algae:
TMRL medium (Tung Kang Marine Res. Lab.)

Potassium nitrate	10 g/100 ml of DW
Sodium orthophosphate	1 g/100 ml of DW
Ferric chloride	0.3 g/100 ml of DW
Sodium silicate	0.1 g/100 ml of DW

The chemicals are kept separately in 100 ml reagent bottle. 1 ml each to 1 l of sterilized seawater is added. This medium can be used for the mass culture of diatom.

Schreiber's medium

Potassium nitrate	0.1 g
Sodium orthophosphate	0.02 g
Soil extract	50 ml
Filtered and sterilized seawater	1

Soil extract is prepared by boiling 1 kg of garden soil in 1 l of fresh water for 1 h. After 24 h, clear water is decanted and kept in a bottle. 50 ml of this soil extract can be added to each liter of sterilized seawater. This can be used as a medium while isolating the nanoplankton.

F/2 medium

$NaNO_3$ (75.0 g/L dH_2O)	1.0 ml
$NaH_2PO_4 \cdot H_2O$ (5.0 g/L dH_2O)	1.0 ml
$Na_2SiO_3 \cdot 9H_2O$ (30.0 g/L dH_2O)	1.0 ml
F/2 trace metal solution	1.0 ml
F/2 vitamin solution	0.5 ml
Filtered seawater to	1.0 L

After all additions, the medium will be autoclaved.

F/2 trace metal solution

$FeCl_3 \cdot 6H_2O$	3.15 g
$Na_2EDTA \cdot 2H_2O$	4.36 g
$CuSO_4 \cdot 5H_2O$ (9.8 g/L dH_2O)	1.0 ml
$Na_2MoO_4 \cdot 2H_2O$ (6.3 g/L dH_2O)	1.0 ml
$ZnSO_4 \cdot 7H_2O$ (22.0 g/L dH_2O)	1.0 ml
$CoCl_2 \cdot 6H_2O$ (10.0 g/L dH_2O)	1.0 ml
$MnCl_2 \cdot 4H_2O$ (180.0 g/L dH_2O)	1.0 ml
Distilled water to	1.0 L

F/2 vitamin solution

Vitamin B_{12} (1.0 g/L dH_2O)	1.0 ml
Biotin (0.1 g/L dH_2O)	10.0 ml
Thiamine HCl	200.0 mg
Distilled water to	1.0 L

Filter sterilizes into plastic vials, and stored in the refrigerator.

Note: F/2 medium contains extensive silica precipitate and should be used only when growing diatoms. For other algal groups, use F/2-Si medium.

Conway's or Walne's Medium

1. *Nutrient solution A per liter of DW*

FeCl3.6H2O	1.3 g
MnCl2.4H2O	0.36 g
H_3BO3	33.6 g
EDTA (disodium salt)	45.0 g
NaH2PO4.2H2O	20.0 g
NaNO3	100.0 g
TMS stock	1.0 ml

2. *Trace metal solution B (TMS) per 100 ml DW*

ZnCl2	2.1 g
CoCl2.6H2O	2.0 g
(NH4)6Mo7O24.4H2O	0.9 g
CuSO4.5H2O	2.0 g

This solution is normally cloudy. Acidify with a few drops of conc. HCl to give a clear solution.

3. *Vitamin solution C per 100 ml*

Cyanocobalamin	10.0 mg
Thiamine	10.0 mg
Biotin	200.0 μg (micrograms)

Medium per liter

Nutrient solution (**A**)	1.0 ml
Trace metal solution (B)	0.5 ml
Vitamin solution (**C**)	0.1 ml
Sterilized seawater	1.0 l

Besides the above mentioned laboratory-prepared chemical which serves as nutrients, commercial fertilizers can be used for the mass culture of diatoms and nanoplankters, in open tanks for economic purpose. The media used for the open culture are:

Urea 46	10 mg/l
Super phosphate	10 mg/l
Ammonium sulfate	100 mg/l

Stock Culture Maintenance

Stock cultures are kept on hand for initial start-up and as a source of "clean" algae when the mass tank cultures or the smaller intermediate cultures become contaminated, die, or are harvested. Stock cultures are usually kept in heat-resistant, transparent test tubes or conical flasks. All algae culture systems require a set of stock cultures, usually of about 250 ml in volume, to provide the reservoir of algal cells from which to start the larger-scale cultures which will be used for feeding. Stock cultures are kept in small flasks, such as 500 ml borosilicate glass, flat-bottomed boiling flasks fitted with cotton-wool bungs. Flasks are available in sizes ranging from 1.7 fl. oz. to 4 quarts (50–4,000 ml). The culturist can choose the size that is most convenient and practical for his/her operation. An adequate volume of stock culture must be available for the next step in vessel size. Air is not bubbled into stock flasks, thus eliminating one potential source of contamination. Each flask is half filled (or less) with media. Partially filled flasks provide a larger surface area for gas exchange than full flasks. Good gas exchange increases the prospect of stable, long-lived, and healthy stock cultures. Many hatcheries keep backup stock cultures. Some are kept on-site in duplicate culture chambers. Others are kept at locations remote from the hatchery. In this way, problems with equipment, power failures, unexplained crashes, and other unforeseen difficulties can be minimized. By following a strict transfer regime, pure algal monocultures can be kept healthy and used indefinitely to seed new cultures.

Subculture

Stock cultures must be subcultured frequently (preferably weekly). Subculturing involves inoculating some cells from an old stock culture into fresh culture medium, so that the cells can continue to grow and divide and remain healthy. If subculturing is not carried out, the algal cells in the stock culture will eventually die. It is important to take precautions to prevent contaminants from the air entering the stock cultures when subculturing. To start a new stock culture, about 20 ml of algae are taken from a stock culture which has been growing for 6–7 days and poured into a flask containing 250 ml of fresh culture medium. After removing the cotton-wool bungs, and before and after pouring, the necks of both flasks should be passed through a gas flame, such as that from a Bunsen burner or spirit lamp, to kill any surface and airborne contaminants, such as bacteria, that might enter the culture. To grow, the new stock culture should be put about 20 cm from a fluorescent lamp that is lit continually. After subculturing, the remainder of the old stock culture can be used to start a batch culture of up to 10 l. This method is described more fully in stock culture maintenance. If the stock culture is not required immediately, it may be kept for up to 3 weeks on a shelf in a north-facing window (away from direct sunlight), but after this time, it should be discarded and the culture should be less than 25 °C. It is noted that stock cultures do not require an air/CO_2 supply.

Mass Cultures

Stock cultures are used to inoculate intermediate-sized cultures. Many hatcheries use five gallon bottle (19 L) cultures as a step between flask stock cultures and large mass tank cultures. These bottles are available in either plastic (polycarbonate) or glass. Glass bottles are heavy and breakable but can be autoclaved without ill effects. Lightweight plastic bottles have seams that could develop leaks, particularly if autoclaved. Other intermediate-sized vessels such as translucent 2.5 gal bottles, plastic one gallon milk jugs, or polyethylene bags can be used. Polyethylene material for constructing bags can be purchased as tubing (diameters from 1.5–58 in.). The ends can be heat sealed to any convenient length. They require external support but are disposable and do not need to be sterilized when first used. Many hatcheries are using this technique to mass culture algae. For a small, laboratory-sized hatchery, these containers may provide enough algae to satisfy production needs. Commercial-sized bivalve hatcheries require mass tank cultures containing 25–5,000 gal (100–20,000 L) or more of quality algae. In these operations, mass tank cultures are typically started from healthy bottle cultures. Tank and bottle cultures should be provided with the most favorable environment possible. Deviating from ideal conditions will reduce algal growth rate. Lower light levels or reliance on natural sunlight could be used to reduce electrical costs but with a concurrent decrease in production. Some hatcheries eliminate aeration to reduce the spread of biological contaminants. Departures from optimal conditions may be necessary for other reasons. For example, if several algal species are cultured in one room, a compromise in environmental conditions may be necessary. A cooler temperature, such as 65 ° F, would increase the time needed to attain peak density for some species but may eliminate a biological contaminant. Inoculum sufficient for a bottle produces a visible tint in the water. For example, one or two pints (500–1,000 ml) would be adequate inoculums for a 5 gal bottle. Under favorable conditions, a newly inoculated *Isochrysis galbana* culture in a 5 gal (19 L) bottle takes 10–14 days to reach peak density (10–12 million cells/ml). The faster-growing *Thalassiosira pseudonana* reaches its peak (four million/ml) in about 3 days. Once maximum density is attained, healthy *I. galbana* bottle cultures are stable for two additional weeks. Cultures of *T. pseudonana* will deteriorate within 5 days once reaching maximum density. Growth rate and culture stability of most other algae fall between these two. Using a larger volume of inoculum will decrease the time for a culture to reach peak density. For example, a half gallon (2 L) inoculum instead of one pint (500 mL) would shorten by 30–50 % the time a 5 gal (19 L) bottle needs to reach peak density. Algal growth could be increased by bubbling in a mixture of air and 10 % carbon dioxide instead of air alone. Fast-growing cultures could be enhanced with carbon dioxide pulsed in several times a day, keeping the pH below 10. Carbon dioxide is available in pressurized bottles from bottled air suppliers in most communities. A dense (10 million cells/ml), healthy 5 gal (19 L) bottle culture contains enough inoculum to seed a 240 gal (900 L) mass tank culture. At lower densities (five million cells/ml, for example), two or more bottles would be necessary to produce the same amount of algal inoculum. Larger tank cultures would benefit from a proportionally larger inoculum. Mass cultures in 240 gal tanks (8′ × 4′ × 1′) should reach peak density in 2–3 days when configured as follows:

Isolating/Obtaining and Maintaining of Cultures

Sterile cultures of microalgae used for aquaculture purposes may be obtained from specialized culture collections. A list of culture collections is provided by Vonshak (1986) and Smith et al. (1993a). Alternatively, the isolation of endemic strains could be considered because of their ability to grow under the local environmental conditions. Isolation of algal species is not simple because of the small cell size and the association with other epiphytic species. Several laboratory techniques are available for isolating individual cells, such as serial dilution culture, successive plating on agar media, and separation using capillary pipettes. Bacteria can be eliminated from the phytoplankton culture by washing or plating in the presence of antibiotics. The sterility of the culture can be checked with a test tube containing seawater with 1 $g.l^{-1}$ Bacto™ Peptone. After sterilization, a drop of the culture to be tested is added, and any residual bacteria will turn the Bacto™ Peptone solution turbid. The collection of algal strains should be carefully protected against contamination during handling and poor temperature regulation. To reduce risks, two series of stocks are often retained, one which supplies the starter cultures for the production system and the other which is only subjected to the handling necessary for maintenance. Stock cultures are kept in test tubes at a light intensity of about 1,000 lux and a temperature of 16–19 °C. Constant illumination is suitable for the maintenance of flagellates but may result in decreased cell size in diatom stock cultures. Stock cultures are maintained for about a month and then transferred to create a new culture line.

Some Types of Culture

Batch Culture

Batch culture is a system where the total culture is harvested and used as food. If required, another culture can be set up to replace it. The batch culture consists of a single inoculation of cells into a container of fertilized seawater followed by a growing period of several days and finally harvesting when the algal population reaches its maximum or near-maximum density. In practice, algae are transferred to larger culture volumes prior to reaching the stationary phase, and the larger culture volumes are then brought to a maximum density and harvested. The following consecutive stages might be utilized: test tubes, 2 l flasks, 5 and 20 l carboys, 160 l cylinders, 500 l indoor tanks, and 5,000–25,000 l outdoor tanks. According to the algal concentration, the volume of the inoculum which generally corresponds with the volume of the preceding stage in the upscaling process amounts to 2–10 % of the final culture volume. Where small amounts of algae are required, one of the simplest types of indoor culture employs 10–20 l glass or plastic carboys which may be kept on shelves backit with fluorescent tubes. Batch culture systems are widely applied because of their simplicity and flexibility, allowing to change species and to remedy defects in the system rapidly. Although often considered as the most reliable method, batch culture is not necessarily the most efficient method. Batch cultures are harvested just prior to the initiation of the stationary phase and must thus always be maintained for a substantial period of time past the maximum specific growth rate. Also, the quality of the harvested cells may be less predictable than that in continuous systems and, for example, varies with the timing of the harvest (time of the day, exact growth phase). Another disadvantage is the need to prevent contamination during the initial inoculation and early growth period. Because the density of the desired phytoplankton is low and the concentration of nutrients is high, any contaminant with a faster growth rate is capable of outgrowing the culture. Batch cultures also require a lot of labor to harvest, clean, sterilize, refill, and inoculate the containers.

Semicontinuous Culture

Semicontinuous culture is a system where part of the culture is harvested and used as food and the amount taken is replaced with fresh culture medium (clean seawater and nutrient salts). After allowing 2–3 days for the remaining cells to grow and divide, the process is repeated. Semicontinuous cultures may be operated for up to 7–8 weeks. The semicontinuous technique prolongs the use of large tank cultures by partial periodic harvesting followed immediately by topping up to the original volume and supplementing with nutrients to achieve the original level of enrichment. The culture is grown up again, partially harvested, etc. Semicontinuous cultures may be indoors or outdoors, but usually their duration is unpredictable. Competitors, predators, and/or contaminants and metabolites eventually build up, rendering the culture unsuitable for further use. Since the culture is not harvested completely, the semicontinuous method yields more algae than the batch method for a given tank size

Continuous Culture

The continuous culture method (i.e., a culture in which a supply of fertilized seawater is continuously pumped into a growth chamber and the excess culture is simultaneously washed out) permits the maintenance of cultures very close to the maximum growth rate. Two categories of continuous cultures can be distinguished:

Turbidostat culture, in which fresh medium is delivered only when the cell density of the culture reaches some predetermined point, as measured by the extinction of light passing through the culture. At this point, fresh medium is added to the culture, and an equal volume of culture is removed. The diluted culture increases in cell density until the process is repeated.

Chemostat culture, in which a flow of fresh medium is introduced into the culture at a steady, predetermined rate. The latter adds a limiting vital nutrient (e.g., nitrate) at a fixed rate, and in this way, the growth rate and not the cell density is kept constant. In a chemostat, the medium addition ultimately determines growth rate and cell density.

Laing (1991) described the construction and operation of a 40 l continuous system suitable for the culture of flagellates, e.g., *Tetraselmis suecica* and *Isochrysis galbana*. The culture vessels consist of internally illuminated polyethylene tubing supported by a metal framework. This turbidostat system produces 30–40 l per day at cell densities giving optimal yield for each flagellate species. A chemostat system that is relatively easy and cheap to construct is utilized by Seasalter Shellfish Co. Ltd, UK. The latter employs vertical 400 l capacity polyethylene bags supported by a frame to grow *Pavlova lutheri*, *Isochrysis galbana*, *Tetraselmis suecica*, *Phaeodactylum tricornutum*, *Dunaliella tertiolecta*, and *Skeletonema costatum*. One drawback of the system is the large diameter of the bags (60 cm) which results in self-shading and hence relatively low algal densities. The disadvantages of the continuous system are its relatively high cost and complexity. The requirements for constant illumination and temperature mostly restrict continuous systems too indoors, and this is only feasible for relatively small production scales. However, continuous cultures have the advantage of producing algae of more predictable quality. Furthermore, they are amenable to technological control and automation, which in turn increases the reliability of the system and reduces the need for labor. In many chemostat continuous culture systems, the nutrient medium is delivered to the culture at a constant rate by a peristaltic pump or solenoid gate system. The rate of media flow can be adjusted and is often set at approximately 20 % of culture volume per day. Air is pumped into the culture vessel through a sterile filter. This bubbling air has three effects: it supplies CO_2 and O_2 to the culture, aids in circulation and agitation of the cultures, and pressurizes the head space of the culture vessel so as to provide the force to "remove" an amount of media (and cells) equal to the volume of inflowing media. The culture may be aseptically sampled by opening the clamp on a sample port. The magnetic stirrer and aeration help to prevent the cells from collecting in the bottom of the culture vessel. A truly continuous culture will have the medium delivered at a constant volume per unit time. However, delivery systems such as peristaltic pumps or solenoid gates are inherently unreliable. In order to deliver exactly the same amounts of medium to several cultures growing at once, a "semicontinuous" approach can be taken. The rate of flow of medium into a continuous culture system is known as the "dilution rate." When the number of cells in the culture vessel remains constant over time, the dilution rate is said to equal the rate of cell division in the culture, as the cells being removed by the outflow of medium are being replaced by an equal number through cell division in the culture. The principal advantage of continuous culture is that the rate of dilution controls the rate of microbial growth via the concentration of the growth-limiting nutrient in the medium. As long as the dilution rate is lower than the maximum growth rate attainable by the algal species, the cell density will increase to a point at which the cell division rate ("birth rate") exactly balances the cell washout rate ("death rate"). This steady-state cell density is also characterized by a constancy of all metabolic and growth parameters. On the other hand, if the dilution rate exceeds the maximum cell division rate, then cells are removed faster than they are produced and total washout of the entire cell population occurs.

Culture of Diatoms

Diatoms are important members of phytoplankton community, and they are sometimes called golden algae, because the characteristic yellow-brown pigment masks the green chlorophyll. Diatoms have radiosymmetry they are round and shaped like pill boxes, while others have bilateral symmetry and are elongated. Round diatoms float better than the elongate forms, therefore the elongate diatoms are often found in the shallow seafloor attached to floating objects, while the round diatoms are more truly planktonic and are found in areas of cold, nutrient-rich water. A hard, rigid transparent cell wall impregnated with silica surrounds each diatom. Pores connect the living position of the cell inside to its outside environment; buoyancy of the diatoms is increased by low density of the interior of the cell and the production of oil as a storage product. The small volume and accompanying relatively large surface area also provide them with greater exposure to sunlight, and water contains the gases and nutrients for photosynthesis and growth. In addition, some diatoms have spines or other projections that increase their ability to float. Diatoms reproduce very rapidly by cell division; when their populations discolor the water, it is known as bloom. Culture of this group of phytoplankton requires specialized media like TMRL medium, PM solution, Takeda medium, etc.

Culture of Dinoflagellates

Dinoflagellates are single celled protists that contain photosynthetic pigment vary in colour from green to brown to red and can exist at lower light levels. Their external walls do not contain silica; some are smooth and flexible, but others are armored with plates of cellulose. Dinoflagellates usually have two whiplike appendages or flagella that beat within grooves in the cell wall, giving the cells limited motility. Under favorable conditions, they multiply even more rapidly than diatoms to form blooms, but they are not as important as the diatoms as a primary ocean food source. Some dinoflagellates are called fire algae, because they glow with bioluminescence at night. Dinoflagellates are hard to grow; certain species such as *Prorocentrum micans* grow better in medium FE, which is prepared by mixing equal volumes of Guillard's medium F with Foyn's Schreiber medium E (Subba Rao 1980) and autoclaved at 125 °C (2 atm pressure) for 20 min; some dinoflagellates such as *Ceratium* grow well in Chan's medium (1978) or MET 44 (Schöne and Schöne 1982) and the toxigenic *Protogonyaulax tamarensis* or *Pyrophacus steinii* in T1 medium (Ogata et al. 1987; Pholpunthin et al. 1999). Oceanic ultraplankton is growing successfully in medium K (Keller et al. 1987).

Determining Algal Densities

Monitoring algal growth in mass tank cultures is essential for successful production; a more accurate measure of growth can be made by comparing actual cell numbers with the previous day's count. At present, there are many kinds of counting procedures available for enumerating phytoplankton, two of the procedures are used when counting algae in mixtures, as from field sampling while the other two are used when counting unialgal samples, such as in growth or bioassay experiments in a laboratory. Some techniques are relatively "low-tech" and can be used also in remote locations; others are "high-tech" and require an expensive instrument. The Sedgewick-Rafter counting chamber (Fig. 2) is a low-tech device routinely used for counting algae in mixed assemblages. This cell limits the volume and area of the sample to enable easier counting and calculation of phytoplankton numbers. It consists of a brass or polystyrene rectangular frame mounted on a heavy glass slide from which a precise internal chamber has been cut; its dimensions are 50 × 20 × 1 mm^3, with an area of 1,000 mm^2 and a volume of 1.0 ml. The base is ruled in a 1 mm grid. When the liquid sample is held in the cell by its large, rectangular glass cover slips, the grid subdivides 1 ml volume into microliters.

The phytoplankton sample placed into the Sedgwick-Rafter counting chamber is allowed to stand on a flat surface for 20 min to enable the phytoplankton to settle. It is then transferred to the stage of an upright light microscope and securely positioned, ready for counting. Counts are done with the 4× or (more usually) the 10× objectives of the compound microscope. A Whipple disk is inserted into one of the ocular lenses in order to provide a sample grid. It is necessary to first determine the area (A) of the Whipple field for each set of ocular and objective lenses used. This is accomplished with a stage micrometer. There are 50 fields in the length and 20 fields in the width of the chamber (comprising a total of 1,000 fields). A horizontal strip corresponds to 50 fields. All cells within randomly selected fields are counted.

"High-tech" methods for counting unialgal samples are the electronic particle counter (e.g., Coulter® counter) and the digital microscopy. In spite of relatively high cost, an

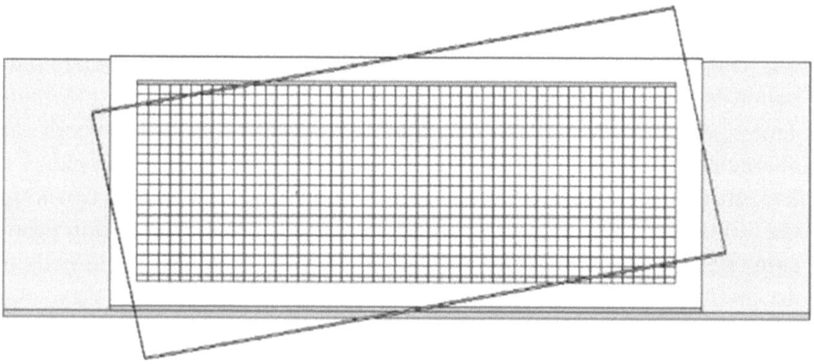

Fig. 2 The Sedgewick-Rafter counting chamber

electronic particle counter is highly recommended for performing growth or bioassay studies that require many counts and high accuracy. In addition, the instrument will provide particle size biovolume distributions. The principle of operation is that particles, suspended in an electrolyte solution, are sized and counted by passing them through an aperture having a particular path of current flow for a given length of time. As the particles displace an equal volume of electrolyte in the aperture, they place resistance in the path of the current, resulting in current and voltage changes. The magnitude of the change is directly proportional to the volume of the particle; the number of changes per unit time is proportional to the number of particles in the sample. When opened, the stopcock introduces vacuum into the system, draws sample through the aperture, and unbalances the mercury in the manometer. The mercury flows past the "start" contact and resets the counter to zero. When the stopcock is closed, the mercury starts to return to its balanced position and draws sample through the aperture. Variously sized aperture tubes are available for use in counting variously sized particles; the aperture size is chosen to match that of particles.

Digital microscopy necessitates of a microscope equipped with a digital TV camera connected to a personal computer running dedicated application software for cell recognition and cell counting. It represents an automatic, reliable, and very fast approach to growth determination.

The number of cells can be determined by studying a sample from a culture under a 400× microscope. The sample is placed on a special slide called a hemacytometer. These slides are available from scientific or medical supply stores and come with instructions. To use the hemacytometer, place a pipette filled with a sample from the culture tank against the "V" groove of the slide (a cover slip must be in place). Withdraw the pipette before the sample runs off the flat grid area, but not before the grid area is completely covered with the algal suspension. Wait about 5 min for the cells to drift onto the grid. The grid area resembles the pound symbol (#) inside a square border so that there are nine equal squares. Each of the four corner squares is further divided into 16 smaller squares. The total number of algal cells in all 16 small squares in the 4 corner areas (64 total small sqs.) is counted. Divide this number by 4, for an average, and multiply by 10,000 (hemacytometer volume factor). The product is equal to the number of cells per ml in the sample. Motile algae like *I. galbana* must be killed before counting. A weak toxin such as acetic acid or ethanol may be used (one drop into 20–30 ml of algal sample). Dense cultures (>3 million) should be diluted 1:10 to facilitate counting and to reduce error. Remember to multiply by 10 (for a 10:1 dilution) or by the appropriate correction factor if a different dilution is made. There are numerous other methods to quantify algal densities; however, using a microscope and hemacytometer will reveal the presence of biological contaminants. Peak or maximum density (for existing conditions) is attained when, on consecutive days, the cell count does not increase by a factor of two. With practice, approximate cell counts can be estimated by observing the gross color of each culture. These estimates are based on the tint of the water caused by the algal cells and comparing it to the actual cell count. Tedious and time-consuming cell counts (or other growth measurements) can be eliminated from the daily routine. Cell counts should be made periodically as a check on the gross estimate. In practice, culture crashes can be caused by a variety of reasons, including the depletion of a nutrient, oxygen deficiency, overheating, pH disturbance, or contamination. The key to the success of algal production is maintaining all cultures in the exponential phase of growth. Moreover, the nutritional value of the produced algae is inferior once the culture is beyond phase 3 due to reduced digestibility, deficient composition, and possible production of toxic metabolites. Cells can be counted either with an electronic particle counter or directly under a microscope, using a hemacytometer. The Coulter® counter and similar instruments need appropriate calibration for each algal species to be counted. Detailed instructions on operation of electronic cell counting can be found in Sheldon and Parsons (1967). The presence of contaminating particles in the same size range as the algae and failure of cells to separate after cell division may be possible sources of erroneous counts. Counting with a microscope has the advantage of allowing control of the quality of the cultures. The major difficulty in microscopic counts is reproducibility, which is a function of the sampling, diluting, and filling of the counting chamber, as well as the choice of the right type of counting chamber and range of cell concentration. A relationship between optical density and cellular concentration, can be established using a spectrometer. However, variations may occur due to the fact that the chlorophyll concentration in the algal cell varies according to the culture conditions and therefore affects this relationship. In this way, a culture under low lighting conditions will be comparatively more pigmented and will eventually result in higher readings for optical density.

Harvesting and Preserving Microalgae

In most cases, it is unnecessary to separate microalgae from the culture fluid. Excess and off-season production may, however, be concentrated and preserved. The various techniques employed to harvest microalgae have been reviewed by Fox (1983) and Barnab (1990). High-density algal cultures can be concentrated by either chemical flocculation or centrifugation. Products such as aluminum sulfate and ferric chloride cause cells to coagulate and precipitate to the bottom or float to the surface. Recovery of the algal biomass is then accomplished by, respectively, siphoning off the supernatant or skimming cells off the surface. Due to the increased particle size, coagulated algae are no longer suitable as food for filter feeders. Centrifugation of large volumes of algal culture is usually performed using a cream separator; the flow rate being adjusted according to the algal species and the centrifugation rate of the separator. Cells are deposited on the walls of the centrifuge head as a thick algal paste, which is then resuspended in a limited volume of water. The resulting slurry may be stored for 1–2 weeks in the refrigerator or frozen. In the latter case, cryoprotective agents (glucose, dimethyl sulfoxide) are added to maintain cell integrity during freezing. However, cell disruption and limited shelf life remain the major disadvantages of long-term preserved algal biomass. Concentrated cultures of *Tetraselmis suecica* kept in darkness at 4 °C maintain their viability, whereas the latter is completely lost upon freezing. Furthermore, cultures stored in hermetically sealed vials lose their viability more rapidly than those kept in cotton-plugged vials.

References

Allen EJ, Nelson EW (1910) On the artificial culture of marine plankton organisms. J Mar Biol Assoc UK 8:421–474

Barnab G (1990) Harvesting micro-algae. In: Barnab G (ed) Aquaculture, vol I. Ellis Horwood, New York, pp 207–212

Beijerinck MW (1890) Culturversuche mit Zoochlorellen. Lichenengonidien und anderen niederen Algen Bot Zeitung 48:725–785

Burlew JS (ed) (1953) Algal culture from laboratory to pilot plant. Carnegie Institution of Washington, Washington, DC, 357 pp

Chan TA (1978) Comparative physiological studies of marine diatoms and dinoflagellates in relation to irradiance and cell size. I. Growth under continuous light. J Phycol 14:396–402

Chu SP (1942) The influence of the mineral composition of the medium on the growth of planktonic algae: part I. Methods and culture media. J Ecol 30(2):284–325

Fox JM (1983) Intensive algal culture techniques. In: McVey JP (ed) Crustacean aquaculture, vol 1. CRC Press, Boca Raton, pp 43–69

Kawaguchi K (1980) Microalgae production systems in Asia. In: Shelef G, Soeder CJ (eds) Algae biomass. Elsevier/North-Holland Biomedical Press, Amsterdam, pp 25–33

Keller MD, Selvin RC, Claus W, Guillard RRL (1987) Media for the culture of oceanic ultraplankton. J Phycol 23:633–638

Laing I (1991) Cultivation of marine unicellular algae. MAFF laboratory leaflet number 67. Directorate of Fisheries Research, Lowestoft, 31 pp

Miquel P (1890–1893) De la Culture artificielle des Diatomees. Le Diatomiste pp. 73, 93, 121, 149, 165; Le Micro. Pieparature, 1897 69.

Ogata T, Ishimaru T, Kodama M (1987) Effect of water temperature and light on growth rate and toxicity change in *Protogonyaulax tamarensis*. Mar Biol 95:217–220

Pholpunthin P, Fukuyo Y, Matsuoka K, Nimura Y (1999) Life history of a marine dinoflagellate *Pyrophacus steinii* (Schiller) Wall et Dale. Bot Mar 42:189–197

Provasoli L, McLaughlin JJA, Droop MR (1957) The development of artificial media for marine algae. Arch Mikrobiol 25:392–425

Schöne HK, Schöne A (1982) MET 44: A weekly enriched-sea water medium for ecological studies on marine plankton algae, and some examples of its application. Bot Mar XXV:117–122

Sheldon RW, Parsons TR (1967) A continuous size spectrum for particulate matter in the sea. J Fish Board Canada 24(5):909–915

Smith LL, Fox JM, Granvil DR (1993) Intensive algae culture techniques. In: McVey JP (ed) CRC handbook of mariculture. Volume 1. Crustacean aquaculture, 2nd edn. CRC Press, Boca Raton, pp 3–13

Subba Rao D (1980) Measurement of primary production in phytoplankton groups by size-fractionation and by germanic acid inhibition techniques. Oceanol Acta 3(1):31–42

Vonshak A (1986) Laboratory techniques for the cultivation of microalgae. In: Richmond A (ed) Handbook of microalgal mass culture. CRC Press, Boca Raton, pp 117–145

Warburg O (1919) Uber die Greschwindiling Keit der Koti; ensarezusammensetzung in Lebendenzellen. Biochemischezeitschrift 100:230–270

Recent Advances in Rotifer Culture and Its Application for Larviculture of Finfishes

M. Kailasam, Thirunavukkarasu A.R., A.G. Ponniah, S. Selvaraj, and P. Stalin

Introduction

Global production of fish from aquaculture has grown substantially in the past decade reaching 52.5 million tonnes in 2008, compared with 32.4 million tonnes in 2000. Aquaculture continues to be the fast-growing animal food-producing sector and currently accounts for nearly half (45.6 %) of the world's food fish consumption compared with 33.8 % in 2000. The value of world aquaculture, excluding aquatic plants, is estimated at US$ 98.4 billion in 2008. A forecast indicates that aquaculture food fish production is expected to be 57.2 % million tonnes in 2010 (FAO 2010). Asian countries are dominating the world in aquaculture production due to the favourable tropical climatic conditions and the availability of suitable sites and natural water resources such as bays, rivers, estuaries and brackishwater lagoons.

Even though India has been blessed with 1.2 million hectares of physical resources for brackishwater aquaculture, only 15 % of resources presently are being utilised for farming purposes. Brackishwater farming in India is synonym with the culture of only shrimp *Penaeus monodon*. The monoculture practices of *P. monodon* in the past have resulted in the occurrence of viral diseases. Under these circumstances, diversification of farming practices to other species is vital and fish farming is one option to diversify culture species in order to reduce the dependence of shrimp culture activity. However, saltwater fish farming is very much limited in India. Aquaculture in coastal ecosystem is restricted to shrimp farming mainly because of high unit cost of the production and the export value. Due to problems encountered in the recent years, the outbreak of uncontrollable diseases, environmental and social issues, etc., and the sustainability of coastal aquaculture are hampered, and to overcome these problems, many options are put forth, and one of the easiest and adoptable options is the diversification to another species farming. In this context, finfish aquaculture assumes greater significance. But, for this, the major inputs like quality seed and feed are important prerequisite. Quality fish seed production in hatcheries needs many inputs, and the most important and critical are live feed organisms such as green algae, rotifers and *Artemia* nauplii.

M. Kailasam (✉) • Thirunavukkarasu A.R.
A.G. Ponniah • S. Selvaraj • P. Stalin
Fish Culture Division, Central Institute of Brackishwater Aquaculture, No. 75, Santhome High Road, R.A. Puram, Chennai 600028, Tamil Nadu, India
e-mail: kailu66@hotmail.com

Importance of Rotifer

Among the live feeds, the rotifers are the most useful organisms for the rearing of fish larvae because of its unique characteristics such as very small in size, relatively slow in motility, rapid reproduction rate, acceptance of algal feed, easy to enrich with required components

S. Perumal et al. (eds.), *Advances in Marine and Brackishwater Aquaculture*,
DOI 10.1007/978-81-322-2271-2_2, © Springer India 2015

and mass culture possibility. These characters contributed their usefulness as good prey for fish and shellfish larvae. Rotifers can tolerate a wide range of salinity and temperatures, because of its euryhaline characteristics, and thus suit the continuous mass culture in outdoor systems. Rotifers are microscopic organisms abundantly found in all the aquatic systems. It thrives in the eutropic condition. Rotifer succeeds normally after the phytoplankton bloom crashed out. It feeds on macroscopic unicellular algae like *Chlorella salina*, *Tetraselmis costatum* and *Nannochloropsis oculata*. Because of their apt size (100–280 μ) for fish larvae ingestion, it is cultured in large scale and used in all the fish hatcheries. Rotifer is considered as an important live food organism.

morphologically and ecologically different groups: "L type" (large) and the other "S type" (small) (Oogami 1976; Rumengan 1990). On the basis of several studies, L- and S-type *B. plicatilis* have been classified into *B. plicatilis* and *B. rotundiformis* (Segers 1995). These two species can be distinguished morphologically by the shape of lorica and spines. *B. plicatilis* has obtuse anterior spines, while *B. rotundiformis* has pointed anterior spines (Fu et al. 1993). Rumengan et al. (1991) reported that the number of chromosomes also differed between the S and L types. Moreover, it was reported that S and L types were reproductively isolated from each other (Fu et al. 1993; Gómez 2005). As a result it was reclassified the S and L types as different species as *B. rotundiformis* and *B. plicatilis.*

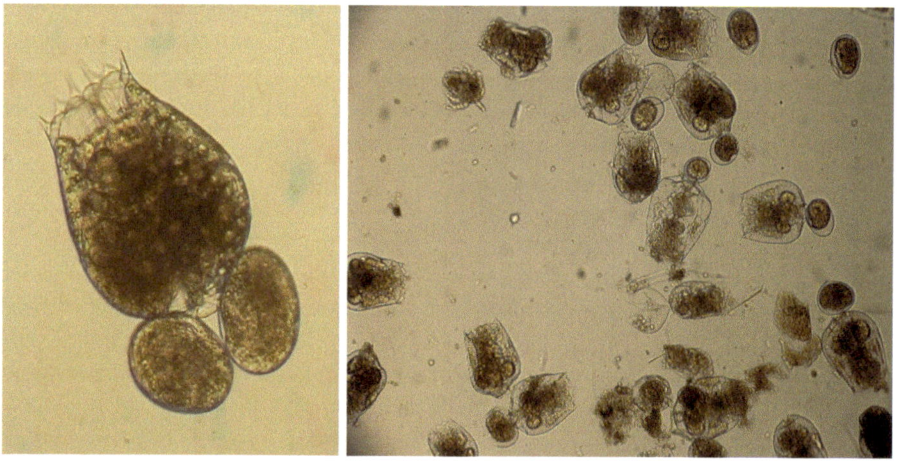

Rotifer *Brachionus plicatilis*

Taxonomy and biology of *Brachionus plicatilis*

Phylum	Nemathelminthes
Class	Rotifera
Order	Monogononta
Suborder	Ploima
Family	Brachionidae
Genus	*Brachionus* Pallas
Species	*plicatilis* Muller

Initially, rotifers were classified into a single species, *B. plicatilis.* However, it was later reported that *B. plicatilis* consists of two

Few studies have reported that some strains could not be classified as either *B. plicatilis* or *B. rotundiformis* (Hagiwara et al. 1995). The small size rotifer strains observed in the tropical regions were designated as SS type by the aquaculturists (Hagiwara et al. 1995). Hagiwara et al. (1995) reported that SS-type rotifers belonged to *B. rotundiformis* as they are morphologically, ecologically and genetically similar and do not have any pre-mating reproductive isolation. In recent years rotifers are classified into three size groups based on the lorica length as SS type (100–140 μm), S type (141–220 μm) and L type (above 220 μm) which is followed by several research groups (Gondol Research Institute for Mariculture, Indonesia; Aslianti 2003).

Brachionus plicatilis is distributed in tropical and subtropical waters. Morphologically, rotifers consist of head bearing wheel organ, body forming lorica and foot consisting of toes. The head carries a corona which is surrounded by cilia. The presence of cilia facilitates locomotory functions. The size of rotifer *B. plicatilis* is usually ranged between 100 and 400 μm depending on the strain. Fish larvae select the rotifers according to their size, generally preferring small in early stages and large as they grow. The growth of rotifer increases with temperature between 15 and 35 °C. However, its lifespan is shorter at high water temperatures. Under various food levels, *B. plicatilis* flexibly changes its reproductive patterns and lifespans (Yoshinaga et al. 2000). In a food-rich environment, *B. plicatilis* produces approximately 30 offspring during its lifespan of approximately 10 days. In contrast, when fed for only a few hours daily, it suppresses active reproduction and produces less than ten offspring, while surviving for nearly a month. Long-lived individuals resulting from reproductive suppression are likely to obtain a second chance for reproduction in the future. Besides life history parameters, offspring quality (starvation resistance) also increases when *B. plicatilis* reproduces in a food poor environment (Yoshinaga et al. 2001). Thus, *B. plicatilis* possesses a reproductive strategy that is able to adapt to various food levels by changing its life history parameters and offspring quality (Yoshinaga et al. 2003).

The rotifer *B. plicatilis* reproduction can be either sexual or asexual. Asexual reproduction of rotifer is useful for mass culture due to its faster rate and also due to the absence of males as it lacks functional digestive tract. In the asexual reproduction, the offspring are clones genetically identical to their mothers (diploid females). Males are only produced when there is a sudden change in the environment conditions during which females produce haploid eggs. The males and females breed and resting eggs are formed.

Types of Rotifers

There are three types of rotifers cultured in the hatcheries and used depending upon the requirements.

1.	SS (super small) type	100–140 μm
2.	S type	141–220 μm
3.	L type	Above 221 μm

Generally the water volume of algal culture is two to five times greater than the volume of rotifer culture. The required daily parameters for rotifer culture are densities of *Nannochloropsis* spp. $> 10 \times 10^6$ cells/ml to be provided for the rotifers and temperature range between 25 and 28 °C. Rotifer starter cultures are drawn from stock cultures maintained. Fibreglass tanks varying in capacity from 1 to 2 tons are used for rotifer culture. Tanks should be preferably elevated at about 3 ft above the ground for easier collection and harvest.

Propagation and Maintenance of Pure Culture of Rotifer

To raise a pure culture of rotifer, initial samples must be collected from stagnant water bodies, using a net of 50–80 μ mesh size. About 50–60 l of pond water then filtered yields sufficiently large numbers of individuals of *Brachionus plicatilis*. Using a fine dropper, individual specimens are isolated and introduced into a glass cavity block containing filtered, sterilised seawater, the pH of which is adjusted to be same as that of the field sample. Using a Pasteur pipette, each of the rotifer species is transferred into the cavity blocks and serially transferred through several cavity blocks to eliminate many associate organisms. Cover the cavity block and place in diffused light. After isolation, the rotifers are maintained feeding with algal cells like *Nannochloropsis oculata* at the density of one million cells/ml. The stock rotifer culture can be continued up to 8–10 days to reach peak density of stationary phase, and afterwards fresh culture has to be carried out following the same procedure.

Mass Culture of Rotifer

The rotifer culture tanks are cleaned and filled with algal water with a cell density >20 × 10^6 cells/ml. Rotifers are introduced with an initial density of 10 ind/ml. Rotifer density estimated by taking 1 ml aliquot with a glass rotifer pipette and number is counted. The culture is allowed for a period of 8–10 days for rotifer density to increase. Using a 48 μ plankton net, rotifers are harvested. Some harvested stock is reserved as starter culture for other tanks and rests as feed for larvae. After each harvest, the tanks were thoroughly washed. Culture of rotifer and algae should be scheduled to ensure daily harvest of rotifer and an uninterrupted production. Culturists prefer rotifers to reproduce asexually, because of the shorter lifespan and better nutritive value of the asexual forms. This is accomplished by regulating feed, water, temperature, salinity and aeration during the culture process. Fertility is a measure of the general health of the rotifer culture. Under normal circumstances greater than 30 % of the rotifers should be carrying eggs 24 h after initial stocking. This value will fall to 10 % at the time of harvesting.

Techniques of High-Density Rotifer Culture

High density of rotifer culture is usually undertaken using commercial product known as Culture Selco. Culture Selco is a dry and complete rotifer diet that does not require algae and is also effective as enrichment medium. The particle size (5–7 μm) and physical characteristics of product ensure an optimal uptake by rotifers. The average daily production of rotifers fed on Culture Selco ranges consistently from 45 to 60 % of the initial rotifer density in the culture system. In addition, rotifers are also enriched with high levels of essential polyunsaturated fatty acids. To prepare Culture Selco medium for use, the amount required for a single meal in tap water (up to 50 g Culture Selco/l) is taken and mixed vigorously for 3–5 min using kitchen or industrial blender. This is distributed four to six meals evenly over the 24 h period. The inoculum of rotifers for undertaking high-density rotifer culture is 300–500 rotifers/ml, and harvesting of rotifers at 2,000–5,000 rotifers/ml in 8–10 days can be achieved using this technique.

Ultra-high-density culture practices are also followed in several places reaching the density from 10,000 nos/ml to 40,000 nos/ml. Algal paste is supplied continuously in the rotifer culture system, and ammonia is being removed by adding various ammonia detoxifying products like ammonax, etc. This high-density rotifer culture tank water is being recirculated continuously with the filter to remove the suspended particles if any. The advantages of the high-density practices are less space and low manpower, through which higher production of rotifers can be achieved.

Nutritive Value of Rotifers

The nutritive value of live rotifer is very important for the survival of fish larvae. Therefore, along with the production of sufficient quantities of rotifers, it is also necessary to ensure proper nutritional quality required for the larvae. The lipid content of rotifer plays an important role in the early development of fish larvae as it forms a part of the cellular membranes of the larvae and, therefore, is crucial in determining the rates of enzymatic processes placed at these sites (Lubzens et al. 1989). In addition, it has been recognised that the nutritional quality of rotifers can be manipulated to ensure that a nutritionally augmented rotifer is provided to fish larvae as a first food (Watanabe et al. 1983). The nutritive value of cultured *B. plicatilis* for larval rearing of finfishes depends on the transfer of dietary components from the algal feed. Though proximate composition of the rotifer and its diet may not be similar, constituent units of dietary macronutrients, particularly fatty acids, are known to be transferred (Watanabe et al. 1983). Fatty acid composition of rotifers is extremely important because polyunsaturated fatty acids, particularly 20:5 n3 and 22:6 n3 acids, are essential for early growth of fish larvae of many marine fish species. Rotifers can synthesise polyunsaturated fatty acids, but tissue levels are insufficient to meet the demands for normal growth and development of fish larvae (Lubzens et al. 1985). Tissue accumulation of these fatty acids by rotifers is highly dependent on exogenous supply of microalgae.

In general, the nutritive value of rotifer depends on the type of microalgae in which it is raised under mass culture conditions. Though several microalgae are used as food for rotifers, it was observed that *Nannochloropsis oculata* is the best feed for rotifers in terms of nutritive content as it contains high lipid content when compared to other algal species.

Rotifer Enrichment

The protein and fatty acid content of the rotifers cultured may not be sufficient for growth of marine and brackishwater fish larvae. Though rotifer can synthesise some n-3 highly unsaturated fatty acids (Lubzens et al. 1989), the quantity may not be sufficient to satisfy the demand of fish larvae. Therefore, rotifers must be fed with these fatty acids. Fatty acid composition of rotifer is largely dependent on the type of food. The polyunsaturated fatty acids, specifically eicosapentaenoic acid (EPA, 20:5n-3) and docosahexaenoic acid (DHA, 22:6n-3), have been shown to be essential in the diet of marine fish larvae (Watanabe et al. 1983). The deficiency of essential fatty acids in the diet results in poor growth, low feed efficiency, anaemia and high mortality. Ascorbic acid or vitamin C is required in larval fish diets. Scoliosis, distorted/twisted gill filaments, short opercula and snout are few of the gross signs of ascorbate deficiency (Dabrowski 1990). Hence, to augment the required levels, enrichment of rotifer can be done.

Rotifer can be enriched using the following methods: (1) emulsified marine oils such as cod liver oil and menhaden oil rich in omega-3 HUFAs, (2) microencapsulated oils containing high concentrations of n-3 HUFAs and (3) live microalgae. Mostly, enrichment is provided by using specially formulated artificial diets such as Protein and DHA Protein Selco. This oil emulsion gives good results in terms of high levels of EPA, DHA and vitamin C. Rotifers can be enriched in their mass culture tanks or after harvesting and placing them in separate containers/tanks. The first method includes enrichment of the tissues since it is continuous along with the culture period. The second method is short-term enrichment or rather gut enrichment in which harvesting and rinsing of the rotifers is carried out in a separate enrichment tank.

Rotifer as First Feed for Fish Larvae

Sea Bass

Asian seabass *Lates calcarifer* larvae are voracious feeders, and feed has to be supplied adequately in the hatchery production (Kailasam et al. 2006). Rotifer is used as initial feed for seabass larvae soon after mouth opening (mouth size, 250 μm) at 48 h of hatching (1.6 ± 0.2 mm). Rotifer is supplied for seabass larvae up to the 9th day post hatch (dph) and from 10th to 15th dph; rotifer is supplied along with *Artemia* nauplii (Dhert et al. 1990). Usually, rotifer is supplied to the seabass larvae at the rate of 20–30 nos/ml in the rearing tank.

Milkfish

In milkfish (*Chanos chanos*), mouth opens 54 h after hatching and the mouth width is 500 μm. Feeding with rotifers is initiated 45–50 h after hatching. If onset of feeding is considered to be at 54 h from hatching, milkfish larvae then have about 66 h to initiate feeding effectively before the yolk is completely resorbed at 120 h. (Bagarino 1986). Rotifer is supplied to milkfish larvae up to 20 days post hatch.

Mullet

The first feeding of rotifers starts 65–70 h after hatching as presence of rotifers in the gut of striped mullet *Mugil cephalus* was observed first 70 h post hatch. Rotifer feeding is continued till the 25th day in larval rearing of *M. cephalus* (Eda et al. 1990).

Groupers

The newly hatched larvae of humpback grouper, *Cromileptes altivelis*, measures 1.52 mm with volume of yolk 1.71 mm^3 and oil globule of 3.76×10^{-2} mm^3 in average. Mouth opens after 2 days and the size of mouth is 180–200 μm and the larvae start first feeding. The larvae consume all of their reserves around 48 h after mouth opening.

The first feeding of estuarine grouper, *Epinephelus tauvina*, larvae starts on day 3 with SS-type rotifer (153 ± 13 μm), and rotifer feeding is continued up to the 15th day (Chen et al. 1977). For the brown-marbled grouper *E. fuscoguttatus*, initial feeding starts on day 3 with S-type rotifer (176 ± 10 μm) (Chao and Lim 1991). Grouper larvae indeed have extremely small mouth size and cannot always be raised with rotifers. In the past, the eggs (55–60 μm) and trochophore larvae (60–80 μm) of green mussel *Perna viridis* were used as initial feed for the greasy grouper *E. tauvina* larvae (Lim 1993). Mouth gap at first feeding of orange-spotted grouper *E. suillus* was reported as 150–180 μm, and first feeding starts on day 2 with rotifer *B. plicatilis* (162.2 ± 18.6 μm) (Duray et al. 1997).

Gilthead Sea Bream

First feeding of sea bream *Sparus aurata* starts on day 3 after hatching. Rotifers (50–100 μm) are introduced on day 3, and rotifer feeding is continued till the 16th day (Moretti et al. 1999).

Rabbit Fishes

Newly hatched larvae of *Siganus guttatus* are very fragile and measure 1.5–2.6 mm in total length. The yolk sac is oval in shape, measures 0.70 × 0.24 mm and has two oil globules (Hara et al., 1986). Larvae of *S. guttatus* grow fast within the first 24 h. The mouth opens at 36 h after hatching, larvae start to feed rotifer after 48 h of hatching and the yolk is completely resorbed at 72 after hatching (Bagarino 1986). First feeding of *S. guttatus* starts on day 2. Feeding is normally done with *B. plicatilis* (<90 μm) at the rate of 10–20 individuals/ml. Rotifer feeding is continued till the 15th day in seed production of *S. guttatus* (Hara et al. 1986).

Spotted Scat *Scatophagus argus*

The larvae of spotted scat starts its first feeding after 48 h of mouth opening, since the larvae of *S. argus* had the yolk and oil droplets volume of 0.108 and 0.00982 mm^2, respectively, at hatching and thereafter the yolk volume reduced on subsequent days and reached to an olw of 0.012 mm^2 at 72 h post hatch. Similarly, the oil droplets reduced to a low of 0.00119 mm^2 at 96 h post hatch and recommended the first feeding with rotifers before 48 h of mouth opening (Kailasam et al. 2011).

Conclusion

The coastal aquaculture of finfishes in ponds and cages is considered as an eco-friendly and sustainable practice of farming and gaining a lot of importance. Some of the candidate fish species such as Asian seabass, grouper, milkfish, grey mullets and sea bream are identified as potential candidate species for brackishwater aquaculture. However, the availability of quality seed of the species in adequate quantity and cost-effective feed are very important inputs for the development of finfish farming. Efforts made for development of seed production of Asian seabass under captive condition have been successful (Thirunavukkarasu et al. 2001). Since, live feed especially rotifer plays significant role in the production of quality seed, the techniques for the mass culture of rotifer with adequate nutritional quality are being addressed in priority.

References

Aslianti T (2003) Live food mass culture. In: Training manual on grouper hatchery production. Gondol Research Institute for Mariculture, Bali, Indonesia, pp 1–9

Bagarino T (1986) Yolk resorption, onset of feeding and survival potential of larvae of three tropical marine fish species reared in the hatchery. Mar Biol 91:449–459

Chao TM, Lim LC (1991) Recent developments in the breeding of grouper (*Epinephelus* spp.) in Singapore. Singap J Prim Ind 19:78–93

Chen FY, Chow M, Chao TM, Lim R (1977) Artificial spawning and larval rearing of the grouper, *Epinephelus tauvina* (Forskal) in Singapore. Singap J Prim Ind 5:1–21

Dabrowski K (1990) Ascorbic acid status in the early life of whitefish (*Coregonus lavaretus* L.). Aquaculture 84:61–70

Dhert PH, Duray M, Lavens P, Sorgelos P (1990) Optimized feeding strategies in the larviculture of the Asian Seabass *Lates calcarifer*. In: Hirano R, Hanyo I (eds) Proc. the second Asian Fisheries Forum, Asian Fisheries Society, Manila, Philippines, pp 319–323

Duray MN, Estudillo CB, Alpasan LG (1997) Larval rearing of the grouper *Epinephelus suillus* under laboratory conditions. Aquaculture 150:63–76

FAO (2010) FishStat fishery statistical collections: aquaculture production (1950–2008; released March 2010). Food and Agriculture Organization of the United Nations, Rome

Eda H, Murashige R, Ooozeki Y, Hagiwara A, Eastham B, Bass P, Tamaru CS, Lee CS (1990) Factors affecting intensive larval rearing of striped mullet, *Mugil cephalus*. Aquaculture 91:281–294

Fu Y, Hagiwara A, Hirayama K (1993) Crossing between seven strains of the rotifer *Brachionus plicatilis*. Nippon Suisan Gakkaishi 59:2009–2016

Gómez A (2005) Molecular ecology of rotifers: from population differentiation to speciation. Hydrobiologia 546:1–17

Hagiwara A, Kotani T, Snell TW, Aree MA, Hirayama K (1995) Morphology, genetics, and mating behavior of small tropical marine *Brachionus strains* (Rorifera). J Exp Mar Biol Ecol 194:25–37

Hara S, Duray MN, Parazo M, Taki Y (1986) Year-round spawning and seed production of the rabbitfish, Siganus guttatus. Aquaculture 59(3):259–272

Kailasam M, Thirunavukkarasu AR, Selvaraj S, Stalin P (2006) Effect of delayed initial feeding on growth and survival of Asian Seabass *Lates calcarifer* (Bloch) larvae. Aquaculture 271:298–306

Kailasam M, Thirunavukkarasu AR, Stalin P, Madhavi M, Sundaray JK, Biswas G, Prem Kumar, Subburaj R, Thiagarajan G (2011) Yolk and oil globule utilization in the larvae of spotted scat *Scatophagus argus* and Asian seabass of *Lates calcarifer*. In: 9th Indian Fisheries Forum, December 19–23, 2011 Abstracts. P.91

Lim LC (1993) Larviculture of the Greasy Grouper *Epinephelus tauvina* F. and the brown marbled grouper *E. fuscoguttatus* F. in Singapore. J World Aquacult Soc 24:262–273

Lubzens E, Tandler A, Minkoff G (1989) Rotifers as food in aquaculture. Hydrobiologia 186(187):387–400

Moretti A, Fernandez-Criado MP, Cittolin G, Guidastri R (1999) Manual on hatchery production of seabass and gilthead seabream. FAO Publications, Rome

Oogami H (1976) On the morphology of *Brachionus plicatilis*. Newsletter from Izu Branch, Shizuoka Prefectural Fisheries Research Center 184: 2–5

Rumengan IFM (1990) Studies on growth characteristics and karyotypes of S and L type rotifers *Brachionus plicatilis*. Doctoral thesis, Nagasaki University, 147 pp

Rumengan IFM, Kyayano H, Hirayama K (1991) Karyotypes of S and L type rotifers *Brachionus plicatilis* O.F. Muller. J Exp Mar Biol Ecol 154:171–176

Segers H (1995) Rotifera 2. The Lecanidae (Monogononta). In: Nogrady T, Dumont HJ (eds) Guides to the identification of the microinvertebrates of the continental waters of the world 6. SPB Academic, The Hague

Thirunavukkarasu AR, Kailasam M, Kishore Chandra P, Shiranee P, Abraham M, Charles AVK, Subburaj R (2001) Captive broodstock development and breeding of sea bass *Lates calcarifer* (Bloch) in India. In: Menon NG, Pillai PP (eds) Perspectives in mariculture, CMFRI, Kochi, India, pp. 111–124

Watanabe T, Tamiya T, Oka A, Hirata M, Kitajima C, Fujita S (1983) Improvement of dietary value of live foods for fish larvae by feeding them on ω3 highly unsaturated fatty acids and fat soluble vitamins. Bull Jpn Soc Fish 49:471–479

Yoshinaga T, Hagiwara A, Tsukamoto K (2000) Effect of periodical starvation on the life history of *Brachionus plicatilis* O.F. Muller (Rotifera): a possible strategy for population stability. J Exp Mar Biol Ecol 253:253–260

Yoshinaga T, Hagiwara A, Tsukamoto K (2001) Effect of periodical starvation on the survival of offspring in the rotifer *Brachionus plicatilis*. Fish Sci 67:373–374

Yoshinaga T, Kaneko G, Kinoshita S, Tsukamoto K, Watabe S (2003) The molecular mechanisms of life history alterations in a rotifer: a novel approach in population dynamics. Comp Biochem Physiol 136:715–722

Production and Preservation of *Artemia*

Joseph Royan

Introduction

Most of the aquatic animal species of commercial interest which are grown in intensive hatchery systems have to be offered a live food throughout their life cycle (Kinne and Rosenthal 1977). Zooplankton, the natural food of fish and shrimp larvae, is very difficult to culture on a commercial level. As a result the efforts of early pioneers to rear marine fish were hampered by inadequate and unsuitable larval food supplies (Shelbourne 1968).

A breakthrough in hatchery aquaculture was made with the discovery by Seale (1933) in the USA and Rollefsen (1939) in Norway that the 0.4 mm/nauplius of *Artemia* constitutes an excellent food source for the newborn fish larvae.

The advantage of using *Artemia* is that one starts from an apparently inert product, namely, the dry cysts. These cysts which inflict inactive embryos are commercially available, can be stored for years and only have to be incubated for 24 h in seawater to produce free-swimming larvae. Furthermore, brine shrimps are very well accepted as a food source. It is not exactly known if this is due to their biochemical composition, a very thin carapace, the fact that they are a moving prey (swimming) or a combination of all these factors.

J. Royan (✉)
National Institute of Oceanography, Dona Paula, Goa 403004, India
e-mail: royan39@yahoo.com

Artemia has been found to be a suitable food for the most diversified group of organisms of the animal kingdom, e.g. foraminifers, coelenterates, flatworms, polychaetes, squids, insects, chaetognaths, many freshwater and marine crustaceans and fishes. More than 85 % of the marine animals which are being cultivated are offered *Artemia* as food source along with other foods or sometimes as a sole diet.

In most cases, brine shrimp *Artemia* are used as freshly hatched nauplii, but adult *Artemia* is reported to be a better diet than nauplii for many predators. Adult *Artemia* are harvested from many saline biotopes as food for the larvae of lobsters and the freshwater prawn *Macrobrachium rosenbergii*.

Exploitation of Natural *Artemia* Habitats

Natural populations of *Artemia* are found throughout the world in salt lakes, saline lagoons and also man-made solar saltworks. In saltworks, *Artemia* is found only in the evaporation ponds of the saltworks especially in ponds having salinity more than 90 ppt. Similarly, the most popular sites of *Artemia* are the Great Salt Lake, San Francisco Bay and Mono Lake California; USA from where most of the *Artemia* products especially the cysts are supplied to the entire world. Iran, Algeria, Egypt, Kenya and Nigeria are also having natural *Artemia*. Tientsin in China is one of the spots located in recent times. In India we

S. Perumal et al. (eds.), *Advances in Marine and Brackishwater Aquaculture*,
DOI 10.1007/978-81-322-2271-2_3, © Springer India 2015

have natural populations in Tuticorin in Tamil Nadu, Bhayander in Mumbai, Salt Pans in Jamnagar, Great Rann of Kutch in Gujarat and also Sambhar Lake which is an inland lake in Rajasthan.

The *Artemia* cysts produced from the Great Salt Lake (GSL) and the San Francisco Bay (SFB) are much superior to other strains in terms of hatching efficiency (>300,000/g), hatching percentage (>90 %), naupliar size (428 pm) and high HUFA levels especially 18:3w3, 20:5w3 and 22:6w3. Aquaculturists throughout the world started preferring both GSL and SFB cysts, and therefore there is always great demand for cysts from these regions.

World Consumption of *Artemia* Cysts

Year	Quantity (mt)
1974	23
1980	46
1992	900
1993	1,500
2000	5,000
2010	>10,000

It is very evident from the above table that there is a tremendous increase in the consumption of *Artemia* since 1974 with more countries getting involved in aquaculture. The heavy demand for cysts also reflected in the sharp rise in the price per kg of cysts. During the early seventies, one kilogram of *Artemia* cysts was sold for 5 US$ in the international market. At present it is sold for nearly 80 US$. This high cost of the feed made aquaculturists around the world to look for natural *Artemia* resources in their own countries or sometimes to take up artificial production in suitable saline biotopes. As a result nearly 600 spots with natural populations of *Artemia* have been located so far and artificial culture taken up in several countries like Brazil, India, the Philippines, Indonesia, Thailand, Sri Lanka and the West Indies.

Prerequisites for the Production of *Artemia*

1.	Site	Suitable sites for *Artemia* culture are saline lakes, lagoons and man-made solar salt pans
2.	Region	Tropical and subtropical regions with less rainfall
3.	Location	Preferably closer to the sea in established salterns
4.	Salinity	Between 100 and 150 ppt
5.	Temperature	35 °C is the upper tolerance limit and can thrive a temperature as low as 5 °C
6.	pH	Between 7 and 8
7.	Food	*Artemia* is a continuous non-selective filter feeder; enough algae and detrital matter should be available
8.	Pond depth	Minimum 0.5 m; less depth will heat up the water column during midday which will kill all *Artemia*
9.	Predators	*Artemia* has no defence mechanism against predators. All ponds should therefore be free of predators
10.	Strain	Preferably bisexual strain of GSL/SFB/Vietnam
11.	Culture period	Starting immediately after monsoon rains (October/November) and ending before summer (May/June)

Advantages of Doing *Artemia* Culture in Existing Large Salt Pans

- There is no need to procure vast lands, as developmental cost of such lands for separate *Artemia* culture would be very high and uneconomical. On the other hand, existing saltworks have well-laid-out ponds (condenser ponds) with varying densities of seawater. Such ponds have strong bunds with no seepage and a desirable water depth of more than 50 cm and sometimes even up to 80 cm, ideal for growing *Artemia*.
- Most of the saltworks powerful pumps for pumping the seawater.
- The water circulation between the ponds is through gravitational force, and therefore no need to install separate pumps to lift the water.
- All ponds are easily accessible.

- Large saltworks have their own analytical laboratory and also routinely maintain meteorological data.
- The presence of *Artemia* in the salt pans is known to improve the quality and quantity of salt, in addition to forming a valuable by-product for the aquaculture industry.

Status of Technology

The technology of *Artemia* culture was successfully demonstrated in India and abroad. It was in Mundra near Bhuj, Gujarat, that first successful culture was carried out. Large-scale commercial culture was tried with success at Jamnagar in Gujarat. This technology developed at the National Institute of Oceanography, Goa, was also transferred to countries like Sri Lanka, Jamaica and Trinidad and Tobago in the West Indies. The technology involves:

- Site selection
- Fertilization of ponds with either organic or inorganic fertilizers
- Selection of suitable strain
- Incubation of cysts and inoculation of first instar nauplii
- Maintenance of salinity, temperature and algal production
- Maintaining viviparous reproduction of the population
- Gradually increasing the salinity level
- Switching over from viviparous to oviparous reproduction
- Harvesting of cysts
- Processing of cysts (washing, quality separation and drying)
- Packing (vacuum packing under nitrogen)
- Quality control on hatching percentage, hatching efficiency, diameter of cysts and finally fatty acid content

Methodology for Semi-intensive Culture of *Artemia* in Salt Farms

The biomass of *Artemia* and their cysts can be produced under intensive as well as semi-intensive and extensive culture conditions; while the intensive culture is carried out in indoor culture systems using tanks, the semi-intensive and extensive cultures are done in open outdoor ponds especially the evaporation or condenser ponds of large salt farms.

The most important requirement for a semi-intensive and extensive culture operation is the availability of high-density brine which is totally devoid of any type of predator. In a typical salt pond, predators are normally found up to 80 or 90 ppt, and hence *Artemia* are found from 100 ppt onwards.

In natural populations, *Artemia* densities are lower mainly due to low levels of nutrients in the water, except in large saltworks which are situated near estuaries or mangrove regions. The salterns with low productivity are best suited for the production of only adult biomass, whereas cyst production is possible in large salt farms with stable ecological conditions.

In a commercial saltworks, seawater is pumped into reservoirs or during high tide water is allowed to enter through automatic sluice gates. Most of the time, the seawater which is pumped in has 35 ppt salinity. After a week's retention time, when the salinity has reached 40 ppt, the water is allowed to flow into evaporation or condenser pans by gravitational force, occasionally also lifted by pumps. During the regular flow into various condenser pans, salts with low solubility get precipitated as carbonates and gypsum (calcium sulphate). When the salinity reaches 200 ppt, brine is charged into crystallizers where pure sodium chloride gets deposited. *Artemia* is found in intermediate ponds having salinity between 100 and 200 ppt.

Very large solar salt industries are situated in climatic zones with low rainfall and high evaporation rate. They have *Artemia* production throughout the year. In contrast to this, several salt units in tropical and subtropical regions operate only for a few months in a year (3–6 months). They are comparatively smaller in size where *Artemia* biomass and cyst production are easily managed along with the environmental parameters.

The basic principles and working procedures are outlined below for semi-intensive production of *Artemia* in seasonal salt ponds.

General Characteristics of a Salt Farm

Salt farms engaged in the production of commercial grade salt are situated mostly along the Gujarat coast and Tamil Nadu coast. They generally range in size from 500 to 5,000 ha. The individual condenser pans are 1–5 ha in size with a water depth ranging from 40 to 80 cm. Seawater to the reservoir is mostly through tidal creeks. The salt operation is stopped during monsoon months from June to September.

Pond Modification

The condenser pans which are identified for *Artemia* culture should have a minimum water depth of 40–50 cm (minimum), whereas 70–100 cm will be most ideal. This is very essential as salt pans with low water levels will experience lethal water temperature for *Artemia* (35 °C and more). Furthermore, low water levels promote phytobenthos, instead of phytoplankton which are larger to be ingested by *Artemia*. Phytobenthos also start floating, preventing evaporation as well as smooth harvest of cysts and biomass. They also interfere with the free-swimming nature of *Artemia*. Higher water depths can easily be achieved by digging an inner perimeter trench and at the same time increasing the height of the builds using the soil from the trenches.

Preparation of Ponds

Prior to *Artemia* inoculation, it is recommended that the ponds are completely emptied and dried for 10–15 days, followed by raking the upper layer of the soil to enhance mineralization of accumulated organic matter. Fish or any other predator left behind may be killed using tea seed cake or by application of lime. The soil acidity limits the availability of plant nutrients and subsequently results in lower phytoplankton and *Artemia* production. Soil pH lower than 6.5 needs to be limed to neutralize the acidity. In a number of cases, soil acidity may be visually observed, i.e. air-exposed soils turn yellow to brownish red.

The amount of lime which should be applied is very important. Inadequate application of lime will only partially reduce acidity, while any overdose might result in a loss of phosphate through the formation of insoluble calcium compounds.

Water Intake to the Culture Ponds

Most of the salt farms draw seawater either by pumping or through automatic sluice gates which allow seawater to enter the main reservoirs during high tides. From the reservoir, the water flows into different evaporation or condenser pans through gravitational force. This is achieved by maintaining a high water depth in the reservoir. The ponds in which *Artemia* are to be introduced should have a salinity of 100 ppt and totally free from predators. The salinity should be maintained through regular intake of water which should be carefully screened to prevent any accidental entry of predators. By regular intake of water, a desired water depth is also maintained.

Fertilizer Application Schedule

Before inoculating the *Artemia* nauplii into the ponds, it should be made sure that enough particulate food is available in the water for a good growth of *Artemia*. The pond water with a green colouration and a transparency of less than 20 cm indicates high concentrations of organic detrital particles or algae which can be easily consumed by the growing *Artemia*. Under this condition, there is no need for fertilizing the pond water before the inoculation of *Artemia*. On the other hand, water with a transparency level of more than 25 cm and a light colour requires fertilization a week before the inoculation.

It is recommended to fertilize the pond when the salinity is around 100 ppt or less. In high-density water, due to the absence of algal species, fertilizers fail to initiate any phytoplankton bloom.

As regards the fertilizers, either organic fertilizers like dried chicken manure or inorganic fertilizers like urea and ammonium phosphates can be used. By experience, a faster phytoplankton growth with inorganic fertilizers has been noticed, whereas the organic fertilizers take time to induce algal growth. Moreover, the salt operators prefer using inorganic fertilizers as they have a fear that organic fertilizers might contaminate the salt.

Dosage for Inorganic Fertilizers

1. Initially, prior to inoculation, 100 kg/ha monoammonium phosphate (NPK ratio 16:20:0) in combination with 50 kg/ha ammonium nitrate (33:0:0) have to be applied. Thereafter, weekly application of 50 and 25 kg/ha, respectively, is recommended.
2. 50 kg/ha diammonium phosphate (18:46:0) and 50 kg/ha urea (44:0:0) at the beginning and a weekly application of 25 and 20 kg/ha, respectively.

Dosage for Organic Fertilizers

The common organic fertilizers are chicken manure, cow dung and goat dung. These have to be dried and finely powdered.

To begin with 0.5–1.0 mt/ha has to be applied, and thereafter 100 to 200 kg/ha is applied once in 3–5 days.

Method of Application

Fertilizers have to be applied evenly on the surface of the water. If the inorganic fertilizers are in the form of pellets/granules, they have to be thoroughly dissolved in plastic bucket before applying to the pond water. In 24–48 h after fertilization, the pond water colour will change to green due to algal blooms (normally the algae are *Dunaliella* sp. and *Spirulina* sp.).

Inoculation Procedures

Before attempting to inoculate the pond, one must make sure that there are absolutely no predators (fish, copepods, etc.). Under normal circumstances, predators are absent in the salinity level of 100 ppt and above. It is sufficient to inoculate only the ponds with 100 ppt water as higher salinity ponds will automatically get populated with *Artemia* escaping out from the inoculated pond. The inoculation density should be restricted to 15–20 nauplii per litre as it is found to be effective and economical.

The quantity of cysts required for obtaining the required numbers of nauplii (taking into consideration approximately 30 % mortality) has to be calculated from the pond volume and the hatching efficiency of the selected strain of cysts.

The following aspects are to be considered before the incubation of the cysts:

- Cyst incubation should be set up as nearer to the pond as possible.
- The incubation should take place preferably under a shaded room to avoid excessive heat from the sun.
- The hatching container should be funnel shaped and also transparent.
- Seawater should have a salinity of 30–35 ppt, filtered with 100 pm bolting cloth.
- $NaHCO_3$ (sodium bicarbonate) at a rate of 1 g/l is to be added to buffer the seawater.
- Cyst density to be hatched must not exceed 1 g/l.
- There should be a strong aeration from the bottom of the funnel-shaped hatching container to keep all the cysts in suspension.
- Hatching container should be illuminated as light stimulus triggers the metabolism and subsequent embryonic development.

It is mandatory to harvest the baby nauplii while they are in the first instar stage for inoculation as elder stages will not survive the salinity shock from 35 ppt to 100 ppt.

After the hatching is completed, nauplii should be collected on a 125 jam filter, washed thoroughly and reintroduced in filtered seawater before transferring to the pond. If the pond is not within a short distance, aeration should be provided from a battery-operated aquarium aerator while transporting.

The ideal time of the day for inoculation is late evening when the pond water temperature is minimum and will continue to drop down till next day morning. This will facilitate the nauplii getting acclimatized to the change in temperature.

Culture Maintenance

The first day after inoculation, it is very difficult to spot the nauplii as they are no longer bright orange in colour, and moreover they prefer to concentrate in deeper parts of the pond. It is only when the nauplii have grown into preadult, one realizes that the inoculation has been successful.

In the fertilized ponds, so long the conditions are normal, i.e. temperature within tolerance limit (35 °C); pH between 7.5 and 8.5; salinity level remaining around 100 + 15 ppt; and sufficient levels of feed in the form of phytoplankton, the *Artemia* population will grow in size reproducing ovoviviparously. This happens from 7–10 days after inoculation resulting in fast increase of the population (the size of the population is determined by pond conditions such as water depth, temperature and food concentration). The most critical parameter out of these is the food availability which is indeed the controlling factor in production success. In order to maintain a healthy algal population, regular fertilizer application should be carried out, along with periodical intake of water.

Inducing Cyst Production

Cyst production (oviparity) is known to be triggered by environmental stress. In culture ponds, cysts appear when the salinity level goes high. Similarly, salinity shocks have also been found effective in switching the population towards cyst production. In addition to salinity, low oxygen concentrations also induce oviparous reproduction in *Artemia*. This may be accomplished by raising the salinity and/or by increasing the rate of fertilization to induce phytoplankton bloom creating extensive diurnal fluctuation in dissolved oxygen. Oviparity should only be induced when the population density is sufficiently high as no further recruitment will take place.

Pond Management

For a good pond management, it is essential to monitor the physicochemical and biological parameters, on a regular basis, which includes:

(a) Dissolved oxygen – this could be estimated either by an oxygen electrode or by titration by Winkler's method.
(b) Salinity could be read by a refractometer or with hydrometers (Baum scale).
(c) pH values using a pH metre.
(d) Air and water temperature at the surface and at the bottom using a maximum-minimum thermometer.
(e) Depth to be read from a graduated staff.
(f) Maintenance of a healthy phytoplankton population is very important for a good *Artemia* production. This is done by monitoring the rate of water intake, regular fertilizer application and periodical harvesting of biomass.
(g) Water turbidity is another parameter to be checked regularly with the help of a Secchi disc which is one way of determining the changes in phytoplankton densities.

Cyst Harvesting

Cysts which are released in the culture ponds float because of the high density of water and tend to accumulate along the windward side of the ponds.

Cysts should be collected from the water surface using a double-screen hand dip net, while they are still afloat which will ensure a good hatching quality. Cysts which are not collected this way will get washed ashore and mixed with sand and mud, making it difficult for harvesting and processing. Cysts which are washed ashore also become dry very soon and will be carried away by strong wind. Cysts not collected on time are also exposed to high temperature (40 °C) which is a lethal temperature for wet (hydrated) cysts.

Cysts could be prevented from being washed ashore by constructing cyst barrier at the windward side of the pond. Whenever there is heavy wind, strong waves are created forming thick foam, in which cysts are trapped and carried away. To avoid such situation, foam barriers can be installed with wooden poles.

Cyst Processing

As soon as the cysts are harvested, it should be cleaned at the site itself using the saturated brine from the pond, over screens with different mesh size (500 μm and 125 μm) in order to remove feathers, twigs, molluscan shells and sometimes even live or dead *Artemia*. Cysts which are smaller will settle on the 125 pm screen. These cysts could be immediately stored in plastic drums or buckets containing saturated brine (>180 ppt). In this, all cysts are in dehydrated condition and will be floating at the surface. It is advisable to stir the cysts once every day to ensure all cysts are properly dehydrated. This treatment in brine will reduce the water content in the cysts to about 20 % and at the same time will clean the cysts from heavy debris which will sink in the concentrated brine.

Storing cysts in saturated brine is purely temporary and should be processed (further cleaning, drying and packing) for the production of good quality cysts with maximum hatching, purity and long shelf life. This is achieved by adopting the following procedures.

The cysts from high-density brine are given a quick wash with freshwater to remove all the salt and reintroduce in a long conical container having freshwater or seawater. In this, all the full and viable cysts will sink, whereas the empty cysts and foreign bodies will float which are scooped off and discarded. This treatment of cysts in freshwater/seawater should not take longer than 15 min in order to avoid higher hydration levels which might trigger the cyst hatching metabolism.

The settled cysts are collected from the bottom of the conical container onto a 125 pm bolting net. Excess water from the cysts should be removed by squeezing or centrifuging the cysts. The wet cysts should be dried immediately to reduce the water level in the cysts to below 10 %. This is necessary to arrest the metabolic activity in the cysts. Drying of cysts below 40 ° C assures optimal cyst quality in terms of hatching efficiency, hatching rate and the energy content. This is done depending on the equipment available and applying the following techniques.

Drying Cysts in Open Air

The cysts are spread thinly on a drying surface consisting of trays made of PVC wire screen covered with muslin cloth. The trays are kept one above the other leaving a gap of 15 cm between the trays for better air circulation.

The wet cysts are never to be kept directly under the sun for drying, as it would increase the temperature inside the cysts to more than 40 ° C, killing the embryos instantly. Drying is continued till a constant weight is reached (below 10 % with a preference between 2 and 5 %). Once the cysts are dried properly, they resemble a free-flowing powder.

Drying Cysts in Oven

Cysts can be dried in temperature-controlled room or in an oven with temperature setting around 35–38 °C with good air exchange and regular redistribution of cysts on the tray. Hygroscopic materials such as silica gel or calcium chloride may be kept for absorbing moisture.

Fluidized Bed and Rotary Drying

Quick and at the same time a uniform drying is obtained when the cysts are held in continuous motion in the drying air. In a fluidized bed dryer, cysts are dried in a continuous air stream which keeps the cysts suspended in the drying chamber. In a rotary drier, cysts are kept under continuous movement by rotation of the container (5 rpm).

To ensure a proper drying, a continuous airflow through rotating cylinder can he maintained by installing a ventilator.

Storage and Packing of Cysts

Dry *Artemia* cysts are highly hygroscopic. Once the moisture is absorbed, the cyst will incidentally hydrate and will reach a breaking stage. In order to avoid such a situation, the cysts should be stored in airtight containers. For a long period of storage, the cysts should preferably be packed in oxygen-free condition. This could be done by vacuum packing in metal cans or aluminium pouches. Cysts also can be packed in oxygen-free condition by nitrogen flushing the cans prior to vacuum packing.

Production Figures of *Artemia* Cysts

Studies on both natural and inoculated populations of *Artemia* especially in solar salt pans indicate an annual production (i.e. during a minimum of 4 to 6 months salt operation period) of 18–50 kg/ha. According to Boone and Baas Becking (1931), Marina Salina in California gave a production of 18 kg/ha (Royan et al. 1978; Royan 1981, 1985), while studying the natural populations at various salterns obtained production figures ranging from 25 to 38 kg/ha. In a 0.22 ha pond, inoculated with San Francisco Bay strain, the harvest was about 21 kg dry cyst/ha. A moderate figure of 10–20 kg/ha was the estimate of Persoone and Sorgeloos (1980).

Production figures vary between strains and locations. Similarly, it differs between extensive and semi-intensive culture ponds and also fertilized ponds. Therefore, it is advisable to take a conservative figure of 20 kg/ha for working the economics of *Artemia* culture.

References

Boone E, Baas Becking LGM (1931) Salt effects on eggs and nauplii of *Anemia solina* L. J Gen Physiol 14:753–763

Kinne O, Rosenthal FI (1977) Commercial cultivation (aquaculture). In: Kinne O (ed) Marine ecology, vol III, Cultivation, Part 3. Wiley, New York, pp 1296–1521

Persoone G, Sorgeloos P (1980) General aspects of the ecology and biogeography of *Anemia*. In: Persoone G, Sorgeloos P, Roels O, Jaspers E (eds) The brine shrimp, *Artemia*, vol 3, Ecology, culturing, use in aquaculture. Universa Press, Wetteren, pp 3–24, 456 p

Rollefsen G (1939) Artificial rearing of fry of seawater fish (Preliminary communication). Rapp Proc Verb Reun Cons Perm Explor Mer 109: 133pp

Royan JP (1981) Brine shrimp culture. In: Status report on coastal aquaculture in India. National Institute of Oceanography, Goa

Royan JP (1985) Estimated cyst production at Hambantota salterns in Sri Lanka. NARAIOCC/85/1, 19–28

Royan JP, Wafar MVM, Vijayaraghavan S (1978) The brine shrimp, Artemia salina and its culture potential in India. Indian J Mar Sci 7(2):116–119

Seale A (1933) Brine shrimp *(Anemia)* as a satisfactory live food for fishes. Trans Am Fish Soc 63:129–130

Shelbourne JE (1968) The culture of marine fish larvae with special reference to the plaice *(Pleuroneetes-platessa)* and the sole *(Solea solea)*. Thesis, London University, 143 p

Intensive Indoor and Outdoor Pilot-Scale Culture of Marine Copepods

Santhanam Perumal, S. Ananth, R. Nandakumar,
T. Jayalakshmi, M. Kaviyarasan, and Perumal Pachiappan

Introduction

Copepods are more abundant than any other group of multicellular animals, including the hyper-abundant insects and nematodes. They consume phytoplankton and microorganisms, and they are inturn preyed upon by higher trophic level, animals including fish and whales. In particular, they serve as primary prey for the larval stages of many fish species of economic importance. In aquaculture, copepods have been proven to be the much preferred and most adequate food for many marine fish larvae (Houde 1973; May et al. 1974; Kraul 1983, 1989, 1993) and are also used for the shrimp larvae (Shamsudin and Saad 1993). Good fish productivity of an aquatic ecosystem is related to the presence of copepods and their role as the main food component (May 1970; Bent 1993). The larvae of many marine fish require prey with size of about 50–100 μm wide at their first feeding stage (Detwyler and Houde 1970; Yufera and Pascual 1984). Even the rotifer of type "S" is too large in many cases (Houde 1973; May et al. 1974; Doi and Singhagraiwan 1993). The results concerning first feeding of commercially important fish on dry food organisms are encouraging (Fernandez-Dıaz and Yufera 1997; Cahu and Zambonino Infante 2001). However, live feeds cannot always be substituted because of biochemical and behavioural constraints of the fish larvae (Drillet et al. 2006).

S. Perumal (✉) • S. Ananth • R. Nandakumar
T. Jayalakshmi • M. Kaviyarasan
Department of Marine Science, Bharathidasan University, Tiruchirappalli 620024, Tamil Nadu, India
e-mail: santhanamcopepod@gmail.com

P. Pachiappan
Department of Biotechnology, Periyar University, Salem 636011, Tamil Nadu, India

Global Status

Since the 1960s, culturing copepods have become increasingly more reliable, and approximately 60 copepod species have been successfully raised (Mauchline et al. 1998). The oldest copepod culture is from the Danish Technical University of Denmark (*Acartia tonsa*) (Stottrup et al. 1986). The World Copepod Culture Database was initiated in 2006 at Roskilde University (http://copepod.ruc.dk/main.htm) in an attempt to supply and share information among copepod scientists, aquaculturists and the public at large. The database contains details on various cultures and up-to-date recent knowledge on cultivation procedures. To date, approximately 30 copepod cultures have been referenced in the database (Tables 1, 2).

Copepods have also been satisfactorily used after a period of using rotifers and before introducing *Artemia* nauplii (Kraul 1993; Alvarez-Lajonchere et al. 1996) and are also offered simultaneously (Leu and Chou 1996).

S. Perumal et al. (eds.), *Advances in Marine and Brackishwater Aquaculture*,
DOI 10.1007/978-81-322-2271-2_4, © Springer India 2015

Table 1 Available reports on copepod culture in foreign countries

Genus/species	Identification type: morphological or genetical	Geographical origin	Cultivation conditions, Temperature, salinity, light regime, food
Acartia grani	Morphological	Barcelona harbour, Spain (NW Mediterranean)	19 °C/38 ppt/12 L:12D/*Rhodomonas salina*
Acartia sinjiensis	Morphological	Townsville channel, QLD (Australia)	27–30 °C/30–35 ppt/ 18 L:6D/ *Tetraselmis chuii* and T-iso
Acartia southwelli	Morphological	Pingtung - Taiwan	25–30 °C/15–20‰/12 L:12D/*Isochrysis galbana*
Acartia tonsa	Morphological	Origin Unknown	?
Acartia tonsa	Morphological and genetics	Oresund (Denmark)	17 °C/30 ppt/0 L:24D/*Rhodomonas salina*
Acartia tonsa	Morphological	Punta del Este, Uruguay	25–30 °C/17 ppt/indirect natural light/*T-ISO-Tetraselmis*
Ameira parvula	Morphological	Kiel Bight, Germany	18 °C, 17 ppt, 12 L:12D, different algae
Amonardia normani	Morphological	Kiel Bight, Germany	18 °C, 17 ppt, 12 L:12D, different algae
Amphiascoides atopus	Morphological	USA	25 °C, 12/12, cultured phytoplankton
Apocyclops royi	Morphological	Pingtung - Taiwan	25–30 °C/15–20‰/12 L:12D/*Isochrysis galbana*
Centropages typicus	Morphological	Gulf of Napoli Italy (W Mediterranea)	19–21 °C/38 ppt/12 L:12D/*Prorocentrum minimum*/*Isochrysis galbana*/*Tetraselmis suecica*
Eurytemora affinis	Morphological	River Seine Estuary (France)	10–15 °C/15 ppt/12 L:12D/fed *Rhodomonas marina*
Eurytemora affinis	Morphological	Gironde Estuary (France)	10–15 °C/15 ppt 12 L:12D/fed *Rhodomonas marina*
Eurytemora affinis	Morphological	Loire Estuary (France)	10–15 °C/15 ppt 12 L:12D/fed *Rhodomonas marina*
Eurytemora affinis	Morphological	Canada	10–15 °C/15 ppt/12 L:12D/*Rhodomonas marina*
Euterpina acutifrons	Morphological	Mediterranea	19 °C/38 ppt/12 L:12D/*Rhodomonas salina*
Eurytemora affinis	Morphological	Gironde Estuary (France)	10–15 °C/15 ppt 12 L:12D/fed *Rhodomonas marina*
Euterpina acutifrons	Morphological	Mediterranea	19 °C/38 ppt/12 L:12D/*Rhodomonas salina*
Eurytemora affinis	Morphological	Gironde Estuary (France)	10–15 °C/15 ppt 12 L:12D/fed *Rhodomonas marina*
Eurytemora affinis	Morphological	Canada	10–15 °C/15 ppt/12 L:12D/*Rhodomonas marina*
Euterpina acutifrons	Morphological	Mediterranea	19 °C/38 ppt/12 L:12D/*Rhodomonas salina*
Gladioferens imparipes	Morphological	Swan River, Perth Western Australia	23–27 °C, 18 ppt, continuous dark, T-Iso and *Chaetoceros muelleri*
Mesocyclops longisetus	Morphological	Florida/USA	http://edis.ifas.ufl.edu/IN490
Microcyclops albidus	Morphological	Florida/USA	http://edis.ifas.ufl.edu/IN490
Oithona davisae	Morphological	Barcelona harbour, Spain (NW Mediterranean)	20 °C/30 ppt/natural light/*Oxhyris*

(continued)

Table 1 (continued)

Genus/species	Identification type: morphological or genetical	Geographical origin	Cultivation conditions, Temperature, salinity, light regime, food
Pseudodiaptomus annandalei	Morphological	Pingtung - Taiwan	25–30 °C/15–20‰/12 L:12D/*Isochrysis galbana*
Tachidius discipes	Morphological	Kiel Bight, Germany	18 °C, 17 ppt, 12 L:12D, different algae
Temora longicornis	Morphological	North Sea	15 °C/30 ppt/0 L:24D/*Thalassiosiera weissflogii, Rhodomonas salina, Heterocapsa, Prorocentrum minimum*
Temora longicornis	Morphological	Plymouth, Devon, UK	Temp according to current sea temperatures, salinity 30–36 ppt, 12 L:12D, fed mixture of *Isochrysis galbana, Rhodomonas* and *Oxyrrhis*
Temora stylifera	Morphological	Gulf of Napoli Italy (W Mediterranea)	19–21 °C/38 ppt/12 L:12D/*Prorocentrum minimum/Isochrysis galbana/Rhodomonas baltica*

Table 2 Available reports on copepod culture in India

Culture species	Culture size	Productivity	Food/conditions	Reference
Oithona rigida	25 L	33,867 ind/L	*Chlorella vulgaris, Coscinodiscus centralis, Chaetoceros affinis and Skeletonema costatum*	Santhanam (2002)
Acartia centrura	25 L	41,603 org/l.	*C. marina, Nannochloropsis salina* and *Isochrysis galbana*	Vengadeshperumal et al. (2010)
Acartia southwelli	25 L	55,103 org/l.	*Chlorella marina, Nannochloropsis salina* and *Isochrysis galbana*	Vengadeshperumal et al. (2010)
Tigriopus japonicus	210 m^3	10–22/ml	Chlorella minutissima, x-yeast, baker's yeast co-culture with rotifers;	Fukusho (1980)
Acartia erythreae	25 L	–	Microalgae	Rajkumar (2006)
Macrosetella gracilis	100 L	2.86 ind/ml	Mixed microalgae	Ananth and Santhanam (2011)

Copepods offer a great variety of sizes, species and qualities (Kinne 1977; Yufera and Pascual 1984; Delbare et al. 1996) and have high levels of protein, highly unsaturated fatty acids (HUFA), carotenoids and other essential compounds. Kraul et al. (1992) and Watanabe et al. (1983) reported that culture media did not influence the copepod chemical composition, although Delbare et al. (1996) reported that copepod n-3 HUFA reflected the culture diet. A positive correlation of the n-3 HUFA levels and highest nauplii production was reported by Norsker and Stottrup (1994). Other good characteristics of copepods are their swimming movements as a larval visual stimulus; their tank-cleaning performance primarily by benthic harpacticoids, which are grazers (Stottrup et al. 1995); their high digestive enzyme content (Delbare et al. 1996); and a possible enhancement of their feeding rates with improved growth and survival (Støttrup and Norsker 1995; 1997).

Indian Scenario

In India, work on copepod culture is very scanty. Merrylal James and Martin Thompson (1986) cultured the copepods belonging to the genera *Cyclops*, *Oithona* and *Pseudodiaptomus* and introduced them in mariculture. Santhanam

(2002) has studied the culture and utilization of *Oithona rigida* in seabass *Lates calcarifer* and tiger shrimp *Penaeus monodon* larval rearing. Rajkumar et al. (2004) cultured the copepod *Acartia clausi* in laboratory condition. Recently, marine copepods, viz., *Nitocra affinis* and *Oithona rigida*, have been successfully cultured with the density of 1,30,000 ind./l and 1,15,000 ind./l, respectively, fed with mixed marine microalgae such as *Chlorella marina*, *Isochrysis* sp., *Dunaliella* sp. and *Nannochloropsis* sp. The initial stocking density of both species ranged between 100 and 200 individuals, and the culture duration lasted for 45 days. Used plastic cans bought from shops served as culture vessels. This is the first successful achievement on *Nitocra affinis* and *Oithona rigida* for high-density culture in low volume of seawater at lower cost. The culture technology has been developed for marine copepods *Nitocra affinis* and *Oithona rigida* in the Marine Planktonology and Aquaculture Laboratory of the Department of Marine Science, Bharathidasan University, Tiruchirappalli, by Dr. P. Santhanan, Assistant Professor, and his team for the first time in India.

Need for Marine Copepods Culture

Aquaculture is considered to be one of the lucrative industries due to the high market price (of shrimp) and the unlimited demand for it in the international market. By virtue of its geographical location in the Indian Ocean, India possesses a rich fishing ground in the sea and offers immense potential for aquafarming (Varghese 1995). For a sustainable growth of the aquaculture industry, regular supply of adequate quantities of quality fish seeds is one of the prerequisites. Quality seeds are those that ensure high growth rate and low mortality and that can withstand stress during culture (Santhanakrishnan and Visvakumar 1995). To produce good quality of adult fish, an effective larval rearing is necessary to produce quality fish fry. A reduction in the use of chemicals and drugs or addition of hormones would increase the natural immunity of fish larvae (Sorgeloos and Leger 1992), though providing sufficient nutrients to the larvae and preventing bacterial infections are still the most important requirement.

Therefore, the importance of live feeds in fish and shrimp culture has been well documented. The use of live feeds for larval fish is well established with brine shrimp, *Artemia* and rotifer, *Brachionus plicatilis* being the most common among them. While brine shrimps are very amenable to commercial culture (Loya-Javellana 1989), difficulties have been reported on the use of rotifers because of their small size, their nutritional variability and their culture susceptibility to crashing (Kovalenko et al. 2002). Although the *Artemia* nauplii have been widely used as live food, by no means it is the optimal live food organism in terms of nutritional requirement of fish and shrimp larvae. The main disadvantages of *Artemia* are marked variation in cost, physical properties and nutritional quality among different sources. Hence, the production of very small, rapidly developing and highly vulnerable larvae remains a bottleneck in the commercially successful culture of many marine fish species (Shields et al. 1999). The bio-enrichment of *Artemia* has been widely adopted to overcome the problem of inferior nutrition supply. But there are still other nutrient deficiencies in the enriched *Artemia* nauplii, such as free amino acids availability (Helland et al. 2003), and the biological composition of *Artemia* is not stable after enrichment (Olsen et al. 2000; Olsen et al. 1999). In addition, common lipid enrichment actually reduces the relative protein content and alters the amino acid profile of *Artemia* nauplii (Helland et al. 2003). Losses of nutrients may take place if the live food is not fed to the fish larvae immediately after enrichment (Olsen et al. 2000). Moreover, enrichment of *Artemia* with commercial emulsifier is increasing the cost of production of fish larvae.

Nutritional compounds such as n-3 fatty acids, essential amino acids (EAA) and protein content of live feeds are critical factors for the survival and optimal growth of larval finfish and crustaceans. Hence, the need for the production of quality copepod gains importance. Copepods are very nutritious larval feeds, containing more EPA and DHA. Copepods also have the highest DHA to EPA ratio (Nanton and Castell 1999; Toledo et al. 1999). On the contrary, rotifers and *Artemia* are poor in polyunsaturated fatty acids (PUFAs) and need to be enriched before feeding, which in turn has its own shortcomings, as mentioned before. Among the very many benefits that application of copepod food could have in aquaculture industries are improved larval survival (Shields et al. 1999), higher growth rates (Stottrup and Norsker 1997), better pigmentation (Spenelli 1979; Støttrup et al. 1998), improved gut development (Luizi et al. 1999) and a source of exogenous enzymes (Munilla-Moran et al. 1990).

The marine copepods are considered to be "nutritionally superior live feeds" for commercially important cultivable species, as they are a valuable source of protein, lipid (especially HUFA, 20:5 n-3 and 22:6 n-3) and enzymes (amylase, protease, exonuclease and esterase), which are essential for larval survival, growth, digestion and metamorphosis (Stottrup 2000; Kleppel 1993). Copepods are known to have greater digestibility (Schipp et al. 1999) and a relatively high weight-specific caloric content (Sun and Fleeger 1995). In addition, the growth stages of copepods from first nauplius to adult form a broad spectrum of prey sizes. This makes them suitable prey for a similarly broad range of developing fish sizes (Schipp et al. 1999). It is well known that the red pigment astaxanthin is one of the strongest antioxidants in nature (Edge et al. 1997) and is abundant in crustaceans (Matsuno 1989; Matsuno et al. 1990). It is well established that the pigmentation of copepods can improve the survival and growth of larvae. Furthermore, carotenoids are important antioxidants and often play other biological functions, such as regulatory effects on intra- and intercellular signalling and gene expression (Sies and Stahl 2005).

HUFAs are essential for marine fish larvae (Watanabe et al. 1983; Witt et al. 1984; Sorgeloos et al. 1988; Watanabe and Kiron 1994; Kanazawa 1995). Docosahexaenoic acid (DHA; 22:6 (n-3)) has a significant influence on larval stress resistance (Kraul et al. 1991, 1993). DHA content is higher in copepods than in even enriched *Artemia* nauplii which give better results in terms of survival, growth and stress resistance in fish larvae (Fujita 1979). Superior larval stress resistance can be achieved with copepods, even when DHA content is less than in enriched *Artemia* nauplii (Kraul et al. 1992). Although *Artemia* enrichment has greatly helped in improving commercial aquaculture, the most advanced rotifer and *Artemia* bioencapsulation techniques have not matched the good results of copepods, and their composition was subsequently taken as the standard for improving enrichment techniques for rotifers and *Artemia* (Kraul et al. 1988, 1992; Sorgeloos and Leger 1992). Most of the large-scale copepod culture systems are based on outdoor semi-controlled polyculture techniques, although several attempts have been made to culture some species in intensive systems (Stottrup et al. 1986; Stottrup and Norsker 1997).

Free-living copepods have been intensively studied because of their impact as key players in the marine pelagic environment. In terms of biomass, copepods can represent up to 80 % of the mesozooplankton (Mauchline et al. 1998). They are an important food source for planktivorous fish and fish larvae in general (e.g. Fox et al. 1999; Mollmann et al. 2004). The life cycles and physiology of copepods have been intensively studied in relation to various environmental conditions on the pelagic ecosystem. Experimental studies such as feeding behaviour, fecundity, developmental biology, nutritional composition of copepods with reference to environmental parameters and feeds have benefited from the small-scale cultivation of copepods. Also, when compared to field-sampled specimens, the advantage is that the history and consistent condition of cultivated copepods are well known and thus can even be manipulated. The data generated during various culture experiments could be applied in a variety of research areas such as copepod genetics, feeding behaviour, population dynamics, parameterization of standing stocks, production rates for ecosystem models, etc.

Microalgal Culture

The amount and quality of algal feeds are essential parameters that enhance the production of copepod. Hence, a series of preliminary trials were carried out to determine a suitable diet for copepods. Our results confirmed that copepods benefit from mixed algal cultures with *Isochrysis* being the important constituent. Unialgal cultures of *Isochrysis* sp., *Chlorella* sp. and *Tetraselmis* sp. were maintained as live feeds for copepods. The algae were maintained at an optimum temperature of 24–28 °C, salinity of 28–30 ppt and with 12:12 h light and dark conditions.

Indoor algal stock culture was maintained in special air-conditioning room using Conway's medium. About 10 ml of the inoculum in the growing phase was transferred to the culture flask provided with 12:12 h. light and dark cycle. Outdoor mass culture was maintained in large volume of FRP tanks using commercial fertilizers, viz., ammonium sulphate/urea/super phosphate, in the ratio of 10:1:1 g.

Water Quality

Copepods were reared in clean, filtered seawater diluted to approximately 24 ppt filtered through filter bags (5 μm). This salinity was maintained at 26–28 ppt during most part of the study. Aeration was achieved via air pumps with air stones inserted from the top of the vessels. Ammonia, nitrite, pH and dissolved oxygen were measured at periodic intervals in all cultures.

Copepods Source

Copepods for culture were collected from Muthupet and Muttukadu estuaries (Southeast coast of India). Copepods were collected using plankton net (158 μm) at dawn. Contamination from other zooplankters was reduced by thorough rinsing and filtering with meshes of various sizes upon return to the laboratory. The majority of copepods collected were found to be calanoids. *Pseudodiaptomus* sp. being positively phototactic was easily separated and isolated in the presence of lamp using a hand net. *Oithona rigida* was the second best in terms of density from our samples followed by *Nitocra affinis*.

Isolation of Copepods

After collection, the zooplankton was screened to isolate the size fraction containing predominantly adult and late-stage copepodids. This was achieved by first removing fish and prawn larvae. Grading was accomplished by using a set of superimposed sieves with varying mesh sizes and with decreasing mesh size from upstream to downstream. Copepod samples were screened coarsely through a 500 μm mesh to remove fish and prawn larvae. Then the samples were screened through 190 μm mesh to remove rotifers, nauplii of copepods and barnacles. After grading, copepods were identified using standard manuals, monographs and text books (Kasturirangan 1963; Perumal et al. 2000; Santhanam and Perumal 2008).

Indoor Stock Culture of Marine Copepods

After the species confirmation, known numbers of gravid females were isolated using capillary tubes, fine brush and needle and were stocked initially in low volume of glass beakers and conical flasks provided with microalgae without aeration. Later, the copepods were subcultured into average volume of plastic containers filled with filtered seawater and were provided with vigorous aeration. The optimum water quality conditions like temperature, salinity, pH and dissolved oxygen were maintained. Copepods were fed with a daily ration of microalgae diet in the constant concentration.

Copepods were generally cultured in seawater of 26–28 ‰ salinity. The stock was maintained mostly at a temperature of 29 °C. Copepod stock cultures were maintained in cylindrical, flat-bottomed, polyethylene 5 or 7 l plastic cans at air-conditioning room. Initially the stock was maintained with 50–80 adult copepods. The microalgae cultures grown non-axenically in 5-l flasks were provided as food.

Outdoor Pilot-Scale Culture of Marine Copepods

Pilot-scale copepods culture was started with clean FRP tanks, algae and a filtered, UV-treated seawater. Tanks were stocked with known numbers of gravid females. Gravid females had released the nauplii within 36–42 h. The detritus in the tanks were removed daily using graded sieves connected with siphon hose, and then the adults and nauplii were introduced back to the tank. The sequential batch cultures were started at 5–7-day intervals for continuous copepods production.

Air Quality Control

The cool, filtered, interior air was used for aeration. The disposable in-line 0.2 μm pore antibacterial air filters were used for providing aeration. Aeration was used to maintain the algae culture in a condition of CO_2 saturation, pH stability and uniform mixing.

Water Quality Control

The seawater conditions like salinity, pH, DO, colour and scent (particularly the "rotten egg" smell of H_2S) were checked prior to collection and treatment. The seawater was serially filtered through 50, 10 and 1 μm mesh bags and then passed through an ultraviolet sterilizer. Filter bags were cleaned and then sanitized overnight in hypochlorite solution once a week under normal usage. Filtered, UV-treated seawater was used directly to culture copepods. The filtered seawater was treated with 10 % commercial hypochlorite solution at 0.2 ml/l and stand for overnight without aeration. After that, seawater was dechlorinated with thiosulfate solution volume for volume (V/V) at 0.2 ml/l. The dechlorinated water was used for filling all wash bottles, stacked-sieve holders, harvest samples, population counts, etc.

System and Equipments Preparation

The fibreglass tank was washed with a low-residue laboratory detergent (e.g. Alconox or Sparkleen) and water followed by thorough rinsing. Then the tank was treated with 100 % muriatic acid (HCl) solution in the outdoors followed by thorough rinsing with filtered seawater. Tanks were leached three times (24 h each time) to remove all water-soluble remnants of the manufacturing process. Tanks were filled to the rim with filtered, UV-treated seawater and adjusted the salinity as needed, and water was chlorinated with 60 ml (0.2 ml/l) commercial 10 % hypochlorite solution per litre. The treated system was allowed to stand for 24 h. Thereafter, the system was dechlorinated with 60 ml of stock thiosulfate solution. The vigorous aeration was started. After an hour, "free chlorine" test strip was dipped and zero "free chlorine" remaining was confirmed. The treated seawater was aerated to at least 6 mg/l DO.

Harvest Stipulations

The stacked-sieve holder and wash bottles were washed with treated seawater at culture tank temperature. The siphon hose were connected to the siphon head and the stacked sieves. The copepods were harvested and filtered onto a wet freestanding sieve. As the tank water level drops, frequent rinsing was done to rinse down copepods stuck to the sidewalls. 30 % of the tank volume was exchanged weekly with new treated seawater.

Conclusion

The fact that copepods are highly valuable source of live food for fish larval rearing is well established. However, despite significant progress in copepod cultivation, their use is still sporadic. This can be attributed to rare use of copepods in commercial settings. Thus, conscious efforts need to be made to upscale the copepod culture to commercial levels in order to ensure production of a reliable, continuous supply of copepods on a large scale. Different species of copepods provide us with different characteristic, i.e. differences in shape, size and movement.

Annual sales of *Artemia* cysts marketed for crustacean and larval fish food have increased to more than 2,000 t (Sorgeloos et al. 1994). This figure may be underestimated considering the report of Stael et al. (1995) that 600 t was used in Thailand just for shrimp culture in 1993. These high consumption levels can lead to a situation where the demand exceeds the supplies (Stael et al. 1995), with consequent costs escalation. It is expected that this situation will not improve much during the next few seasons due to rather unfavourable conditions for *Artemia* production in the lake as well as due to increasing hatchery demand (Lavens and Sorgeloos 1998). This is known as the *Artemia* crisis (Sorgeloos 2000). Hence, the use of copepods at a commercial level needs to be addressed urgently.

Acknowledgment The authors are thankful to the Department of Biotechnology (DBT), Govt. of India, New Delhi, India, for financial support to establish marine copepods culture facility and authorities of Bharathidasan University, Tiruchirappalli, India, for the facilities provided.

References

Alvarez-Lajonchere L, Perez Sanchez L, Hernandez Molejon OG, Torres Gomez E (1996) Mass production of striped patao *Eugerres brasilianus* juveniles in Cuba. J World Aquac Soc 27:347–352

Ananth S, Santhanam P (2011) Laboratory culture and biochemical profile of marine copepod, *Macrosetella gracilis* (Dana). Aquaculture 12:49–55

Bent U (1993) Methods for production of turbot fry based on copepods as food organisms. In: Carrillo M, Dahle L, Morales J, Sorgeloos P, Svennevig N, Wyban J (eds) From discovery to commercialization, osteden. Eur. Aquac. Soc. Spec. Publ., vol. 19. European Aquaculture Society, Oostende, p 609

Cahu C, Zambonino Infante JZ (2001) Substitution of live food by formulated diets in marine fish larvae. Aquaculture 200:161–80

Delbare D, Dhert P, Lavens P (1996) Zooplankton. In: Lavens P, Sorgeloos P (eds) Manual on the production and use of live food for aquaculture, vol 361, FAO Fish. Tech. Pap. Rome, FAO, pp 252–282

Detwyler P, Houde ED (1970) Food selection by laboratory-reared larvae of the scaled sardine *Harengula pensacolae* (Pisces, Clupeidae) and the bay anchovy *Anchoa mitchilli* (Pisces, Engraulidae). Mar Biol 7:214–222

Doi M, Singhagraiwan T (1993) Biology and culture of the red snapper, *Lutjanus argentimaculatus*. The Research Project of Fishery Resource Development in the Kingdom of Thailand, Rayong, Thailand, 51 pp

Drillet G, Jorgensen NOG, Sorensen TF, Ramlov H, Hansen BW (2006) Biochemical and technical observations supporting the use of copepods as live feed organisms in marine larviculture. Aquac Res 37:756–772

Edge R, McGarvey DJ, Truscott TG (1997) The carotenoids as antioxidants: a review. J Photochem Photobiol B 41:189–200

Fernandez-Dıaz C, Yufera M (1997) Detecting growth in gilthead seabream, *Sparus aurata* L., larvae fed microcapsules. Aquaculture 153:93–102

Fox CJ, Harrop R, Wimpenny A (1999) Feeding ecology of herring (*Clupea harengus*) Baltic Sea herring and sprat. J Fish Biol 65:1563–1581

Fujita S (1979) Culture of red sea bream, *Pagrus major* and its food. Spec. Publ.-EMS 14: 183–197

Fukusho K (1980) Mass production of a copepod, *Tigriopus japonicus* in combination culture with a rotifer *Brachionus plicatilis*, fed omega-yeast as a food source. Bull Jpn Soc Sci Fish 46(5):625–629

Helland S, Terjesen BF, Berg L (2003) Free amino acid and protein content in the planktonic copepod *Temora longicornis* compared to *Artemia franciscana*. Aquaculture 215:213–218

Houde ED (1973) Some recent advances and unsolved problems in the culture of marine fish larvae. Proc World Maric Soc 3:83–112

Kanazawa A (1995) Effects of docosahexaenoic acid and phospholipids on stress tolerance of fish larvae. In: Lavens P, Jaspers E, Roelants I (eds) Larvi'95 fish and shellfish larviculture symposium. Spec. Publ., Eur Aquac Soc, vol 24. European Aquaculture Society, Ghent, p 105

Kasturirangan LR (1963) A key for the identification of the more common planktonic copepoda of Indian coastal waters. Council of Scientific and Industrial Research, New Delhi (Publ Indian National Committee on Oceanic Research, No. 2), 87 pp

Kinne O (1977) Marine ecology, vol III, Cultivation: Part 2. Wiley, New York, 1923 pp

Kleppel GS (1993) On the diets of calanoid copepods. Mar Ecol Prog Ser 99:183–195

Kovalenko E, D'Abramo L, Ohs C, Buddington R (2002) A successful microbound diet for the larval culture of freshwater prawn *Macrobrachium rosenbergii*. Aquaculture 210:385–395

Kraul S (1983) Results and hypotheses for the propagation of the grey mullet, *Mugil cephalus* L. Aquaculture 30:273–284

Kraul S (1989) Production of live prey for marine fish larvae. Actes Colloq IFREMER 9:595–607

Kraul S (1993) Larviculture of the Mahimahi *Coryphaena hippurus* in Hawaii, USA. J World Aquac Soc 24:410–421

Kraul S, Nelson A, Brittain K, Wenzel D (1988) Feeding and handling requirements for hatchery culture of the mahimahi *Coryphaena hippurus*. Q Univ Hawaii Sea Grant Prog 10:1–6

Kraul S, Ako H, Brittain K, Ogasawara A, Cantrell R, Nagao T (1991) Comparison of copepods and enriched Artemia as feeds for larval mahimahi, *Coryphaena hippurus*. In: Lavens P, Sorgeloos P, Jaspers E, Ollevier F (eds) Larvi'91 fish and crustacean larviculture symposium. Eur Aquac Soc Spec Publ, vol. 15. European Aquaculture Society, Ghent, pp 45–47

Kraul S, Ako H, Nelson A, Brittain K, Ogasawara A (1992) Evaluation of live feeds for larval and postlarval mahimahi, *Coryphaena hippurus*. J World Aquac Soc 23:299–306

Kraul S, Brittain K, Cantrell R, Nagao T, Ako H, Ogasawara A, Kitagawa H (1993) Nutritional factors affecting stress resistance in the larval mahimahi *Coryphaena hippurus*. J World Aquac Soc 24:186–193

Lavens P, Sorgeloos P (1998) Present status and prospects of the use of *Artemia* cysts and biomass in shrimp farming. Anais do Aquicultura Brasil'98, vol. 1, Recife, 2 a 6 de Novembro de 1998. Latin American Chapter of the World Aquaculture Society, Pernambuco, Brazil, pp 147–162

Leu MM, Chou YH (1996) Induced spawning and larval rearing of captive yellowfin porgy, *Acanthopagrus latus* (Houttuyn). Aquaculture 143:155–166

Loya-Javellana GN (1989) Ingestion saturation and growth responses of *Penaeus monodon* larvae to food density. Aquaculture 81:329–336

Luizi FS, Gara B, Shields RJ, Bromage NR (1999) Further description of the development of the digestive organs in Atlantic halibut (*Hippoglossus hippoglossus*) larvae, with notes on differential absorption of copepod and *Artemia* prey. Aquaculture 176:101–116

Matsuno T (1989) Animal carotenoid. In: Krinsky NI, Mathews-Roth MM, Taylor F (eds) Carotenoids chemistry and biology. Plenum Press, New York, pp 59–74

Matsuno T, Watanabe T, Maoka T, Takemura Y (1990) Carotenoids of crustacea-VII. Carotenoids in the sea louse *Ligia exotica* (Crustacea: Isopoda). Comp Biochem Physiol 95:759–761

Mauchline J, Blaxter JHS, Southward AJ, Tyler PA (1998) The biology of calanoid copepods-introduction. Elsevier Academic Press, California, 710 pp

May RC (1970) Feeding larval marine fishes in the laboratory: a review. Calif Mar Res Commun Cal COFI Rep 14:76–83

May RC, Popper D, McVey JP (1974) Rearing and larval development of *Siganus canaliculatus* (Park) (Pisces: Siganidae). Micronesica 10:285–298

Merrylal James C, Martin Thompson PK (1986) Production of copepods in outdoor culture tank. In: Symposium on coastal aquaculture, 12–18, January 1980, Cochin, India, pp 1275–1280

Mollmann C, Kornilovs G, Fetter M, Koster FW (2004) Feeding ecology of central Baltic Sea herring and sprat. J Fish Biol 65:1563–1581

Munilla-Moran R, Stark JR, Barbour A (1990) The role of exogenous enzymes on the digestion of the cultured turbot larvae, *Scophthalmus maximus* L. Aquaculture 88:337–350

Nanton DA, Castell JD (1999) The effects of temperature and dietary fatty acids on the fatty acid composition of harpacticoid copepods, for use as a live food for marine fish larvae. Aquaculture 175:167–181

Norsker NH, Stottrup JG (1994) The importance of dietary HUFAs for fecundity and HUFA content in the harpacticoid, *Tisbe holothuriae* Humes. Aquaculture 125:155–166

Olsen Y, Evjemo JO, Olsen A (1999) Status of the cultivation technology for production of Atlantic Ž.halibut *Hippoglossus hippoglossus* juveniles in Norway Europe. Flatfish culture. Flatfish Symposium, January 1998, Las Vegas, Nevada, USA. Aquaculture 176:1–13

Olsen EM, Jorstad T, Kaartvedt S (2000) The feeding strategies of two large marine copepods. J Plankton Res 22:1513–1528

Perumal P, Ashok Prabu V, Nedumaran T, Santhanam P (2000) Studies on behaviour and survival rate of *Oithona rigida* Giesbrecht (Copepoda: Cyclopoida) fed with *Coscinodiscus centralis* Ehrenberg and *Skeletonema costatum* Cleve. Seaweed Res Util 22 (1&2):135–137

Rajkumar M, Rajasekar M, Santhanam P, Perumal P (2004) Length-weight relationship in *Siganus javus* L. (Pisces: Siganidae) from Parangipettai coast, southeast coast of India. J Aquat Biol 19:125–126

Rajkumar M (2006) Studies on ecology, experimental biology and live feed suitability of Copepods, *Acartia erythraea* Giesbrecht and *Oithona brevicornis* Giesbrecht from Coleroon Estuary (India). Ph.D. thesis, Annamalai University, India, p 320

Santhanakrishnan G, Visvakumar M (1995) Present status of marine shrimp seed production of future prospectus to meet the demand with suitable Technological development, 43–48. In: INDAQUA'95, MPEDA publication, Cochin, India 134 p

Santhanam P (2002) Studies on the ecology, experimental biology and live food suitability of copepod, *Oithona rigida* Giesbrecht from Parangipettai coastal Environments (India), PhD thesis, Annamalai University, pp 163

Santhanam P, Perumal P (2008) Marine plankton in Indian waters. In: Milton J (ed) Training manual on GIS and marine biodiversity. Lyola College Pub, India, 1–12 pp

Schipp GR, Bosmans JM, Marshall AJ (1999) A method for hatchery culture of tropical calanoid copepods *Acartia* spp. Aquaculture 174:81–88

Shamsudin LS, Saad CR (1993) Live food organisms used in Malaysia for mass propagation of marine shrimp larvae, *Penaeus monodon*. In: Carrillo M, Dahle L, Morales J, Sorgeloos P, Svennecvcig N, Wyban J (eds) From discovery to commercialization, Oosteden. Eur. Aquac. Soc. Spec. Publ, vol 19. European Aquaculture Society, Oostende, p 170

Shields RJ, Bell JG, Luizi FS, Gara B, Bromage NR, Sargent JR (1999) Natural copepods are superior to enriched Artemia nauplii as feed for halibut larvae (Hippoglossus hippoglossus) in terms of survival, pigmentation and retinal morphology: relation to dietary essential fatty acids. J Nutr 129:1186–1194

Sies H, Stahl W (2005) New horizons in carotenoid research. In: Packer L, Obermüller-Jevic U, Kraemer K, Sies H et al. (eds) Carotenoids and retinoids: molecular aspects and health issues. AOCS Press, Urbana, pp. 315–320

Sorgeloos P (2000) President's column. World Aquac 31:3–71

Sorgeloos P, Leger P (1992) Improved larviculture outputs of marine fish, shrimp and prawn. J World Aquac Soc 23:251–264

Sorgeloos P, Leger P, Lavens P (1988) Improved larval rearing of European and Asian seabass, seabream, mahi-mahi, siganid and milkfish using enrichment diets for *Brachionus* and *Artemia*. World Aquac 19:78–79

Sorgeloos P, Dehasque KM, Dhert P, Lavens P (1994) Larviculture of marine finfish: the current status. Infofish Int 4(94):49–54

Spenelli J (1979) Preparation of salmonid diets containing zooplankton and their effect on organoleptic properties of pen-reared salmonids. In: Halver JE, Thiews K (ed) Proceedings of the world symposium on finfish nutrition and fish feed technology, Hamburg, 20–23 June 1978, Heenemann, Berlin, Germany, pp. 383–392

Stael M, Sanygontanagtit T, Van Ballaer E, Puwapanieh N, Tungsutapanich A, Lavens P (1995) Decapsulated cysts and *Artemia* flakes as alternative food sources for the culture of *Penaeus monodon* postlarvae. In: Lavens P, Jaspers E, Roelants I (eds) Larvi'95 fish and shellfish larviculture symposium. Eur. Aquac. Soc. Spec. Publ., vol. 24. European Aquaculture Society, Ghent, pp 342–345

Stottrup JG (2000) The elusive copepods: their production and suitability in marine aquaculture. Aquac Res 31 (89):703–711

Stottrup JG, Norsker NH (1995) Production and use of copepods in marine fish larviculture. In: Lavens P, Jaspers E, Roelants I (eds) Larvi'95 fish and shellfish larviculture symposium. Eur. Aquac. Soc. Spec. Publ., vol. 24. European Aquaculture Society, Ghent, p 320

Stottrup JG, Norsker NH (1997) Production and use of copepods in marine fish larviculture. Aquaculture 155:231–247

Stottrup JG, Richardson K, Kirkegaard E, Pihl NJ (1986) The cultivation of *Acartia tonsa dana* for use as a live food source for marine fish larvae. Aquaculture 52:87–96

Stottrup JG, Gravningen K, Norsker NH (1995) The role of different algae in the growth and survival of turbot larvae *(Scophthalmus maximus* L.) in intensive rearing systems. ICES Mar Sci Symp 201:173–186

Støttrup JG, Shields R, Gillespie M, Gara MB, Sargent JR, Bell JG, Henderson RJ, Tocher DR, Sutherland R, Næss T, Mangor-Jensen A, van der Naas K, Meeren T, Harboe T, Sanchez FJ, Sorgeloos P, Dhert P, Fitzgerald R (1998) The production and use of copepods in larval rearing of halibut, turbot and cod. Bull Aquac Assoc Canada 4:41–45

Sun B, Fleeger JW (1995) Sustained mass culture of *Amphiascoides atopus* a marine harpacticoid copepod in a recirculating system. Aquaculture 136(3):313–321

Toledo JD, Golez MS, Doi M, Ohno A (1999) Use of copepod nauplii during early feeding stage of Grouper *Epinephelus coioides*. Fish Sci 65:390–397

Varghese PU (1995) Aquaculture production to augment export of marine products from India 1–6. In: INDAQUA'95. MPEDA, Publication, Cochin, 134 p

Vengadeshperumal N, Rajkumar PDM, Perumal P, Vijalakshimi S, Balasubramanian T (2010) Laboratory culture and biochemical characterization of the calanoid copepod, *Acartia southwelli* Sewell, 1914 and *Acartia centrura* Giesbretch, 1889. Adv Biol Res 4:97–107

Watanabe T, Kiron V (1994) Prospects in larval fish dietetics. Aquaculture 124:223–251

Watanabe T, Kitajima C, Fujita S (1983) Nutritional value of live organisms used in Japan for mass propagation of fish: a review. Aquaculture 34:115–143

Witt U, Quantz G, Kuhlmann D, Kattner G (1984) Survival and growth of turbot larvae *Scophthalmus maximus* L. reared on different food organisms with special regard to long-chain polyunsaturated fatty acids. Aquac Eng 3:190–1771

Yufera M, Pascual E (1984) La produccion de organismos zooplanctonicos para la alimentacio´n larvaria enacuicultura marina. Inf Tec Inst Investig Pesq 119:3–27

Culture of Marine Polychaetes

J. Sesh Serebiah

Introduction

The *Polychaeta* or *polychaetes* are a class of annelid worms, generally marine. Each body segment has a pair of fleshy protrusions called parapodia that bear many bristles, called chaetae, which are made of chitin. Indeed, polychaetes are sometimes referred to as bristle worms. More than 10,000 species are described in this class. Common representatives include the lugworm (*Arenicola marina*) and the sandworm or clam worm *Nereis*. Polychaetes as a class are robust and widespread, with species that live in the coldest ocean temperatures of the abyssal plain, to forms which tolerate the extreme high temperatures near hydrothermal vents.

They are segmented worms, generally less than 10 cm (3.9 in.) in length, although ranging at the extremes from 1 mm (0.039 in.) to 3 m (9.8 ft). They are often brightly colored and may be iridescent or even luminescent. Each segment bears a pair of paddle-like and highly vascularized parapodia, which are used for movement and, in many species, act as the worm's primary respiratory surfaces. Bundles of bristles, called setae, project from the parapodia. However, polychaetes vary widely from this generalized pattern and can display a range of different body forms. The most generalized polychaetes are those that crawl along the bottom, but others have adapted to many different ecological niches, including burrowing, swimming, pelagic life, tube-dwelling or boring, commensalism, and parasitism, requiring various modifications to their body structure. The head, or prostomium, is relatively well developed, compared with other annelids. It projects forward over the mouth, which therefore lies on the animal's underside. The head normally includes two to four pairs of eyes, although there are some blind species. These are typically fairly simple structures, capable of distinguishing only light and dark, although some species have large eyes with lenses that may be capable of more sophisticated vision. The head also includes a pair of antennae, tentacle-like palps, and a pair of pits lined with cilia, known as "nuchal organs." These latter appear to be chemoreceptors and help the worm to seek out food. The outer surface of the body wall consists of a simple columnar epithelium covered by a thin cuticle. Underneath this, in order, there are a thin layer of connective tissue, a layer of circular muscle, a layer of longitudinal muscle, and a peritoneum surrounding the body cavity. Additional oblique muscles move the parapodia. In most species, the body cavity is divided into separate compartments by sheets of peritoneum between each segment, but in some species, it is more continuous. The mouth of polychaetes varies in form depending on their diet, since the group includes predators, herbivores, filter feeders, scavengers, and

J.S. Serebiah (✉)
Department of Marine Studies and Coastal Resources Management, Madras Christian College, Tambaram, Chennai, 600054, Tamil Nadu, India
e-mail: seshserebiah@yahoo.com

S. Perumal et al. (eds.), *Advances in Marine and Brackishwater Aquaculture*,
DOI 10.1007/978-81-322-2271-2_5, © Springer India 2015

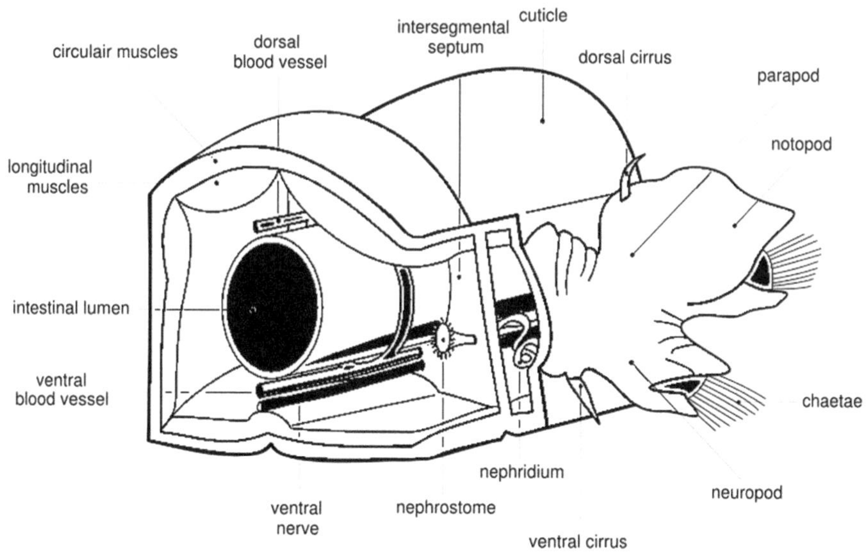

Fig. 1 Parts and cross-sectional diagram of polychaete

parasites. In general, however, it possesses a pair of jaws and a pharynx that can be rapidly everted, allowing the worm to grab food and pull it into the mouth. In some species, the pharynx is modified into a lengthy proboscis. The digestive tract is a simple tube, usually with a stomach partway along. The smallest species and those adapted to burrowing lack gills, breathing only through their body surface. Most other species, however, have external gills, generally, although not always, associated with the parapodia. There is usually a well-developed, if simple, circulatory system. There are two main blood vessels, with smaller vessels to supply the parapodia and the gut. Blood flows forward in the dorsal vessel, above the gut, and returns back down the body in the ventral vessel, beneath the gut. The blood vessels themselves are contractile, helping to push the blood along, so most species have no need of a heart. In a few cases, however, muscular pumps analogous to a heart are found in various parts of the system. Conversely, some species have little or no circulatory system at all, transporting oxygen in the coelomic fluid that fills their body cavity. The blood itself may be colorless or have any of three different respiratory pigments. The most common of these is hemoglobin, but some groups have hemerythrin or the green-colored chlorocruorin instead (Fig. 1).

The nervous system consists of a single or double ventral nerve cord running the length of the body, with ganglia and a series of small nerves in each segment. The brain is relatively large, compared with that of other annelids, and lies in the upper part of the head. An endocrine gland is attached to the ventral posterior surface of the brain and appears to be involved in reproductive activity. In addition to the sensory organs on the head, there may also be photosensitive eyespots on the body, statocysts, and numerous additional sensory nerve endings, most likely involved with the sense of touch. Polychaetes have a varying number of protonephridia or metanephridia for excreting waste, which in some cases can be relatively complex in structure. The body also contains greenish "chloragogen" tissue, similar to that found in oligochaetes, which appears to function in metabolism, in a similar fashion to that of the vertebrate liver. Their cuticle is constructed from cross-linked fibers of collagen and may be 200-nm to 13-mm thick. Their jaws are formed from sclerotized collagen and their setae from sclerotized chitin.

Fig. 2 Polychaete scoop

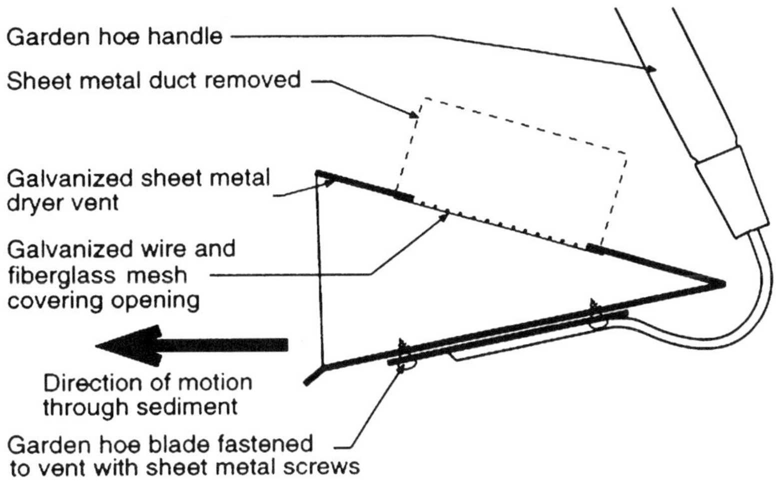

Sampling and Identification

Polychaetes in sandy/muddy sediments are sampled using corers. They are collected by "polychaete scoop" (Fig. 2). This is made from a modified galvanized sheet metal dryer vent (available at any good building supply store) fastened to a garden hoe, as a handle. The scoop is dragged along the sediment catching the top polychaete-containing layer. The vent has a hole covered with galvanized wire mesh, which in turn is covered with fiberglass window screening adhered with silicone caulk. This hole allows water to pass through the scoop while the mesh catches any polychaetes that swim out of the sediment. The full scoop of sediment is then sieved through two screens, an upper one with a 500-μm mesh and a lower one constructed of fiberglass window screening (approximately 64-μm mesh size). The upper sieve catches larger stones and macrobenthic organisms, which retained the polychaetes. This operation is repeated numerous times, resulting in a bucket of concentrated polychaetes containing the desired species. The sieving operation must be performed gently, with the lower sieve submerged in seawater so that the interstitial organisms are not ground up in the sand or against the mesh. Additional sediment is sieved only through the coarse mesh, to remove rocks and larger meiofauna. Once a sufficient quantity of sediment has been sieved, the concentrated polychaetes must be taken back to the laboratory to separate the desired species from the other interstitial species present. Identification of polychaetes is performed by Day (1967) and Kristian Fauchald (1977). This is important to avoid the isolation of larvae of different species, which might be difficult to distinguish from those of the study species. In addition, it is necessary to remove polychaetes that might feed on the study species or that might outcompete the study species for food. Transfer a small amount of sediment to a shallow petri dish for examination under a low-power stereo microscope. The worms hide in their sand-mucus tubes, but they can be captured by agitating the tube with a wide-bore Pasteur pipette until the worm swims out and then sucking it into the pipette for transfer to another dish. The worms are checked to make sure they are the correct species before being added to the culture vessel. The brief identification chart and keys up to family are given herewith for classification and identification of polychaetes (Fig. 3 and Table 1).

Significance of Polychaetes Culture

The importance of polychaetes in aquaculture industry is being understood now and caused

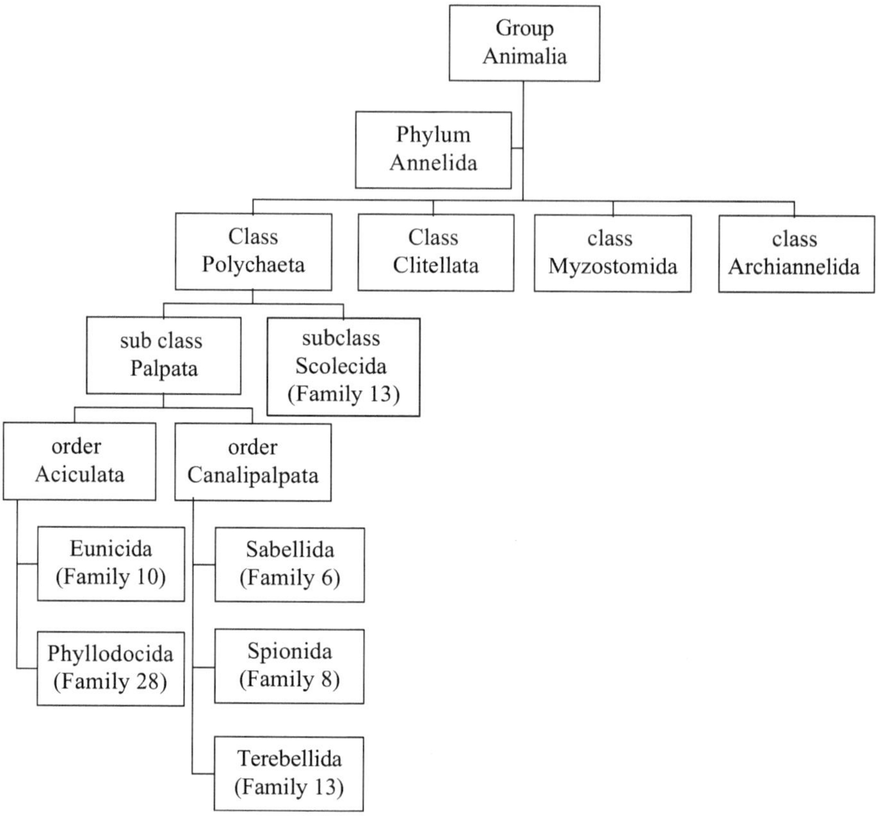

Fig. 3 Flowchart of polychaetes' classification up to order

Table 1 Showing polychaetes identification key up to subclass and family

<table>
<tr><td colspan="2">Head without pairedfeeding tentacles, without jaws. Each segment with or without a pair of gills</td><td>Scolecida: A group of subsurface sediment eaters. Mostly found on soft shores. Few of this group lives on rocky shores.</td></tr>
<tr><td colspan="2">Head with one pair of flexiable grooved feeding tentacles without jaws. Anterior segment often with a pair of gills</td><td>Spionida: Tube or burrow dwelling surface particle pickers. Mostly found on soft shores or subtidally, Few of this group lives on rocky shores.</td></tr>
<tr><td colspan="2">Head with many pairs of feeding flexiable tentacles, without jaws.</td><td>Terebellida: A group of tube-dwelling particle pickers. Mostly found on soft shores or in sediment subtidally, and some of this group is common in rock crevices.</td></tr>
<tr><td colspan="2">Head usually with a terminal funnel like fan of inflexiable 'tentacles', without jaws</td><td>Sabellida: A group of tube dwelling particle filterers. Some of this group lives in colonial group on rocky shores, some on soft shores. Most live subtidally.</td></tr>
<tr><td rowspan="2">Head with chitinous jaws, usually (not glyceridae, Goniadidae) also conspicuous eyes, and short sensory tentacles.</td><td>Jaws up to two pincer-like pairs terminal on extensible proboscis. Each segment without a pair of gills.</td><td>Phyllodocida: A group of mostly surface wandering food graspers occurs everywhere. Some of this group is well adapted to rock crevices.</td></tr>
<tr><td>Jaws only one pincer like pair, barely extensible, but grouped with other toothed plates. Mid body segments often with pair of gills.</td><td>Eunicida: A group of mostly burrowing food graspers , mostly subtidal, in sediment or on rock coral. A few of this group are well adapted to rock crevices. Few soft species.</td></tr>
</table>

rapid demand as live feed for varieties of cultivable species. Hence the technique on polychaete culture need to be standardized for potential polychaete species. Polychaetes as a live feed can stimulate gonad maturation during spawning in hatchery-reared species, e.g., *Solea vulgaris*, *Solea senegalensis* (Dinis 1986), *Penaeus kerathurus* (Luis 1989), and *Penaeus vannamei* (Lytle et al. 1990). The other main point of the culture is to reduce the substrate-harvesting disturbance and the great biogeochemical and benthic community impact (Gambi et al. 1994) as they are used as commercial bait.

Culture of Polychaetes

Collections of Brood Stocks, Breeding, and Development

Adult worms are collected from the habitat. Adults are gravid from April or May to October. The worms can be kept in room temperature aerated aquaria in 32-ppt artificial seawater and adults sexed according to the color of their gametogenic parapodia. Adult male and female worms are best kept in separate tanks to prevent premature spawning if possible. Separation of adult male and female are easier in tube dwelling polychaetes. The adults were kept inside their tubes at all times, and access to the worm was achieved by making a small longitudinal cut in the tube. Longitudinal cuts are repaired quickly by the worm. Transverse or chevron-shaped cuts cause irreparable damage to the tube, decreasing the ability of the adult to feed and survive. Oocytes are obtained by cutting off a single parapodium from a gravid female and emptying the contents (several thousand eggs) into a 150-μm screen immersed in 0.2-μm filtered artificial seawater (FASW) at 32-ppt salinity in a 35-mm petri dish. Oocytes are washed through this screen; retained mucous and parapodial fragments are discarded. The sieved oocytes are washed three more times by transferring them into 100-mm petri dishes of fresh FASW and then incubated in FASW for about 40 min before fertilization to allow for germinal vesicle breakdown. Sperm is prepared by cutting off the parapodium of a ripe male without allowing the contents to mix with seawater. The "dry sperm" is extracted with a Pasteur pipette and diluted 1:1,000 with FASW. This sperm suspension is examined through a stereo microscope, and when the sperm are seen to be motile, 1–3 drops are added to the eggs in a 100-mm petri dish. Knowing the optimal amount of sperm suspension to use to obtain complete fertilization without excessive polyspermy is a matter of experience, because the concentration of dry sperm varies between animals and over time in the same animal. Swimming ciliated blastulae develop by 12 h at room temperature (21–23 ° C). These blastulae may then be transferred to the culture apparatus. In many interstitial polychaetes, fertilization is internal and its embryos are brooded inside the mother's tube. This makes obtaining early embryonic stages difficult, because gametes are not easily harvested for in vitro fertilization. Most of the tube-dwelling polychaetes have early swimming larvae in the development stages. They are normally swimming in the water. Early-swimming larvae may then be extracted from the culture water.

Simulated Beach

Simulated beach is developed for harboring adult worms for culture. The tube-dwelling worms construct vertical tubes in sand or clay, which are barely visible to the naked eye, and epibenthos can be seen on the surface. Population peaks occur in late post-monsoon and early summer (Zajac 1991), making this the best time to collect the adults. The "simulated beach" is made by depositing about 3 cm of sediment from the collection site into a wide shallow vessel, deep enough to allow for 6–8 cm of water on top (Fig. 4). The sediment should first be frozen to kill any fauna remaining and then rinsed several times with freshwater to remove some of the dead organic material, which could produce unwanted products of decomposition. After deposition of the sediment, natural or artificial

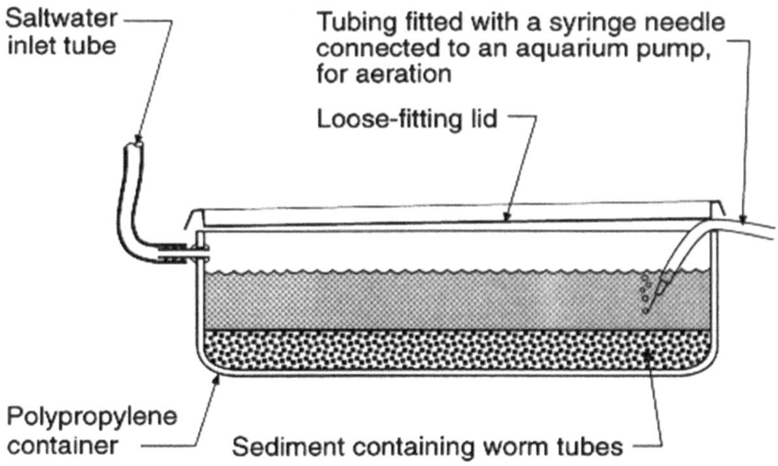

Fig. 4 Simulated beach

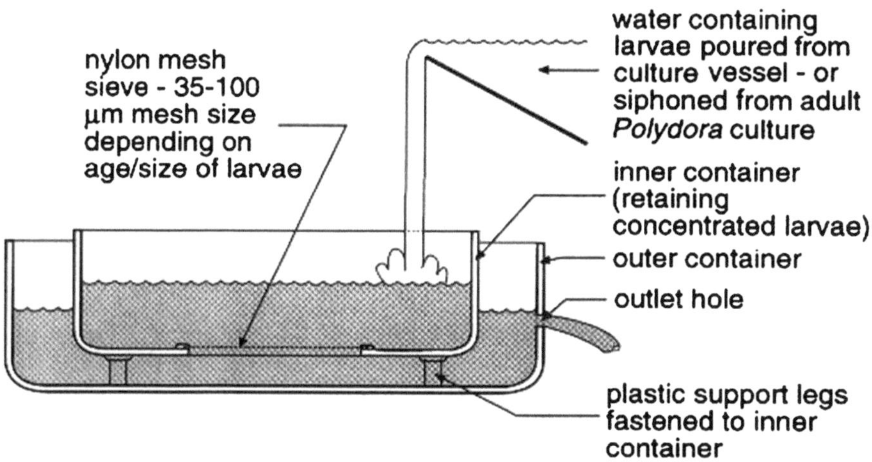

Fig. 5 Ideal method for the changing of water from simulated beach set up

seawater (ASW) is added. The seawater should be changed at least twice to dilute out any remaining freshwater and to remove some of the silty organic debris, which will wash out the newly placed sediment. Washed autoclaved sand may be layered on top if the sediment is too silty. Once the substrate has settled, the sorted worms can be scattered over the surface. They will colonize the sediment soon. Two operations are necessary to maintain the culture long enough to produce larvae. First, adult animals must be fed. For the 30-cm culture vessel, we use 4 ml of concentrated liquid invertebrate food diluted in some ASW and broadcast evenly over the sediment once per week. Overfeeding should be avoided to prevent excessive bacterial growth due to decomposing food matter. Second, the culture water must be changed regularly to prevent waste products, bacteria, and unwanted volunteer organisms from building up and to retrieve any larval offspring that may have been released into the seawater. Changing water by siphoning the old water into a 63-μm sieve (Fig. 5) so that any larvae will be caught can drop once again to culture system. If any other meiofauna are captured in the sieve, they may be removed before the desired polychaete larvae are transferred to the culture apparatus by gently backwashing the sieve with ASW from a squeeze bottle. Two changes of water per week should be sufficient to maintain healthy cultures. The inlet of the siphon tube may be passed back and forth through the culture vessel to ensure that as many larvae as possible are caught. Fresh seawater is added through a side inlet so that the disturbance of the sediments is minimized. If natural seawater is used, it may be necessary to filter out suspended fauna, such as small crustaceans, that might invade the colony. In our experience, larvae begin to appear in the culture systems around the 40th day after beginning the culture. However, the number of larvae retrieved varies over time, with a periodicity of around 30 days, probably due to the breeding and life history characteristics of the worms. In our hands, culture life has been 3 months or roughly 2 months of larval production. After this time the adult population declines and larval collections drop off.

Food for Culture Organisms

Initially, each individual can feed by once in every 3 days with commonly available commercial food used for tropical fish. This food type had been used successfully with polychaete species (Garwood and Olive 1981; Bartels-Hardege and Zeeck 1990). After 20–25 days, they have to feed daily, because of cannibalism activity. Attempts to use commercially available aquarium food, such as the liquid invertebrate food or dried rotifers, resulted in excessive bacterial growth in the colonies and poor survival of the polychaete. Several types of algae can be used, but *Dunaliella salina*, a green alga, and *Isochrysis galbana*, a golden-brown alga, give adequate nutrition, and they are easy to maintain in culture. Autoclaved half-liter widemouthed glass bottles with cotton stoppers are used as culture vessels to avoid bacterial or fungal contamination for algal culture. The supplemented 0.2-μm filtered artificial seawater (FASW) with F/2 medium at standard concentration (Strathmann 1987) can act as good culture medium.

References

Bartels-Hardege HD, Zeeck E (1990) Reproductive behavior of *Nereis diversicolor* (Annelida: Polychaeta). Mar Biol 106:409–412

Day JH (1967) A monograph on the polychaeta of southern Africa part 1. Errantia and part 2. Sedentaria. The Trustees of the British Museum (Natural History), London. Pub No. 656, 948 pp

Dinis MT (1986) Quatre Soleidae de l'estuaire du Tage. Reproduction etcroissance. Essaid'elevage de Soleasenegalensis Kaup. These Doctorat, Universit Bretagne Occidantale, Brest. 357 pp

Fauchald K (1977) The polychaete worms. Definitions and keys to the orders, families and genera. Natural History Museum of Los Angeles County Science Series 28:1–188

Gambi MC, Castelli A, Giangrande A, Lanera P, Prevedelli D, Vandini RZ (1994) Polychaetes of commercial and applied interest in Italy: an overview. In: Dauvin J-C, Laubier L, Reish DJ (eds) Actes de la 4ème Confrence Internationale des Polychètes, vol 162, Memoires du Musum National d' Histoire Naturelle. Musum National d'Histoire Naturele, Paris, pp 593–603

Garwood PR, Olive PJW (1981) The influence of environmental factors on the growth of oocytes in *Nereis diversicolor* (Annelida; Polychaeta). Bull Soc Zool France 106:399–402

Luis OJ (1989) Contribuiçãopara o conhecimento da nutrição de Palaemonserratuse Penaeuskerathurus (Crustacea, Decapoda), com ênfasepara o papel dos lípidos no crescimento e reprodução. Doctoral thesis. Universidade de Lisboa, 374 pp

Lytle JS, Lytle TF, Ogle JT (1990) Polyunsaturated fatty acid profiles as a comparative tool in assessing maturation diets of *Penaeus vannamei*. Aquaculture 89:287–299

Strathmann MF (1987) Reproduction and development of marine invertebrates of the Northern Pacific Coast: data and methods for the study of eggs, embryos, and larvae. University of Washington Press, Seattle

Zajac RN (1991) Population ecology of *Polydora ligni* (Polychaeta: Spionidae). I. Seasonal variation in population characteristics and reproductive activity. Mar Ecol Prog Ser 77:197–206

Probiotics: Live Boon to Aquaculture

Ramasamy Thirumurugan and Venkatasamy Vignesh

Introduction

The term "probiotics" comes from the Greek words "pro" and "bios" meaning "for life". It is opposed to the term "antibiotic" meaning "against life" (Hamilton-Miller 2003). The concept of microfloral manipulation was first appreciated by Metchnikoff (1907) who examined the consumption of yoghurt and found an effect on the longevity of Bulgarian peasants. In "The Prolongation of Life", Metchnikoff (1907) was probably the first to define the health benefits of lactic acid bacteria in yoghurt, playing defence mechanism against harmful bacteria. A beneficial effect on applying certain beneficial bacteria in human, pig, cattle and poultry nutrition has been recorded (Gilliland 1979; Conway 1989; Jong 1993). On the other hand, the use of such probiotics in aquaculture is a relatively new concept (Kozasa 1986). The interest is also being increased in the search of beneficial microorganisms which may contribute to fish, because it has become clear that the microflora plays an important role with reference to the wellbeing and health of fish (Trust and Sparrow 1974).

The term "probiotics" was most likely first proposed by Werner Kollath (1953). He suggested the term as all organic and inorganic food complexes, in contrast to antibiotics, in order to upgrade those food complexes as supplements. He defined probiotics as an "active substance that is essential for a healthy development of life". A few years later, the term probiotics was employed in the perspective of animal feeds by Parker (1974) as "Organisms and substances that have a beneficial role on the host by contributing to its intestinal microbial balance".

A well-defined quotation was given by Fuller (1989) who defined probiotics as "A live microbial feed supplement which beneficially affects the host animal by improving its intestinal microbial balance". The appropriate definition was published by an expert consultation at a meeting convened by the FAO and WHO in October 2001: "Live microorganisms which when administered in adequate amounts confer a health benefits on the host". Probiotics belong to the inherited microflora of the intestinal system with less or no pathogenic nature and having features which are important for the host health and its wellbeing. Therefore, the maintenance of this ecological microflora is essential to prevent diseases, especially infections of the gastrointestinal (GI) tract.

Probiotics are attractive biological stuff with extreme characteristics, and their usage in animal and human nutrition is well-documented (Denev 1996; Fioramonti et al. 2003; Goktepe et al. 2006). Nowadays, probiotics are quite common for promoting health by means of

R. Thirumurugan (✉) • V. Vignesh
Department of Animal Science, Laboratory of Aquabiotics/Nanoscience, School of Life Sciences, Bharathidasan University, Tiruchirappalli 620024, Tamil Nadu, India
e-mail: ramthiru72@gmail.com

S. Perumal et al. (eds.), *Advances in Marine and Brackishwater Aquaculture*,
DOI 10.1007/978-81-322-2271-2_6,

having "functional foods" in humans and natural and ecological alternative of antibiotic growth promoters (AGP) in animal production (Ouwehand et al. 2002; Sullivan and Nord 2002; Senok et al. 2005; Denev 2008).

Aquaculture is mainly considered with aquatic organisms and manipulation of aquatic organisms' life before harvest in order to increase its production. It has become a very important aspect in most of the countries. During large-scale production, aquatic organisms are exposed to stressful conditions; problems related to diseases and deterioration of environmental condition often happen and result in serious economic losses. Prevention and control of diseases through the use of antibiotics and vaccines have been in practice throughout the world. However, the utility of antimicrobial agents as a preventive measure is doubtful, giving extensive documentation of the evaluation of antimicrobial resistance among pathogenic bacteria. The problems from the use of antibiotics have attracted a global concern in terms of development of resistance among the pathogenic bacteria.

There is an urgent need in aquaculture development for microbial control strategies. One alternative to antibiotics in disease control in aquaculture could be the use of probiotic bacteria. The use of probiotics or beneficial bacteria which control pathogen through a variety of mechanisms is increasingly viewed as an alternative to antibiotic treatment. The use of probiotics in human (Anukam et al. 2004, 2006a, b; Anukam and Reid 2007) in animal nutrition (Fuller 1992) is well-documented, and recently they have begun to be applied in aquaculture (Gatesoupe 1999; Gomez-Gil et al. 2000; Verschuere et al. 2000; Irianto and Austin 2002a, b).

Probiotics refer to healthy microbiome possessing the calibre to extend their impacts chronically and promoting the host while taking as feed additives or in water. Probiotic strains enhance the population of healthy gut microbiome which are present in the stomach. The cognizance towards awareness of probiotic strains has been increased sequentially over the past decades since the impacts of probiotic strains in promoting health status of aquatic organisms especially shrimp and fish. The present review reveals the types and forms of probiotics and their applications in aquaculture.

Aquatic Probiotics

Aquatic probiotics are the novel one. While considering application for aquatic purposes, it is most important to care specific features which are basically different from probiotics of terrestrial background. The major difference between aquatic and terrestrial organisms lies in the level of interaction of intestinal microbiome and environment.

Gatesoupe (1999) defined probiotics as "Microbial cells that are administered in such a way as to enter the GI tract and to be kept alive, with the aim of improving health" for aquaculture. The definition focused on the delivery of the probiotic orally and its calibre to improve the health status of the host as a consequence of its presence in the digestive tract. Verschuere et al. (2000) defined aquatic probiotics as "Live microorganisms that have a beneficial effect on the host by modifying the microbial community, associated with the host, by ensuring improved use of the feed or enhancing its nutritional value, by enhancing the host response towards disease, or by improving the quality of its ambient environment".

There are a number of probiotics used in aquaculture industries like a wide range of taxa – from *Lactobacillus*, *Bifidobacterium*, *Pediococcus*, *Streptococcus* and *Carnobacterium* spp. to *Bacillus*, *Flavobacterium*, *Cytophaga*, *Pseudomonas*, *Alteromonas*, *Aeromonas*, *Enterococcus*, *Nitrosomonas*, *Nitrobacter* and *Vibrio* spp. – yeast (*Saccharomyces*, *Debaryomyces*), etc. (Irianto and Austin 2002a, b; Burr et al. 2005; Sahu et al. 2008).

Types

Two types were employed in aquaculture industries: (1) Gut probiotics are being mixed with feed and administrated orally to sustain and enhance the useful microbiome of the gut. Gut probiotics are being mainly used in finfish aquaculture. (2) Water probiotics are being proliferated in water and excluded the pathogenic strains by consuming all available nutrients. Thus, the pathogenic strains can be eliminated through starvation (Nageswara and Babu 2006; Sahu et al. 2008). Water-based probiotics are currently used in shrimp aquaculture. Generally, it is proposed that probiotics act as biological control agents in fish which can be applied in the feed or as a water additive supplement.

Forms of Probiotics

Aquatic probiotics are available in two forms: (1) Dry forms: the dry probiotics are sold out in pockets. They can be given with feed or mixed with water which possess many benefits like safety, easy using, longer shelf life, etc. (Decamp and Moriarty 2007). (2) Liquid forms: the hatcheries generally use liquid forms which are live and ready to act. These liquid forms are directly added to hatchery tanks or blended with farm feed.

Liquid forms show positive results in short duration compared to dry and spore form bacteria, though they are lower in density (Nageswara and Babu 2006). There are no reports of any harmful effect of probiotics, but on applying probiotics, it may be temporarily elevated in the biological oxygen demand. It is advisable to have a minimum dissolved oxygen level of 3 % during probiotic treatment.

Selection Criteria

The ultimate and major need of engaging probiotics are to establish an amicable relationship between beneficial microbiome and pathogenic strains in order to persist the constituents of intestinal or skin mucus of freshwater fish. Probiotics must be nonpathogenic and non-toxic in order to neglect unwanted consequences while administering to fish. Screening of probiotics for antagonism, adhesion and challenge tests *in vitro* etc., are essential things (Fig. 1). There are a few basic and specific properties leading to the beneficial aspects of the host.

Antagonism

Defence against pathogenic strains is one of the primary properties of probiotics to employ and screen their further properties in freshwater fish. Defence mechanism is mainly due to secretion of antimicrobial substances like bacitracin, bacillin, difficidin, mycobacillin, oxydifficidin, polymyxin, subtilin, gramicidin or bacillomycin B in both conditions (in vivo and in vitro) (Korzybski et al. 1978; Zimmerman et al. 1987). There are studies reporting that various probiotics exhibited antibacterial activities against several common fish pathogens, including *Enterococcus durans*, *Escherichia coli*, *Micrococcus luteus* and *Pseudomonas aeruginosa* (Biomin 2009).

Competitive Exclusion

Adhesion and colonisation on the mucosal surfaces are the most feasible shielding mechanisms against pathogenic strains by means of competition in order to get binding sites and nutrients (Westerdahl et al. 1991). There are different *Lactobacilli* reported that they compete and reduce the adhesion of *Aeromonas salmonicida*, *Carnobacterium piscicola* and *Yersinia ruckeri* to intestinal mucus from rainbow trout (Balcázar et al. 2006). Adhesion is one of the most important selection criteria for probiotic bacteria, since it is a prerequisite for colonisation (Beachey 1981).

Enhancement of the Immune Responses

Non-specific immune system can be stimulated by probiotics. Immunostimulants vary according to their mode of action and the way they are

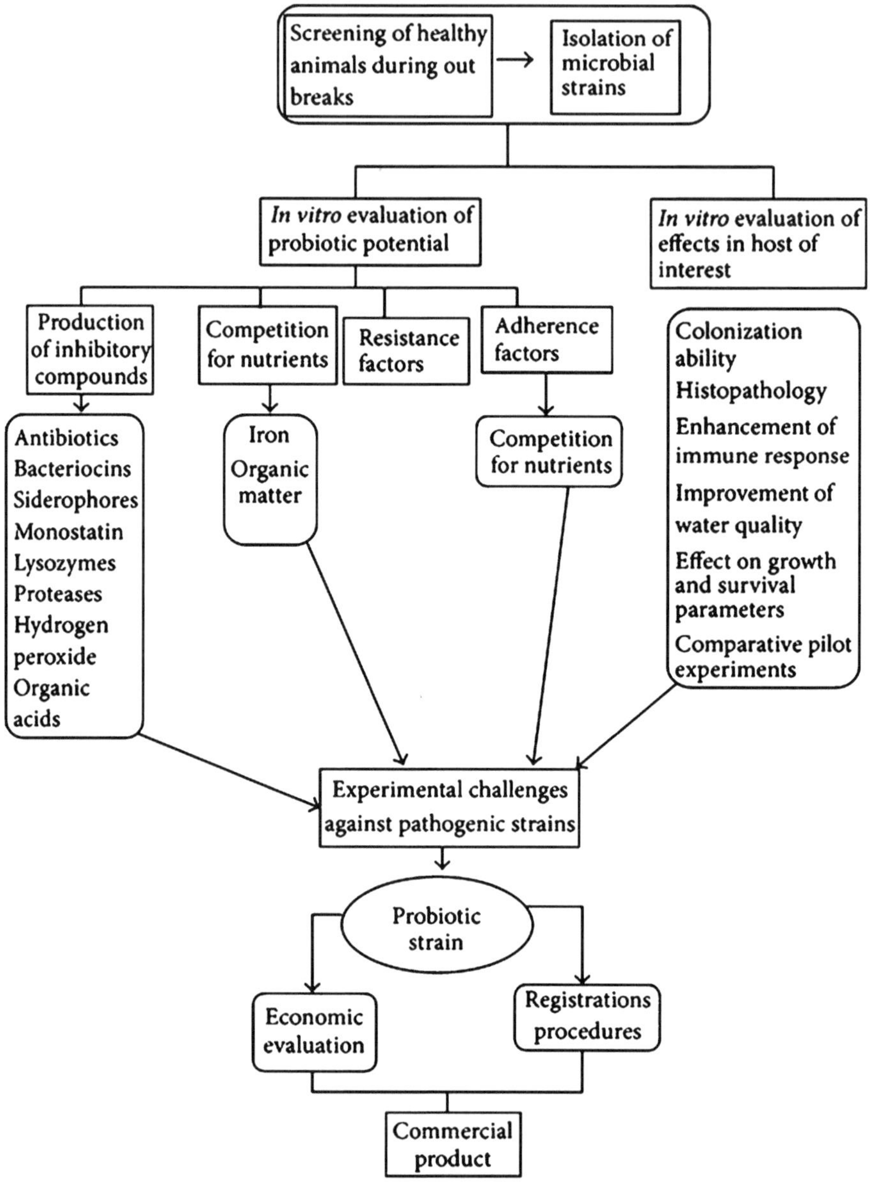

Fig. 1 Procedure for evaluation of probiotic potential of microbial strain for shrimp aquaculture (Bestha Lakshmi et al. 2013)

being used. Certain derivates, such as polysaccharides, lipoproteins, nucleotides and ß-glucans, have the capability to increase phagocytic abilities by activating macrophages. Rengpipat et al. (2000) indicated that the use of *Bacillus* sp. provided disease protection by activating both cellular and humoral immune defences in tiger shrimp (*Penaeus monodon*).

Applications of Probiotics in Aquaculture

These organisms can be administered in aquaculture through feeding, injection or immersion of the probiotic bacteria (Irianto and Austin 2002a, b).

Feed

Probiotics are applied with the feed and a binder (egg or cod liver oil), and most commercial preparation contain either *Lactobacillus* sp. or *Saccharomyces cerevisiae* (Abidi 2003). Regular use of probiotics in feed of fish in various countries has been reported to have several health benefits.

The perception that fermented milk yoghurt is beneficial and already widespread within so many regions because traditionally these products have been used by local healers for the treatment of diverse condition, such as skin allergies, stomach upset especially diarrhoea and vaginal discharges.

According to FAO and WHO guidelines, probiotic organisms used in food must be capable of surviving passages through the gut, i.e. they must have the ability to resist gastric juices and exposure to bile (Senok et al. 2005). Furthermore, they must be able to proliferate and colonise the digestive tract, and they must be safe, effective, potency and maintain their effectiveness for the duration of the shelf life of the product (Senok et al. 2005).

There is more health benefits reported due to regular use of probiotics in feed of fish in most of the countries (Cerrato 2000). In aquaculture, probiotics can also be encapsulated in feed (Ramos et al. 2005) or live feed like rotifers and *Artemia* (Mahdhi et al. 2011). Bioencapsulation/infusion is another efficient application of probiotics to aquatic animals. Preparation of probiotic diets has been demonstrated by Yassir et al. (2002).

Culture Water/Pond

The water probiotics contain multiple strains of bacteria like *Bacillus acidophilus*, *B. subtilis*, *B. licheniformis*, *Nitrobacter* sp., *Aerobacter* sp. and *Saccharomyces cerevisiae*. Application of probiotics through water in tanks and ponds may also have an effect on fish health by improving several qualities of water, since they modify the bacteria composition of the water and sediments (Ashraf Ali 2000; Venkateswara 2007). When probiotics are applied in culture water, they multiply and overgrow the pathogenic organisms present in the water.

Today many researchers attempt to use some kind of probiotics in aquaculture water (Xiang-Hong et al. 2003).

Injection

Application of probiotics by injection is a possibility. Austin et al. (1995) suggested the possibility of freeze-drying the probiont like vaccine and applying it either through bathing or injection. Except for experimental purposes, application of probiotics by injection has not been widely reported. In addition, one of the benefits of probiotics in fish and animal is to boost the immunity, meaning that there is the stimulation of antibodies of the host. Probiotics can therefore confer the best immunity through injection. However, Gram et al. (1999) stated that vaccination by injection which sometimes the only effective route of administration is impracticable when applied to small fish or larger numbers. Yassir et al. (2002) have demonstrated the experimental administration of probiotic *Micrococcus luteus* to *Oreochromis niloticus* by injection through intraperitoneal route which had only 25 % mortality as against 90 % with *Pseudomonas* using the same route. According to Yassir et al. (2002), the use of probiotics stimulates *Rainbow trout* immunity by stimulating phagocytes' activity, complement-mediated bacterial killing and immunoglobulin production. When probiotics are evaluated in freeze-dried form, they can be applied through injection (Table 1).

Benefits of Probiotics

Although some of the effects of probiotics have been documented clearly, research is still ongoing in the area with so many questions on the reality of some of the benefit remaining unanswered. However, it is crucial to remember that different probiotic strains are associated with different health benefits (Senok et al. 2005).

Table 1 Probiotics used in fish nutrition and their effects (Denev et al. 2009)

Identity of the probiotics	Species/method of application	Effects	References
Bacillus subtilis and *Bacillus licheniformis*	Rainbow trout/feed (*Oncorhynchus mykiss*)	Increased resistance to *Yersinia ruckeri*	Raida et al. (2003)
Bacillus subtilis *Lactobacillus delbrueckii*	Gilthead sea bream/ feed	Stimulated cellular innate immune response	Salinas et al. (2005)
Bacillus spp. photosynthetic bacteria	Common carp/feed (*Cyprinus carpio*)	Better digestive enzyme activities; better growth performance and feed efficiency	Yanbo and Zirong (2006)
Bacillus spp.	Rainbow trout/feed *Oncorhynchus mykiss*	Better growth performance and survival	Bagheri et al. (2008)
Bacillus subtilis (ATCC 6633) *Lactobacillus acidophilus*	Nile tilapia (*Oreochromis niloticus*)	Stimulated the gut immune system; enhanced the immune and health status; increased the survival rate and the body weight gain	Mesalhy et al. (2008)
Carnobacterium spp.	Atlantic salmon/feed (*Salmo salar* L.)	Inhibited *A. salmonicida*, *V. ordalii* and *Y. ruckeri*; reduced disease	Robertson et al. (2000)
Carnobacterium maltaromaticum B26 *Carnobacterium divergens*	Rainbow trout/feed (*Oncorhynchus mykiss*)	Enhanced the cellular and humoral immune responses	Kim and Austin (2006a)
Carnobacterium divergens 6251	Atlantic salmon/feed (*Salmo salar* L.)	*Carnobacterium divergens* is able to prevent to some extent pathogen-induced damage in the foregut	Ringø et al. (2007)
Enterococcus faecium ZJ4	Nile tilapia/water (*Oreochromis niloticus*)	Increased growth performance; improved immune response	Wang et al. (2008a)
Lactobacillus rhamnosus	Rainbow trout/feed (*Oncorhynchus mykiss*)	Increased resistance to *Aeromonas salmonicida* ssp. *Salmonicida*; reduced mortality from furunculosis	Nikoskelainen et al. (2001)
Lactobacillus rhamnosus (ATCC 53103)	Rainbow trout/feed (*Oncorhynchus mykiss*)	Enhanced immune parameters; stimulated immune response	Nikoskelainen et al. (2003)
Lactobacillus rhamnosus (JCM 1136)	Rainbow trout/feed (*Oncorhynchus mykiss*)	Increased the serum lysozyme and complement activities	Panigrahi et al. (2004)
Lactobacillus rhamnosus	Rainbow trout/feed (*Oncorhynchus mykiss*)	Stimulated immune response	Panigrahi et al. (2005)
Lactobacillus delbrueckii subsp. *delbrueckii* (AS13B)	European sea bass/feed (*Dicentrarchus labrax* L.)	Positive effects on welfare and growth; increased body weight	Carnevali et al. (2006)
Lactobacillus rhamnosus GG	Tilapia/feed (*Oreochromis niloticus*)	Enhanced the growth performance and immunity	Pirarat et al. (2008)
Lactobacillus delbrueckii subsp. *bulgaricus*	Rainbow trout/feed *Oncorhynchus mykiss*	Enhanced humoral immune response	Tukmechi et al. (2007)
Lactobacillus rhamnosus (ATCC 53103); *Bacillus subtilis* *Enterococcus faecium*	Rainbow trout/feed *Oncorhynchus mykiss*	Modulated cytokine production; stimulated immune response	Panigrahi et al. (2007)
Lactococcus lactis subsp. *lactis*; *Lactobacillus sakei*; *Leuconostoc mesenteroides*	Rainbow trout/Feed *Oncorhynchus mykiss*	Stimulated phagocytosis; enhanced the non-specific immunity	Balcázar et al. (2006a)

(continued)

Table 1 (continued)

Identity of the probiotics	Species/method of application	Effects	References
Lactococcus lactis ssp. *lactis*; *Lactobacillus sake*	Brown trout/feed (*Salmo trutta*)	Modified the intestinal microbiota; stimulated the humoral immune response	Balcázar et al. (2007a)
Leuconostoc mesenteroides			
Leuconostoc mesenteroides CLFP 196; *Lactobacillus plantarum* CLFP 238	Rainbow trout/feed *Oncorhynchus mykiss*	Reduced fish mortality	Vendrell et al. (2008)
Micrococcus luteus	Nile tilapia/feed (*Oreochromis niloticus*)	Enhanced the non-specific immune parameters; improved resistance against *Edwardsiella tarda* infection	Taoka et al. (2006a)
Micrococcus luteus *Pseudomonas* spp.	Nile tilapia/feed (*Oreochromis niloticus*)	Higher growth performance, survival rate and feed utilisation; enhanced fish resistance against *Aeromonas hydrophila* infection	Abd El-Rhman et al. (2009)
Streptococcus faecium *Lactobacillus acidophilus*	Nile tilapia/feed (*Oreochromis niloticus*)	Better growth performance and feed efficiency	Lara-Flores et al. (2003)
Saccharomyces cerevisiae	Nile tilapia/feed (*Oreochromis niloticus*)	Better growth performance and feed efficiency	Lara-Flores et al. (2003)
Live yeasts	European sea bass/feed (*Dicentrarchus labrax*)	Better growth performance and feed efficiency	Tovar-Ramırez et al. (2004)
Saccharomyces cerevisiae strain NCYC Sc 47 (Biosaf® Sc 47)	Rainbow trout/feed *Oncorhynchus mykiss*	No significant effect on enzyme activities	Waché et al. (2006)
Saccharomyces cerevisiae var. *boulardii* CNCM I-1079 (Levucell® SB20)	Rainbow trout/feed *Oncorhynchus mykiss*	Stimulated enzyme activities	Waché et al. (2006)
Saccharomyces cerevisiae (DVAQUA®)	Hybrid tilapia/feed (*Oreochromis niloticus* ♀ × *O. aureus* ♂)	Inhibited potential harmful bacteria; stimulated beneficial bacteria; enhanced the non-specific immunity; no significant effects on growth performance and feed efficiency	He et al. (2009)

Water Qualities

According to Venkateswara (2007), probiotics have been reported to regulate microflora; control pathogenic ones; enhance the decomposition of the undesirable organic substance; improve ecological environment by minimising the toxic gases like NH_3, N_2O, H_2O_2, methane, etc.; increase population of food organism in the water; increase nutritional level of the aquatic host; and improve their immunity in the culture water. In several studies, improved water quality has been recorded during the addition of probiotics especially with *Bacillus* sp. (Verschuere et al. 2000). The rationale is that Gram-positive *Bacillus* spp. are generally more effective in converting organic matter back to CO_2 than Gram-negative bacteria which could convert a greater percentage of organic carbon to bacterial biomass or slime.

Growth Promoters

One of the activities of probiotic bacteria is expected to have a direct growth promoting effect of fish either by direct involvement in nutrient uptake or by providing nutrients or vitamins. However, it has been demonstrated experimentally that probiotics indeed may enhance the growth of fish. The ability of organisms to outgrow the pathogens in favour of the host or to improve the growth and yet no side effect on the host made. Yassir et al. (2002)

made an attempt to use probiotic bacteria as growth promoter on tilapia (*Oreochromis niloticus*) and identified that the highest growth performance was recorded with *Micrococcus luteus*, a probiotic bacterium, and the best feed conversion ratio was observed with the same organism. So *M. luteus* may be considered as a growth promoter in fish aquaculture. Lactic acid bacteria also had an effect as a growth promoter on the growth rate in juvenile carp though not in sea bass. Also *Enterococcus faecium* had been used to improve growth when applied in feed to fish. Irianto and Austin (2002a, b) reported that probiotics may stimulate appetite and improve nutrition by producing vitamins and detoxification of compounds in the diet and breakdown of indigestible components. *Streptococcus faecium* improved the growth and feed efficiency of *Israeli carp*. Probiotics therefore can be regarded as growth promoters in aquaculture organisms in addition to various benefits.

Disease Prevention

Probiotics or their products for health benefits to the host have been found useful in aquaculture, in terrestrial animals and in human disease control. These include microbial adjunct that prevents pathogens from proliferating in the intestinal tract, on the superficial surfaces and in culture environment of the culture species (Verschuere et al. 2000). The effect of these beneficial organisms is achieved through optimising the immune system of culture organism, increasing their resistance to disease or producing inhibitory substance that prevents the pathogenic organisms from establishing disease in the host.

Conclusion

This review mainly deals with the strategies in production of more effective fish feed for aquaculture species to improve the physical, chemical and nutritional quality of feed and their respective ingredients. Probiotics as an alternative to antibiotics in aquaculture becomes imperative. The most highly researched probiotic bacteria are *Lactobacillus acidophilus*, *L. bulgarium*, *Bifidobacterium longum*, *B. infantis* and *Bacillus* spp.

Acknowledgment The authors are thankful to the UGC for providing financial support through Major research project (F.No: 38-239/2009 (SR)); thankful to the Department of Zoology, The Madura College (Autonomous), Madurai and Department of Animal Science, School of Life Sciences, Bharathidasan University, Tiruchirappalli, Tamil Nadu.

References

Abidi R (2003) Use of probiotics in larval rearing of new candidate species. Aquac Asia 2:15–16

Abdel-Rhman AM, Khattab YAE, Adel ME, Shalaby AME (2009) Micrococcus luteus and Pseudomonas species as probiotics for promoting the growth performance and health of Nile tilapia, Oreochromis niloticus. Fish Shellfish Immunol 27:175–180

Anukam KC, Reid G (2007) Organisms associated with bacterial vaginosis in Nigerian women as determined by PCR-DGGE and 16S rRNA gene sequence. Afr Health Sci 7(2):68–72

Anukam KC, Osazuwa EO, Reid G, Katsivo MN (2004) Receptivity for probiotic products among pre-menopausal female students in an African university. Sex Transmit Dis 31:460–464

Anukam KC, Osazuwa EO, Ahonkhai I, Reid G (2006a) Lactobacillus vaginal microbiota of women attending a reproductive health care service in Benin City, Nigeria. Sex Transmit Dis 33:59–62

Anukam KC, Osazuwa E, Osemene GI, Ehigiagbe F, Bruce AW, Reid G (2006b) Clinical study comparing probiotic Lactobacillus GR-1 and RC-14 with metronidazole vaginal gel to treat symptomatic bacterial vaginosis. Microbes Infect 8:2772–2776

Ashraf A (2000) Probiotic in fish farming, evaluation of a candidate bacterial mixture. Vattenbruks Inst Rep 19:1–15.

Austin B, Stukey LF, Robertton PA, Effendi I, Griffith DRW (1995) A probiotic strain of vibrio alginolyticus effective in reducing disease caused by *Aeromonas salmonicida, Vibrio anguillarum and Vibrio ordalii.* J Fish Dis 18:93–96

Bagheri T, Hedayati S, Yavari V, Alizade M, Farzanfar A (2008) Growth, survival and gut microbial load of rainbow trout (*Onchorhynchus mykiss*) fry given diet supplemented with probiotic during the two months of first feeding. Turkish J Fish Aquat Sci 8:43–48

Balcázar JL, Vendrell D, De Blas I, Ruiz-Zarzuela I, Girones O, Muzquiz JL (2006a) Immune modulation by probiotic strains: Quantification of phagocytosis of Aeromonas salmonicida by leukocytes isolated from gut of rainbow trout (*Oncorhynchus mykiss*) using a radiolabelling assay. Comp Immunol, Microbiol Infect Dis 29:335–343

Balcázar JL, De Blas I, Ruiz-Zazuela I, Cunningham D, Vandrell D, Muzquiz JL (2006b) The role of probiotics in aquaculture. Vet Microbiol 114:173–186

Balcázar JL, De Blas I, Ruiz-Zarzuela I, Vendrell D, Calvo AC, Marquez I, Girones O, Muzquiz JL (2007) Changes in intestinal microbiota and humoral immune response following probiotic administration in brown trout (*Salmo trutta*). Br J Nutr 97:522–527

Beachey EH (1981) Bacterial adherence: adhesin–receptor interactions mediating the attachment of bacteria to mucosal surfaces. J Infect Dis 143:325–345

Bestha Lakshmi, Buddolla Viswanath, Sai Gopal DVR (2013) Probiotics as antiviral agents in shrimp aquaculture. J Pathog 2013:13 (Article ID 424123)

Biomin (2009) Probiotics in aquaculture operations. Biomin Newsl 9:1–4

Burr G, Gatlin GM, Ricke S (2005) Microbial ecology of the gastrointestinal tract of fish and the potential application of prebiotics and probiotics in finfish aquaculture. J World Aquacult Soc 36:425–436

Carnevali O, de Vivo L, Sulpizio R, Gioacchini G, Olivotto I, Silvi S, Cresci A (2006) Growth improvement by probiotic in European sea bass juveniles (Dicentrarchus labrax, L.), with particular attention to IGF-1, myostatin and cortisol gene expression. Aquaculture 258:430–438

Cerrato RM (2000) What fish biologists should know about bivalve growth? Fish Res 46:39–49

Conway PL (1989) Lactobacilli: fact and fiction. In: Grubb R (ed) The regulatory and protective role of the normal microflora. Macmillan, Basingstoke, pp 263–281

Decamp O, Moriarty D (2007) Aquaculture species profit from probiotics. Feed Mix 15:20–23

Denev SA (1996) Probiotics-past, present and future. Bulg J Agric Sci 2:445–474

Denev SA (2008) Ecological alternatives of antibiotic growth promoters in the animal husbandry and aquaculture. D.Sc. thesis, Department of Biochemistry Microbiology, Trakia University, Stara Zagora, Bulgaria, 294 pp

Denev S, Staykov Y, Moutafchieva R, Beev G (2009) Microbial ecology of the gastrointestinal tract of fish and the potential application of probiotics and prebiotics in finfish aquaculture. Int Aquat Res 1:1–29

FAO/WHO (2001) Report of a joint FAO/WHO expert consultation on evaluation of health and nutritional properties of probiotics in food including powder milk with live lactic acid bacteria. Cordoba, Argentina

Fioramonti J, Theodorou V, Bueno L (2003) Probiotics: what are they? What are their effects on gut physiology? Best Pract Res Clin Gastroenterol 17:711–724

Fuller R (1989) Probiotics in man and animals. J Appl Bacteriol 66:365–378

Fuller R (1992) Probiotics: the scientific basis. Chapman & Hall, London

Gatesoupe FJ (1999) The use of probiotics in aquaculture. Aquaculture 180:147–165

Gilliland SE (1979) Beneficial inter-relationships between certain microorganisms and humans: candidate microorganisms for use as dietary adjuncts. J Food Prot 42:164–167

Goktepe I, Juneja VK, Ahmedna M (2006) Probiotics in food safety and human health. CRC Press, Taylor & Francis, 6000 Broken Sound Parkway NW, Suite 300, Boca Raton, FL 33487–2742 p 494

Gomez-Gil B, Roque A, Turnbull JF (2000) The use and selection of probiotic bacteria for use in the culture of larval aquatic organisms. Aquaculture 191:259–270

Gram L, Melchiorison J, Spanggerard B, Hubber I, Nielsen TF (1999) Inhibition of *Vibrio anguillarum* by *Pseudomonas fluorescens* A H2 G possible probiotic treatment of fish. Appl Environ Microbiol 65:969–973

Hamilton-Miller JM (2003) Probiotics and prebiotics in the elderly. Post Grad Med J 80:47–57

He S, Zhou Z, Liu Y, Shi P, Yao B, Ringø E, Yoon I (2009) Effects of dietary Saccharomyces cerevisiae fermentation product (DVAQUA®) on growth performance, intestinal autochthonous bacterial community and non-specific immunity of hybrid tilapia (Oreochromis niloticus ♀×O. aureus ♂) cultured in cages. Aquaculture 294:99–107

Irianto A, Austin B (2002a) Probiotics in aquaculture. J Feed Dis 25:1–10

Irianto A, Austin B (2002b) Use of probiotics to control furunculosis in rainbow trout, *Oncorhynchus mykiss* (Walbaum). J Fish Dis 25:333–342

Jong SC (1993) Probiotics for humans and animals. ATCC Q Newsl 13:1–4

Kim DH, Austin B (2006) Innate immune responses in rainbow trout (Oncorhynchus mykiss, Walbaum) induced by probiotics. Fish Shellfish Immunol 21: 513–524

Kollath W (1953) Nutrition and tooth system: general review with special reference to vitamins. Dtsch Zahnarztl Z 8:7–16

Korzybski T, Kowszyk-Gindifer Z, Kurylowicz W (1978) Antibiotics isolated from the genus Bacillus (Bacillaceae). In: Antibiotics – origin, nature and properties, vol III. American Society for Microbiology, Washington, DC, pp 1529–1661

Kozasa M (1986) Toyocerin (*Bacillus toyoi*) as growth promoter for animal feeding. Microbiol Aliment Nutr 4:121–125

Lara-Flores M, Olvera-Novoa MA, Guzman-Mendez BE, Lopez-Madrid W (2003) Use of the bacteria Streptococcus faecium and Lactobacillus acidophilus, and the yeast Saccharomyces cerevisiae as growth promoters in Nile tilapia (Oreochromis niloticus). Aquaculture 216:193–201

Mahdhi A, Amoun F, Bakhrouf A (2011) Inhibitory activity and adhesive ability of potential probiotics *Bacillus* sp. To confer protection for *Artemia gnotobiotic* culture against pathogenic Vibrio sp. Turk J Vet Anim Sci 35:227–233

Mesalhy ASM, Yousef AGA, Ghareeb AAA, Mohamed MF (2008) Studies on Bacillus subtilis and Lactobacillus acidophilus, as potential probiotics, on the immune response and resistance of Tilapia nilotica (Oreochromis niloticus) to challenge infections. Fish Shellfish Immunol 25:128–136

Metchnikoff E (1907) The prolongation of life. Optimistic studies. William Heinemann, London

Nageswara PV, Babu DE (2006) Probiotics as an alternative therapy to minimize or avoid antibiotics use in aquaculture. Fish Chimes 26:112–114

Nikoskelainen S, Ouwehand A, Salminen S, Bylund G (2001) Protection of rainbow trout (Oncorhvnchus mykiss) from furunculosis by Luctohacillus rhamnosus. Aquaculture 198:229–236

Nikoskelainen S, Ouwehand AC, Bylund G, Salminen S, Lilius EM (2003) Immune enhancement in rainbow trout (Oncorhynchus mykiss) by potential probiotic bacteria (Lactobacillus rhamnosus). Fish Shellfish Immunol 15:443–452

Ouwehand AC, Salminen S, Isolauri E (2002) Probiotics: an overview of beneficial effects. Antonie van Leewenhoek 82:279–289

Panigrahi A, Kiron V, Kobayashi T, Puangkaew J, Satoh S, Sugita S (2004) Immune responses in rainbow trout Oncorhynchus mykiss induced by a potential probiotic bacteria Lactobacillus rhamnosus JCM 1136. Vet Immunol Immunopathol 102:379–388.

Panigrahi A, Kiron V, Puangkaew J, Kobayashi T, Satoh S, Sugita H (2005) The viability of probiotic bacteria as a factor influencing the immune response in rainbow trout Oncorhynchus mykiss. Aquaculture 243:241–254

Panigrahi A, Kiron V, Satoh S, Hirono I, Kobayashi T, Sugita H, Puangkaew J, Aoki T (2007) Immune modulation and expression of cytokine genes in rainbow trout Oncorhynchus mykiss upon probiotic feeding. Dev Comp Immunol 31:372–382

Pirarat N, Ponpornpisit A, Rodkhum C, Chansue N (2008) Lactobacillus rhamnosus GG, a potential human-derived probiotic candidate for Tilapia (Oreochromis niloticus) culture. In: Proceedings the 15th congress of FAVA, 27–30 October, Bangkok, Thailand, pp 141–142

Parker RB (1974) Probiotics, the other half of the antibiotic story. Anim Nutr Health 29:4–8

Raida MK, Larsen JL, Nielsen ME, Buchmann K (2003) Enhanced resistance of rainbow trout, Oncorhynchus mykiss (Walbaum), against Yersinia ruckeri challenge following oral administration of Bacillus subtilis and B. licheniformis (BioPlus2B). J Fish Dis 26:495–498

Robertson PAW, O'Dowd C, Burrells C, Williams P, Austin B (2000) Use of Carnobacterium sp. as a probiotic for Atlantic salmon (Salmo salar L.) and rainbow trout (Oncorhynchus mykiss, Walbaum). Aquaculture 185:235–243

Ramos EA, Relucio JL, Torres-Villanueva CA (2005) Gene expression in tilapia following oral delivery of chitosan encapsulated plasmid DNA incorporated into fish feeds. Mar Biotechnol 7:89–94

Rengpipat S, Runkpratanporn S, Piyatiratitivorakul S, Menasaveta P (2000) Immunity enhancement on black tiger shrimp (*Penaeus monodon*) by a probiont bacterium (*Bacillus* S11). Aquaculture 191:271–288

Ringø E, Salinas I, Olsen RE, Nyhaug A, Myklebust R, Mayhew TM (2007) Histological changes in intestine of Atlantic salmon (Salmo salar L.) following in vitro exposure to pathogenic and probiotic bacterial strains. Cell Tissue Res 328:109–116

Sahu MK, Swarnakumar NS, Sivakumar K, Thangaradjou T, Kannan L (2008) Probiotics in aquaculture: importance and future perspectives. Indian J Microbiol 48:299–308

Salinas I, Cuesta A, Esteban MA, Meseguer J (2005) Dietary administration of Lactobacillus delbriieckii and Bacillus substilis, single or combined, on gilthead seabream cellular innate immune responses. Fish Shellfish Immunol 19:67–77

Senok AC, Ismaeel AY, Botta GA (2005) Probiotics: facts and myths. Clin Microbiol Infect 11:958–966

Sullivan A, Nord CE (2002) The place of probiotics in human intestinal infections. Int J Antimicrob Agent 20:313–319

Taoka Y, Maeda H, Jo JY, Kim S, Park S, Yoshikawa T, Sakata T (2006a) Use of live and dead probiotic cells in tilapia Oreochromis niloticus. Fish Sci 72(4):755–767

Tovar-Ramırez D, Zambonino IJ, Cahu C, Gatesoupe FJ, Vazquez-Juarez R (2004) Influence of dietary live yeast on European sea bass (Dicentrarchus labrax) larval development. Aquaculture 234:415–427

Tukmechi A, Morshedi A, Delirezh N (2007) Changes in intestinal microflora and humoral immune response following probiotic administration in rainbow trout (Oncorhynchus mykiss). J Animal Vet Adv 6(10):1183–1189

Trust TJ, Sparrow RAH (1974) The bacterial flora in the alimentary tract of freshwater salmonid fishes. Can J Microbiol 20:1219–1228

Vendrell D, Balcazar JL, De Blas I, Ruiz-Zarzuela I, Girones O, Muzquiz JL (2008) Protection of rainbow trout (Oncorhynchus mykiss) from lactococcosis by probiotic bacteria. Comp Immunol Microbiol Infec Dis 31:337–345

Venkateswara AR (2007) Bioremediation to restore the health of aquaculture. Pond Ecosystem, Hyderabad, pp 1–12

Verschuere L, Rombaut G, Sorgeloos P, Verstraete W (2000) Probiotic bacteria as biological control agents in aquaculture. Microbiol Mol Biol Rev 64:655–671

Waché Y, Auffray F, Gatesoupe FJ, Zambonino J, Gayet V, Labbé L, Quentel C (2006) Cross effects of the strain of dietary Saccharomyces cerevisiae and rearing conditions on the onset of intestinal microbiota and digestive enzymes in rainbow trout (Onchorhynchus mykiss, fry). Aquaculture 258:470–478

Wang YB, Tian ZO, Yao JT, Li W (2008a) Effect of probiotics, Enterococcus faecium, on tilapia (Oreochromis niloticus) growth performance and immune response. Aquaculture 277:203–207

Westerdahl A, Olsson JC, Kjelleberg S, Conway PL (1991) Isolation and characterization of turbot (*Scophthalmus maximus*)-associated bacteria with inhibitory effects against Vibrio anguillarum. Appl Environ Microbiol 57:2223–2228

Xiang-Hong W, Jun L, Wei-shang J, Huai-shu X (2003) Application of probiotics in aquaculture. Ocean University of Qungo, Qingdao. Online pp l–10

Yanbo W, Zirong X (2006) Effect of probiotics for common carp (Cyprinus carpio) based on growth performance and digestive enzyme activities. Animal Feed Sci Technol 127:283–292

Yassir AL, Adel ME, Azze A (2002) Use of probiology bacteria as growth promoters, antibacterial and the effect on physiological parameters of *Oreochromis niloticus*. J Fish Dis 25:633–642

Zimmerman SB, Schwartz CD, Monaghan RL, Pleak BA, Weissberger B, Gilfillan EC, Mochales S, Hernandez S, Currie SA, Tejera E, Stapley EO (1987) Difficidin and oxydifficidin: novel broad spectrum antibacterial antibiotics produced by *Bacillus subtilis*. J Antibiot 40:1677–1681

Sponge-Associated Microbes: A Potential Probiotics for the Management of Fish Diseases

S. Ravikumar

Introduction

A probiotic is a "live microbial feed supplement which beneficially affects the host animal by improving its intestinal balance". Probiotics, which are microorganisms or their products with benefit to the host, have found use in aquaculture as a means of disease control, supplementing or even in some cases replacing the use of antimicrobial compounds. A wide range of yeast and Gram-positive and Gram-negative bacteria have been evaluated. However, the mode of action of the probiotics is rarely investigated, but possibilities include competitive exclusion, i.e. the probiotics actively inhibit the colonization of potential pathogens in the digestive tract by antibiosis or by competition for nutrient and/or space, alteration of microbial metabolism and/or by the stimulation of host immunity. Probiotics may stimulate appetite and improve nutrition by the production of vitamins, detoxification of compounds in the diet and by the breakdown of indigestible components. There is accumulating evidence that probiotics are effective at inhibiting a wide range of fish pathogens, but the reasons for the inhibitions are often unstated.

S. Ravikumar (✉)
Department of Oceanography and Coastal Area Studies, School of Marine Sciences, Alagappa University, Thondi Campus, Thondi 623409, Tamil Nadu, India
e-mail: ravibiotech201320@yahoo.com

Marine organisms are well known to have the specific relationship with numerous microorganisms, and sponges are no exception to this. The sponge–microbe association is a topic of research since a long time (Vacelet 1975). It is interesting to note that the bacterial density in sponges is attributed to the temporal variations in the surrounding environment (Thakur 2001). Some bacteria permanently reside in the spongy mesophyll, pointing to a close interaction between the host and associated bacteria (Webster and Hill 2001; Friedrich et al. 2001). It has been estimated that in some sponge species, as much as 40 % of the animal biomass must be attributed to the bacteria; the amount exceeds the population of seawater by two to four orders of magnitude (Friedrich et al. 2001).

The association of sponges and microorganisms have been described for different taxonomic groups like Archaea (Fuerst et al. 1998), Bacteria (Ravikumar et al. 2011a, b, c, d), Cyanobacteria (Diaz and Ward 1999), microalgae (Hill 1996), heterotrophic eukaryotes (Rinkevich 1999), Fungi (Sponga et al. 1999) and actinomycetes (Ravikumar et al. 2011a, b, c, d). The diversity in sponge-associated microbes already indicated that a large variety of metabolic relationships between sponge and symbionts are likely to exist. The biology of

S. Perumal et al. (eds.), *Advances in Marine and Brackishwater Aquaculture*,
DOI 10.1007/978-81-322-2271-2_7, © Springer India 2015

microorganisms–sponge relationship has elucidated considerable interest among researchers. The role of sponge-associated microorganisms in the synthesis of compounds of biological interest is the subject of scientific debate (Imhoff and Stohr 2003). The present study made an attempt to find out the potential sponge-associated microbes for the treatment of fish diseases through feed supplement.

Materials and Methods

Collection of Sponges

Marine sponge samples were collected at every month for the period of 1 year by catch at Thondi Coast of Palk Strait was collected from the selected sites. Normally, the collection was carried out around 8 AM in the morning. All the samples were sealed separately in a presterile sip-lap plastic bag and stored in an icebox for the isolation of total heterotrophic bacteria, actinomycetes and cyanobacteria. The isolation was carried out six hours after samplings.

Isolation of Sponge-Associated Microbes

Isolation and Culture Condition of THB and Actinomycetes

One gram of sponge samples was cut into small pieces using sterile scissors and put into 99 ml of presterilized 50 % seawater blank. This set-up was kept over the shaker at 100 rpm for 1 h for the proper detachment of adhering propagules from the sponge tissues. Serial dilutions were carried out and plated in triplicates with Zobell Marine Agar 2216e medium for THB and starch casein agar for actinomycetes by following the pour plate method. After inoculation, the plates were incubated in an inverted position for 24 h at 37 ± 2 °C and actinomycetes for 7–10 days at 28 ± 2 °C. After incubation, the colonies were counted and recorded. Based on the colour and ornamentation (forms, elevation, margin and colour of the colony), strains were selected and restreaked thrice in a nutrient agar medium for THB and ISP2 medium for actinomycetes. The selected culture is grown on nutrient agar plates and ISP2 medium and stored in refrigerator for further identification and antibacterial studies.

Isolation of Sponge-Associated Cyanobacteria

To isolate the sponge-associated cyanobacterial symbionts, sponge tissues (1 cm^3) were excised from different sponge species by using sterile scissors. The excised tissue was transferred to a 250 ml Erlenmeyer flask containing sterilized BG11 medium (Stanier et al. 1971). Simultaneously, excised sponge tissues (1 cm^3) from each sponge sample were homogenized with sterile BG11 medium by using an electric homogenizer (REMI, Mumbai) and inoculated into a sterile BG11 medium [solution I ($g.l^{-1}$) $NaNO_3$, 300.0; K_2HPO_4, 8.0; $MgSO_4.7H_2O$, 15.0; $CaCl_2.2H_2O$, 7.2; $Na_2EDTA.2H_2O$, 0.2; Na_2CO_3, 4.0 (solution II ($g.l^{-1}$) H_3BO_3, 2.86; $MnCl_2.4H_2O$, 1.81; $NaMoO_4.2H_2O$, 0.39; $CuSO_4.5H_2O$, 0.079; $Co(NO)_3.6H_2O$, 0.0494].

One ml of solution II and 5 ml of solution I were mixed and made up to 1 l for the preparation of BG11 media. The prepared media was further dispensed into suitable glassware and sterilized in an autoclave prior to inoculation. 1.5 % agar was added for the preparation of the BG11 agar medium.

All the inoculated flasks were incubated in a sterile condition for 30 days under 3,000 lux light intensity with 12 h light/12 h dark photoperiod. Triplicates were maintained for each sample. After incubation, cyanobacterial growth in the sides, bottom of the flasks and top of the medium was taken for identification by using a phase contrast microscope (40×). Mixed cultures of cyanobacteria were further purified by standard serial dilution and plating method. The axenic culture of cyanobacterial symbionts were identified at the species level at the National Facility for Marine Cyanobacteria (NFMC), Bharathidasan University, Tiruchirappalli, Tamil Nadu, India, and also by following the method of Desikachary (1959).

Screening of Total Heterotrophic Bacteria (THB) and Actinomycetes Against Fish Pathogens

The antibacterial activities of isolates were tested preliminarily by the cross-streak method. Total heterotrophic bacteria (THB) were streaked across the diameter on Mueller–Hinton agar plates. After that, 24 h-old cultures of fish pathogenic bacteria, viz. *Bacillus subtilis*, *Serratia* sp., *Aeromonas hydrophila*, *Vibrio harveyi* and *Vibrio parahaemolyticus*, were streaked perpendicular to the central strip of total heterotrophic bacteria (THB) culture. All the plates were incubated at 30 °C for 24 h and the zone of inhibition was measured. Similarly, actinomycetes isolates were streaked across the diameter on sterile Mueller–Hinton agar plates and incubated at 28 ° C for 6 days. After incubation, 24 h-old cultures of fish bacterial pathogens, viz. *Bacillus subtilis*, *Serratia* sp., *Aeromonas hydrophila*, *Vibrio harveyi* and *Vibrio parahaemolyticus*, were streaked perpendicular to the central strip of actinomycetes culture. All the plates were incubated at 30 °C for 24 h and the zone of inhibition was measured.

Secondary Screening

Mass Production of Sponge Microbes

Total Heterotrophic Bacteria

A loopful inoculum of the THB-20 strain which showed potential antibacterial sensitivity against isolated fish pathogens was further inoculated into a 500 ml conical flask containing 100 ml of nutrient broth (pH 7.2) prepared with 50 % of aged seawater and kept at 37 ± 2 °C for 24 h with continuous shaking. Twenty millilitre of the broth culture was then transformed to 1,000 ml of nutrient broth prepared with 50 % of aged seawater and incubated for 4–5 days under continuous shaking.

Actinomycetes

A loopful inoculum of the five actinomycetes which showed potential antibacterial sensitivity against isolated fish pathogens was inoculated further into a 500 ml conical flask containing 100 ml of yeast extract–malt extract broth (ISP2) and kept at 28 ± 2 °C for 72 h with continuous shaking. Twenty millilitres of the broth culture was then transformed to 1,000 ml of glycerol–asparagine broth (ISP5) and incubated at 28 ± 2 °C for 7 days under continuous shaking.

Cyanobacteria

Mass cultivation of three species of cyanobacterial symbionts, viz. *Ph. angustissimum*, *O. amphibia* and *C. minor*, was carried out under field conditions for the mass-scale production of bioactive compounds. To achieve this, 1 g of three in vitro grown cyanobacterial symbionts were inoculated into 15 l of filtered seawater (sterilized by UV) with the addition of BG11 ingredients as per the standard concentrations. After that, 150 ml of sponge homogenate (E_2) prepared from three sponge species [*Sigmadocia carnosa* (MSUSR3), *Colloclathria* sp. and *Callyspongia fibrosa*] which were already proved to enhance the maximum number of growth parameters in cyanobacteria were added. The water level in each trough was maintained daily by pouring additional sterilized filtered tap water to avoid loss due to evaporation. The salinity was maintained at 35 ± 2 ppt. The troughs were closed with a single polythene sheet, capable of preventing about 25 % of sunlight. After 30 days, biomass from each treatment was harvested by filtration by using cheesecloth and then shade-dried for the extraction of bioactive compounds. Triplicates were maintained for each treatment.

Extraction of Bioactive Compounds from Sponge Microbes

THB and Actinomycetes (Ravikumar et al. 2011a)

The pH of the mass-cultivated broth (pH 7.2) culture of THB-20 and five actinomycetes (*Streptomyces albogroseolus*, *Streptomyces aureocirculatus*, *Streptomyces raceochromogenes*, *Streptomyces achromogenes* and *Streptomyces furlongus*) was adjusted to pH 5.0 using 1 N hydrochloric acid and filtered by using muslin cloth to remove the cell biomass. The filtrate was mixed with an equal volume of ethyl acetate. After vigorous shaking, the flask was kept undisturbed until two separate layers were obtained (aqueous and organic). The lower aqueous phase was removed and the upper solvent phase was concentrated in a vacuum evaporator at 40 °C for 24 h and the crude extract was obtained. This process was repeated three times to obtain complete extraction of active principles. This crude extract was used for the antibacterial screening against fish pathogens.

Cyanobacteria (Ravikumar et al. 2011a)

The shade-dried cyanobacterial biomass of three species (*Ph. angustissimum*, *O. amphibian* and *C. minor*) was soaked in a chloroform–methanol mixture (3:1) for a week (percolation). After percolation, the extracts were filtered separately and evaporated by using Rotary Flash Evaporator (Buchi, Japan) to obtain dried powder without the solvent. This was further used for the antibacterial screening assay.

Antibacterial Sensitivity Against Fish Pathogens by Disc Diffusion Assay (Ravikumar et al. 2011b)

To find out the effect of the microbial extract (THB, actinomycetes and cyanobacteria) on the sensitivity pattern of chosen fish bacterial pathogens, 100 mg of extracts was dissolved in the same solvent and impregnated into Whatman No. 1 filter paper disc (6 mm dia). The control disc was impregnated with solvent alone. After that, the discs were placed on sterilized Muller–Hinton agar (HiMedia, Mumbai) plates previously seeded with 24 h-old test organisms. After incubation at room temperature for 24 h, the zone of inhibition was measured and expressed as millimetre in diameter.

Determination of Minimum Inhibitory Concentration (MIC) (Ravikumar et al. 2011c)

Minimum inhibitory concentration is defined as the lowest concentration of bioactive compounds that inhibit the growth of a particular microorganism in broth dilution method. 0.1 ml of 24 h bacterial broth culture of identified fish pathogens *Bacillus subtilis*, *Serratia* sp., *Aeromonas hydrophila*, *Vibrio harveyi* and *Vibrio parahaemolyticus* was added to the tubes containing 0.5 ml of varying concentrations of the extracts (125, 250, 500, 1,000, 1,500 μg. ml^{-1}) obtained from sponge-associated bacteria THB-20 and five actinomycetes strains, and further the total volume is made up to 1 ml by adding 0.4 ml of sterilized nutrient broth. Nutrient broth alone served as negative control. The whole set-up in triplicate was incubated at 37 °C for 48 h in a thermostat shaker. After incubation, the tubes were then examined for microbial growth by turbidity observations.

Determination of Minimum Bactericidal Concentration (MBC) (Ravikumar and Suganthi 2011)

To determine the minimum bactericidal concentration, a loopful inoculum of broth culture from each MIC tube was streaked onto a nutrient agar in triplicates and then incubated at 37 °C for 24 h. After incubation, the concentration at which no visible growth was noted is the minimum bactericidal concentration.

Result and Discussion

The present study made an attempt to find out the antibacterial activity of isolated total heterotrophic bacteria (THB), actinomycetes and cyanobacteria isolates against isolated 5 fish pathogens. Of the selected 22 THB and 63 actinomycetes strains, only one strain THB-20 and five actinomycetes isolates showed positive antibacterial activities

Table 1 Antibacterial activity of sponge-associated total heterotrophic bacteria (THB) and actinomycetes against fish pathogens

Sponge-associated microbes	*Bacillus subtilis*	*Serratia* sp.	*Aeromonas hydrophila*	*Vibrio harveyi*	*Vibrio parahaemolyticus*
THB-01	–	+	++	+	–
THB-02	–	+	–	–	–
THB-03	+	++	–	–	+
THB-04	++	++	–	–	+
THB-05	+	+	–	–	+
THB-06	–	–	+	–	+
THB-07	–	+	–	–	+
THB-08	+	–	–	–	–
THB-09	–	–	–	+	–
THB-10	–	–	–	+	+
THB-11	–	–	–	+	–
THB-12	–	–	–	+	+
THB-13	+	+	+	–	+
THB-14	+	–	+	–	+
THB-15	–	+	+	–	+
THB-16	–	+	–	–	–
THB-17	+++	–	++	–	+
THB-18	–	+	–	–	–
THB-19	–	–	–	+	–
THB-20	+++	++	++	+++	++
THB-21	–	–	+	–	+
THB-22	–	–	+	–	–
	Actinomycetes				
ACT1	++	+	+	–	–
ACT2	–	+	–	+	+
ACT3	+	+	–	–	–
ACT4	–	–	–	–	–
ACT5	+	+	++	+	+
ACT6	–	+	+	–	–
ACT7	–	+	+	–	–
ACT8	++	+	+	++	++
ACT9	–	–	+	+	+
ACT10	++	+	++	+	+
ACT11	+	–	–	+	+
ACT12	–	+	+	+	+
ACT13	+	+	–	+	++
ACT14	–	–	–	+	–
ACT15	–	+	–	–	+
ACT16	–	–	–	–	–
ACT17	–	–	–	–	–
ACT18	–	–	–	–	–
ACT19	+	–	–	+	+
ACT20	++	+	+	–	+
ACT21	–	–	–	–	–
ACT22	+	++	++	+	+
ACT23	–	–	–	–	–

(continued)

Table 1 (continued)

Sponge-associated microbes	*Bacillus subtilis*	*Serratia* sp.	*Aeromonas hydrophila*	*Vibrio harveyi*	*Vibrio parahaemolyticus*
ACT24	–	+	+	–	+
ACT25	+	–	+	+	+
ACT26	–	–	–	–	–
ACT27	++	+	+	+	+
ACT28	–	–	–	–	–
ACT29	++	++	++	+++	+++
ACT30	+	+	–	+	–
ACT31	+	–	+	+	+
ACT32	–	–	–	–	+
ACT33	+	–	–	–	+
ACT34	–	–	–	+	–
ACT35	+	–	+	+	–
ACT36	+	+	+	+	+
ACT37	++	+	+	+	+
ACT38	+	+	+	+	+
ACT39	+	+++	+	++	+++
ACT40	+	+	+	+	+
ACT41	+	+	+	+	+
ACT42	–	–	+	+	+
ACT43	–	–	–	–	–
ACT44	+++	+++	+++	+++	+++
ACT45	++	+	+	++	+
ACT46	+++	+++	++	++	++
ACT47	+	+	+	+	+
ACT48	+	+	+	++	+
ACT49	+	+	+	+	+
ACT50	–	–	–	–	–
ACT51	+	+	+	+	+
ACT52	–	–	–	–	–
ACT53	+	+	+	+	–
ACT54	+	+	+	+	–
ACT55	+	+	+	+	+
ACT56	+	++	+	+	+
ACT57	++	++	++	++	++
ACT58	–	–	–	–	–
ACT59	–	+	+	+	–
ACT60	–	–	–	–	–
ACT61	+	–	+	–	+
ACT62	+	+	+	+	+
ACT63	++	+	+	++	+

"+++" = high, "++" = medium, "+" = low, "–" = absence of inhibition

against the isolated fish pathogens with low, medium and high inhibitory activities (Table 1). Similarly, actinomycetes isolated from sponges were also tested for their sensitivity against five fish bacterial pathogens. Of the selected actinomycetes, only five isolates showed sensitivity against fish pathogens. Moreover, the present study also made an attempt to find out the MIC and MBC values of the most promising strains of total heterotrophic bacteria (THB) and actinomycetes

(THB-20 and five actinomycetes strains), and the results showed that the THB and actinomycetes showed concentration-dependent MIC values between the range of 125 and 1,500 μg.ml^{-1}. Moreover, THB-20 extract showed minimum inhibitory concentration values of 500 μg.ml^{-1} against *Bacillus subtilis* and *Serratia* sp. and 1,000 μg.ml^{-1}. Similarly, the extract of actinomycetes showed 500 μg.ml^{-1} concentration of inhibitory activity against all fish pathogens (Table 2). The minimum bactericidal concentration (MBC) test was also carried out by the present study, and it shows that the most promising strains of THB and actinomycetes which confirm the minimum bactericidal concentration against five tested pathogens, and the results are represented in Tables 3 and 4. It clearly reveals that THB-20 and *S. aureocirculatus* showed concentration-dependent MBC values ranged from 500 to 1,500 μg.ml^{-1}. The THB-20 strain showed that the MBC value at the concentration of 1,000 inhibit the *Aeromonas hydrophila*, *Vibrio harveyi and V. parahaemolyticus* and at the concentration level of 1,500 μg.ml^{-1} inhibit all fish pathogens. The actinomycetes extracts showed various MBC concentration values of 500–1,500 μg.ml^{-1} was recorded against all the bacterial fish pathogens *B. subtilis*, *Serratia* sp., *A. hydrophila*, *V. harveyi and V. parahaemolyticus*. Gandhimathi et al. (2008) reported that the minimum inhibitory concentration and minimum bactericidal concentration of *Nocardiopsis dassonvillei* showed active against tested pathogens at 300–600 μg.ml^{-1}. Bhosale and Gadre (2002) and Ravikumar et al. (2011d) reported that the extracts from *Cymodocea rotundata* showed sensitivity against several *Bacillus* species. In addition, Ravikumar et al. (2011a) reported that the ethanolic root extracts of seagrass *Syringodium isoetifolium* exhibited an MIC and MBC value of 1,000 μg. μl^{-1} against three bacterial fish pathogens, viz. *Aeromonas hydrophila*, *Bacillus subtilis* and *Serratia* sp. The results confirmed that invertebrate-associated microorganisms are potential resources of bioactive natural products due to their competition for nutrition and light (Jayanth et al. 2002). The antibacterial sensitivity of ethyl acetate extracts of chosen actinomycetes strains against fish pathogens revealed that the maximum zone of inhibition was noticed from the *S. raceochromogenes* (10 mm dia) against all the tested pathogens, viz. *Vibrio parahaemolyticus*, *V. harveyi*, *Serratia* sp., *A. hydrophila* and *B. subtilis*. The mean zone of inhibition among *Streptomyces* species indicated that *Serratia* sp. was found more sensitive (9.2 mm dia) than the other pathogens. Likewise, the mean zone of inhibition among bacterial pathogens revealed that *S. raceochromogenes* extract showed maximum sensitivity (10 mm dia) than the other actinomycetes species (Table 4). Similarly, the maximum zone of inhibition was noticed from the THB-20 (10 mm dia) against *B. subtilis*, *Serratia* sp. and *A. hydrophila*.

The antibacterial sensitivities of three sponge-associated cyanobacterial species, namely, *Ph. angustissimum*, *C. minor* and *O. amphibia*, were tested against five fish pathogens. This indicates that the antibacterial sensitivity of sponge-associated cyanobacteria isolated from three sponge species showed sensitivity against all the tested fish pathogens (Table 5). The zone of inhibition among the sponge-associated cyanobacterial species reveals that *Ph. angustissimum* mass cultivated in *Colloclathria* sponge extract showed maximum sensitivity (10 mm dia.) against fish pathogen *V. harveyi* (Table 5). It is also reported by the present study that among the three sponge-associated cyanobacterial species, *Ph. angustissimum* showed maximum sensitivity against tested human and fish pathogens. Gopalakrishnan et al. (1998) screened 14 cyanobacterial species against 8 human bacterial pathogens. Of these, *Ph. angustissimum* only showed sensitivity against three tested pathogens than the other cyanobacterial species. It is important to note that the same species isolated from the sponge tissue showed promising sensitivity against the tested fish pathogens, when compared with the free-living forms. It is also interesting to note that all the antagonistic actinomycetes isolated in this study were identified as bacteria, *Streptomyces* and cyanobacteria. The results of the present study are in agreement with the earlier findings that bacteria and *Streptomyces* are mainly found

Table 2 Minimum inhibitory concentration of crude extract obtained from sponge-associated THB-20 and actinomycetes

Sponge-associated extract	Concentration (μg.ml^{-1})																								
	Bacillus subtilis					*Serratia* sp.					*Aeromonas hydrophila*					*Vibrio harveyi*					*V. parahaemolyticus*				
	125	250	500	1,000	1,500	125	250	500	1,000	1,500	125	250	500	1,000	1,500	125	250	500	1,000	1,500	125	250	500	1,000	1,500
THB-20	+++	++	+	–	–	+++	++	–	–	–	+++	+++	+	–	–	+++	+++	+	–	–	+++	+++	++	–	–
S. albogroseolus	+++	+++	++	+	–	+++	+++	+	–	–	++	++	++	+	–	+++	++	++	+	–	+++	++	+	–	–
S. aureocirculatus	+++	++	+	–	–	+++	++	+	–	–	+++	++	+	–	–	+++	++	+	–	–	+++	++	+	–	–
S. raceochromogenes	+++	++	+	–	–	+++	++	+	–	–	++	++	++	+	–	+++	++	++	+	–	+++	++	+	–	–
S. achromogenes	+++	++	+	–	–	+++	++	+	–	–	+++	++	+	–	–	+++	++	+	–	–	+++	++	+	–	–
S. furlongus	+++	++	+	–	–	+++	++	–	–	–	+++	+++	+	–	–	+++	+++	+	–	–	+++	+++	++	–	–

"+++" = high, "++" = medium, "+" = low, "–" = absence of inhibition

Table 3 Minimum bactericidal concentration of crude extract obtained from sponge-associated THB-20 and actinomycetes

	Concentration (μg.ml^{-1})																								
	Bacillus subtilis					*Serratia* sp.					*Aeromonas hydrophila*					*Vibrio harveyi*					*V. parahaemolyticus*				
Sponge-associated extract	125	250	500	1,000	1,500	125	250	500	1,000	1,500	125	250	500	1,000	1,500	125	250	500	1,000	1,500	125	250	500	1,000	1,500
THB-20	+++	+++	++	+	–	+++	++	++	++	–	+++	++	+	–	–	+++	+++	+	–	–	+++	+++	++	–	–
S. albogroseolus	+++	+++	++	+	–	+++	+++	+	–	–	++	++	++	+	–	+++	++	++	+	–	+++	++	+	–	–
S. aureocirculatus	+++	++	++	+	–	+++	+	+–	+	–	+++	++	+	–	–	+++	+	+	–	–	+++	++	++	+	–
S. raceochromogenes	+++	++	+	–	–	+++	++	+	–	–	++	++	++	+	–	+++	++	++	+	–	+++	++	+	–	–
S. achromogenes	+++	++	+	–	–	+++	++	+	–	–	+++	++	+	–	–	+++	++	+	–	–	+++	++	+	+	–
S. furlongus	+++	+++	++	+	–	+++	++	++		–	+++	++	–	–	–	+++	+++	+	–	–	+++	+++	++	–	–

"+++" = high, "++" = medium, "+" = low, "–" = absence of inhibition

Table 4 Antimicrobial susceptibility of antagonistic THB-29 and actinomycetes against fish pathogens

	Zone of inhibition in mm diameter					
Test pathogens	THB-20	*Streptomyces albogroseolus*	*Streptomyces aureocirculatus*	*Streptomyces raceochromogenes*	*Streptomyces achromogenes*	*Streptomyces furlongus*
Bacillus subtilis	10	9	10	10	7	8
Serratia sp.	10	9	8	10	8	8
Aeromonas hydrophila	10	8	10	10	10	8
Vibrio harveyi	9	8	–	10	8	8
V. parahaemolyticus	9	8	–	10	10	8

– = no activity

Table 5 Antibacterial sensitivity of sponge-associated cyanobacterial species against fish pathogens

		Zone of inhibition (mm dia)			
Sponge species	Bacterial pathogens	*Ph. angustissimum* (MSUSRC1)	*C. minor*	*O. amphibia*	Average
Colloclathria sp.	*Serratia* sp.	9	8.5	7	8.2
	Aeromonas hydrophila	8.2	7	6	7.0
	Vibrio harveyi	10	9	6.5	8.5
	Vibrio parahaemolyticus	7	7.8	8	7.6
	Average	8.5	8.0	6.8	
Callyspongia fibrosa	*Serratia* sp.	8.4	7	9	7.1
	Aeromonas hydrophila	7.6	7	8	7.5
	Vibrio harveyi	9	6.6	7.3	7.6
	Vibrio parahaemolyticus	6	6	7	6.6
	Average	7.7	6.6	7.8	
Sigmadocia carnosa	*Serratia* sp.	8	7	8.1	7.7
	Aeromonas hydrophila	7	7.2	9	7.7
	Vibrio harveyi	8	6	6.4	6.8
	Vibrio parahaemolyticus	6.2	6.9	6.9	6.6
	Average	7.3	6.7	7.6	

in shelf and shallow areas, when compared to the other genera of actinomycetes (Weyland and Helmke 1988). The results of the present study clearly suggest that the antagonistic sponge-associated actinomycetes could be used as a potential source for antibiotics. Further studies are continued to isolate, purify and characterize the antibacterial compounds from the sponge-associated microbes and to test their possible use as the alternative chemotherapeutic drugs on a commercial scale.

References

Bhosale P, Gadre RV (2002) Manipulation of temperature and illumination conditions for enhanced β-carotene production by mutant 32 of *Rhodotorula glutinis*. Lett Appl Microbiol 34:349–353

Desikachary TV (1959) Cyanophyta. Indian Council of Agricultural Research, New Delhi, 686 pp

Diaz MC, Ward BB (1999) Perspectives on sponge-cyanobacterial symbioses. In: Hooper JNA (eds) Proceedings of 5th international sponge symposium. Queensland Museum, Brisbane, vol 44, p 154

Friedrich AB, Fischer I, Proksch P, Hacker J, Hentschel U (2001) Temporal variation of the microbial community associated with the Mediterranean sponge, *Aplysinaaerophoba*. FEMS Microbiol Ecol 38:105–113

Fuerst JA, Webb RI, Garson MJ, Hardy L, Reiswig HM (1998) Membrane bounded nucleoids in microbial symbionts of marine sponges. FEMS Microbiol Lett 166:29–34

Gandhimathi RMA, Kumar J, Selvin T, Thangavelu S, Sivaramakrishnan GS, Kiran SS, Natarajasreenivasan K (2008) Antimicrobial potential of sponge associated marine actinomycetes. J Med Mycol 18:16–22

Gopalakrishnan V, Pramod NP, Sundararaman M, Subramaniam G, Thyagarajan SP (1998) Antimicrobial properties of marine cyanobacteria. In: Cyanobacterial biotechnology. Proceedings of the international symposium, Oxford and IBH Publishers, New Delhi, pp 456–458

Hill MS (1996) Symbiotic zooxanthellae enhance boring and growth rate of the tropical sponge *Anthosigmellavarians*. Mar Biol 125:649–654

Imhoff JF, Stohr R (2003) In: Muller WEG (ed) Marine molecular biotechnology. Springer, Berlin, pp 35–57

Jayanth K, Jeyasekaran G, YeyaShakila R, Shanmugam SA (2002) Isolation of marine bacteria inhibitory to shrimp bacterial pathogens. Indian J Fish 49:235–239

Ravikumar S, Suganthi P (2011) Biodiversity of total heterotrophic bacteria and actinomycetes in Thondi coast. World Appl Sci J 9:1189–1192

Ravikumar S, Jacob Inbaneson S, Uthiraselvam M, Kaleeswari R, Ramu A, Margaret Beula B, Rajasekar J (2011a) Antibacterial activity of heterotrophic endophytes from Karangkadu mangrove ecosystem, India. J Pharm Res 4:195–198

Ravikumar S, Jacob Inbaneson S, Uthiraselvam M, RajiniPriya S, Ramu A, Margaret Beula B (2011b) Diversity of endophytic actinomycetes from Karangkadu mangrove ecosystem and its antibacterial potential against bacterial pathogens. J Pharm Res 4:294–296

Ravikumar S, Syed Ali M, Ramu A, Ferosekhan M (2011c) Antibacterial activity of chosen mangrove plants against bacterial specified pathogens. World Appl Sci J 14:1198–1202

Ravikumar S, Syed Ali M, Anandh P, Ajmalkhan M, Dhinakaraj M (2011d) Antibacterial activity of *Cymodocea serrulata* root extracts against chosen poultry pathogens. Indian J Sci Technol 4:98–100

Rinkevich B (1999) Cell cultures from marine invertebrates: new approaches and recent improvements. J Biotechnol 70:133–153

Sponga F, Cavaletti L, Lazzarini A, Borghi A, Ciciliato I, Losi D, Marinelli F (1999) Biodiversity and potentials of marine derived microorganisms. J Biotechnol 70:65–69

Stanier RY, Kunisawa R, Mandel M, Cohen-Bazire G (1971) Purification and properties of unicellular blue-green algae (order Chroococcales). Bacteriol Rev 35 (2):171

Thakur NL (2001) Studies on some bioactivity aspects of selected marine organisms. PhD thesis, Goa University

Vacelet J (1975) Etude en microscopie electronique de l' association entre bacteries et spongiaires du genre verongia (dictyoceratida). J Microsc Biol Cell 23:271–288

Webster NS, Hill RT (2001) The culturable microbial community of the Great Barrier reef sponge *Rhopaloeides odorabile* is dominated by an ά-proteobacterium. Mar Biol 138:843–851

Weyland H, Helmke E (1988) Actinomycetes in the marine environment in biology of actinomycetes. In: Okami Y, Beppu T, Ogawara H (eds) Biology of actinomycetes. Japan Scientific Societies Press, Tokyo, pp 294–299

Controlled Breeding, Seed Production and Culture of Brackishwater Fishes

Thirunavukkarasu A.R., M. Kailasam, J.K. Sundaray, G. Biswas, Prem Kumar, R. Subburaj, and G. Thiagarajan

Introduction

Aquaculture in marine/brackishwater ecosystems in coastal ponds, open sea cages and pens is assuming greater significance in recent years. Out of the total fish production of about 147 million tonnes, the contribution through culture has become half the mark, sharing annual increase between 8 and 10 %. Considering the capture fisheries' stagnations and the growing demand for fish as animal protein source, aquaculture plays an important role in augmenting production. In India, the contribution through aquaculture is 80 % in the freshwater sector and around 160,000 tonnes through coastal aquaculture. Aquaculture has developed rapidly over the last three decades and has become an important growing industry for generating the revenue, providing employment and nutritional security for the millions of people. The ever-increasing population and the rising demand for animal protein are causing pressure on fishery development globally. Fish and fishery products contribute around 15 % of the animal protein supporting the nutritional security. In the global fish production, Asian countries occupy the top eight, where India stands second position after China. China has produced 70 % of the global production which formed 50 % in value, whereas India is a distant second with 5 % production and 4 % in value. The global average per capita consumption of fish is around 15 kg. The present average per capita consumption in India is around 9 kg. In countries like Japan and some of the Southeast Asian countries, the average per capita consumption is more than 100 kg. Reaching the global average of 15 kg, taking into consideration that 50 % of the Indian population will be fish consumers, by 2020 the domestic requirement itself will be in the order of 9 million tonnes. By 2020, the coastal aquaculture in India is expected to support the tune of around 350,000 tonnes, from the current production of around 150,000 tonnes. This implies that a quantum jump has to be made in the ensuing years. Out of this, shrimp is expected to contribute around 250,000 tonnes and the rest has to come through fishes and other nonconventional groups.

Development of aquaculture has become imperative for the following reasons:

- Means of protein-rich fish production for 'nutritional security'
- Generation of employment – livelihood security
- Economic status and social upliftment – social security
- Reduce pressure on wild stock – conservation
- Biological indicator for water quality
- Culture of nutrient utilizers like seaweeds and molluscs improve water quality – environment security

Thirunavukkarasu A.R. (✉) • M. Kailasam
J.K. Sundaray • G. Biswas • P. Kumar
R. Subburaj • G. Thiagarajan
Fish Culture Division, Central Institute of Brackishwater Aquaculture, No. 75, Santhome High Road, R.A. Puram, Chennai 600028, Tamil Nadu, India
e-mail: artarasu@yahoo.com

S. Perumal et al. (eds.), *Advances in Marine and Brackishwater Aquaculture*,
DOI 10.1007/978-81-322-2271-2_8, © Springer India 2015

- Integrated farming like paddy cum fish culture – reduce other outputs

India is bestowed with a coastal line of 8,129 km; estuaries of 3.50 million ha; backwater of 3.90 million ha; mangrove of 0.40 million ha; potential brackishwater area suitable for aquaculture, about 1.19 million ha; freshwater reservoirs of 3.15 million ha; ponds and tank of 2.25 million ha; beels and oxbow lake of 0.82 million ha; medium and large reservoirs of 2.04 million ha; and irrigation canals of 146,000 km. These aquatic systems are either underutilized or unutilized. These areas can be brought into the aquaculture, and production and productivity can be increased.

Status of Coastal Aquaculture

Coastal aquaculture is a traditional practice in India. In the low-lying fields of Kerala (pokkali), West Bengal (bheries and gheries), Orissa, Goa (khzan) and Karnataka (kar) which experiences an influx of salt water, traditional farming of fish/shrimp was practiced. The practice is just allowing juveniles of fish/shrimp in the fields, allowing them to grow, feeding without any supplementary, facilitating water exchange through tidal waters and harvesting periodically at 3–4 months. With the improvement of technologies and realizing the importance of aquaculture, these practices were improved with the supplementary stocking of feeding with water quality management with higher production. The technology improvement made in the aquaculture sector opened new areas for the scientific farming which is called as semi-intensive and intensive farming following all the protocols for farming with production as much as 10 tonnes per ha per culture period of 4–5 months, mainly shrimp in the coastal area. The technology advancement helped in the establishment of more than 380 hatcheries with a production capacity of 5–300 million seeds totalling around 20 billion, and more and more areas were brought under shrimp farming. The present area of operation in the coast line is around 160,000 ha and producing around 160,000 tonnes of shrimp.

The coastal aquaculture witnessed a phenomenal growth during the 1980s and in the beginning of the 1990s. But the growth has not progressed as visualized from the later part of the 1990s due to socioeconomic environmental issues coupled with the outbreak of uncontrollable diseases like WSSV on shrimp. One of the reasons attributed for this is the unregulated development and unforeseen disease outbreaks. The coastal aquaculture in India was also solely dependent on a single group, the shrimp. The effect of this has brought the pronounced impact on the coastal farming sector questioning the very sustainability of the coastal aquaculture.

Diversification in the Coastal Aquaculture

For the sustainable eco-friendly aquaculture practice, diversification to other species is considered as one of the important components. Fishes like the Asian seabass (*Lates calcarifer*), grouper (*Epinephelus tauvina*) and snappers (*Lutjanus* sp.) which are high-value carnivorous fishes and grey mullet (*Mugil cephalus*), milkfish (*Chanos chanos*), pearlspot (*Etroplus suratensis*) and rabbitfish (*Siganus* sp.) which are herbivorous/omnivorous and farmed in the coastal ecosystem are available. The species like cobia (*Rachycentron canadum*) and silver pomfret are being considered as candidate species for farming. Efforts have been made to develop comprehensive technology packages for seed production under controlled conditions and farming for these candidate species. Technologies have been developed elsewhere in the world.

In the Indian scenario, successful technology has been developed by the Central Institute of Brackishwater Aquaculture for the seed production of Asian seabass, *Lates calcarifer*, under controlled conditions and farming. The controlled breeding of groupers *Epinephelus tauvina*, grey mullets *Mugil cephalus* and pearlspot *Etroplus suratensis* has also been successful. The development of broodstock for the captive seed production of milkfish is in progress. Cobia and silver pomfret have been taken up as priority species owing to their high value in the domestic and international markets.

Technology Development for Fish Culture in Coastal Waters

The following discussion will be on the recent technologies developed on the seed production of Asian seabass *Lates calcarifer* which can be a model for the production of marine finfish seed under controlled conditions.

Seed Production Technology for Asian Seabass, *Lates Calcarifer*

Successful seed production in the hatchery depends upon the availability of healthy matured fishes. For selecting potential breeders, viable broodstock under captive conditions has to be developed. Adult and subadult seabass can be procured from wild catch or from farm-reared stock. The fish procured for broodstock should be devoid of external injuries or internal haemorrhage. The fish should be healthy and free from any parasitic infection. The fish can be treated with acriflavine (1 ppm) for 10 min and later with antibiotic, furazolidone (10 ppm), for 1 h as prophylactic treatment to avoid infection due to minor injuries if any during collection and transportation. The fish should be kept under hatchery condition for 3–5 days for close observation before shifting to broodstock-holding facility for further maintenance. Fishes can be maintained at 1 kg/m^3 in the broodstock tank. In the broodstock holding tanks, fishes are fed with trash fish at 5 % of the body weight in frozen form. Fresh low-cost fish like tilapia (*Oreochromis mossambicus*), sardines (*Sardinella* sp.), horse mackerel (*Decapterus* sp.), etc. can be given to the fish. The broodstock tanks are to be disinfected once in 3 months to avoid contamination. From a well-maintained broodstock of fishes under a controlled condition providing good water quality and feed, healthy gravid fishes can be obtained in 6–8 months.

Asian seabass (*Lates calcarifer*)

Captive Maturation and Spawning

The Asian seabass, *Lates calcarifer*, can be made to breed under controlled conditions both spontaneously (natural spawning) and by induced spawning with exogenous hormone administration.

Natural Spawning

This can be achieved by the manipulation of some of the important water quality parameters like salinity, temperature, pH, etc. required for the maturation process, stimulating the conditions prevailing in the marine environment with a flow-through arrangement wherein the seawater pumped into the broodstock maturation tanks is recycled using the biological and pressure sand filters so that the water conditions are stable. With this process, the fish could be made to spawn spontaneously throughout the year, even beyond the normal spawning seasons. This has paved way for the production of seed under controlled conditions throughout the year.

Induced Spawning

Selection of Spawners and Sex Ratio

Matured female fishes will have ova with a diameter more than 450 μ. Males will ooze milt if the abdomen is gently pressed. The gonadal condition is assessed by ovarian biopsy (Szentes et al. 2012). Brood fishes selected for induction of spawning should be active, free from disease, wounds or injuries. Female fishes will be around 4–7 kg and males will be 2.0–3.0 kg. Since seabass spawning is found to have lunar periodicity, days of the new moon or full moon or 1 or 2 days prior or after these days are preferred for inducing the spawning. Female seabass are generally larger (more than 4 kg.), and the males are smaller (in the size of 2.0–3.0 kg) (Davis 1982). To ensure proper fertilization normally two males are introduced for one female in the spawning tank.

Induced Spawning by Hormone Injection

The commonly used hormones in the finfish hatcheries for induced spawning are LHRHa (luteinizing hormone-releasing hormone analogue, available with Sigma Chemicals, USA, and Argent Chemicals, USA), HCG (human chorionic gonadotropins, available in Pharmacy-Medical shops), OVAPRIM, Puberogen, carp pituitary gland extract and pimozide. After selecting the gravid fishes, the requirement of hormone to be injected is assessed. The dosage level has been standardized as LHRHa at 60–70 μg/kg body weight for females and 30–35 μg/kg body weight for males. Since the spawning normally occurs in the late evening hours, when the temperature is cool, the hormone is injected normally in the early hours of the day between 0700 and 0800 h.

Spawning

For fishes injected with the LHRHa hormone, response for spawning was after 30–36 h of injection. Spawning normally occurs late in the evening hours (1900–2000 h). At the time of spawning the fishes will be moving very fast, and in the water surface a milky white substance will be seen. There will be a fishy odour which can be smelled few metres away. Prior to the spawning activity, the males and the female will be moving together exhibiting courtship. The spawning activity in seabass coincides with lunar periodicity. During full moon or new moon days, the activity is found to be in peak. However, with the recirculation system CIBA's seabass hatchery could achieve spawning in any time of the month. Hence, induced spawning is done during new moon/full moon or 1 or 2 days prior or after these days. Seabass has high fecundity. It is a protracted intermittent spawner, and in one spawning the fish may release 1.0–3.0 million eggs (Guiguen et al. 1994). The process of spawning will follow during subsequent days also. If the condition is good, both female and male respond simultaneously resulting to spontaneous natural spawning, and fertilization is affected.

Fertilization

Fertilization is external. In natural spawning of seabass in good maturity condition, fertilization will be 70–90 %. The size of the fertilized eggs will be around 0.75–0.80 mm. The fertilized eggs will be floating on the surface and will be transparent. The unfertilized eggs will be opaque and slowly sink to the bottom. After spawning and fertilization, the water level in the spawning tanks can be increased and allowed to overflow through the overflow outlet. The eggs will be pushed by the water flow. Below the overflow pipe a trough covered with bolting cloth of mesh size 150–200 μm is kept. The water with the egg is allowed to pass through. The eggs are collected in the next bolting cloth washed and transferred to the incubation tanks. Since fertilized eggs will be floating on the surface, a bolting net cloth of 150–200 μm mesh size can be used for collecting the eggs from the surface. The cloth is stretched as net and towed along the water surface. The collected eggs after washing are transferred to the incubation tanks. The water in the spawning tank is siphoned into a small tank covered with collection net cloth through which the water will be allowed to pass through. The eggs collected in the net cloth are transferred periodically to incubation tanks.

Incubation and Hatching

The eggs collected from the spawning tank are washed to remove the debris that would have adhered to and transferred to the hatching tanks for incubation and hatching. The hatching incubation tanks can be 200–250 l capacity cylindroconical tanks. The eggs are kept at 100–200 nos./l density. Continuous aeration is provided. A temperature of 27–28 °C is desirable. The eggs will hatch out in 17–18 h after fertilization. After hatching the larvae are transferred to larval rearing tanks. The unfertilized eggs in the incubation tank can be removed by siphoning (Patnaik and Jena 1976). The larvae are scooped gently using a scoop net and transferred into buckets of known volume. After taking random sample counting depending upon the number required to be kept in the rearing tanks, the larvae will be transferred to rearing tanks.

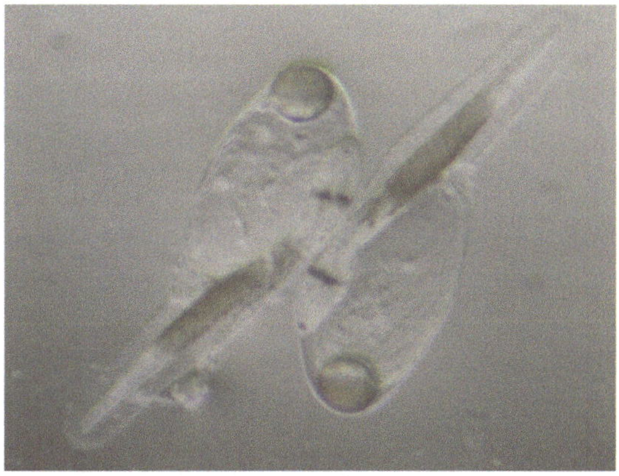

One day old Larvae

Larval Rearing tank

Larval Rearing for Finfishes in India

Rearing of hatchlings through various developmental stages providing required environmental parameters and feed is the most important phase in the seed production technology. This is still more significant in marine fishes like seabass, grey mullet, grouper, cobia, etc. The seabass larval phase extends up to 21 days, and during this time, the feed requirement, type of feed, quantity, etc. also vary with every stage.

Larval Rearing Tanks and Stocking Density

Tanks in the size of 4–5 tonne capacity are preferable for operational convenience. Freshly hatched larvae from the incubation tanks are transferred carefully to the rearing tanks. Larvae are stocked initially at 40–50 nos/l. Depending upon the age and size, the larval density is reduced to 20–25 nos/l on the tenth day or later, and after 15 days, the density is maintained around 10–15 nos/l.

Feed and Feeding During Larval Rearing

Green unicellular algae like the *Chlorella* sp., *Tetraselmis* sp., *Nannochloropsis* sp. or *Isochrysis* sp. are needed for feeding the live feed (zooplankton), rotifer, and for adding to seabass larval rearing tanks for water quality maintenance. Either the rotifer *Brachionus plicatilis* or *B. rotundiformis* is the most preferred diet for the fish larvae in their early stages. The size of the rotifers varies from 50 to 250 μm. The early stage larvae (up to 7 days) are fed with small-sized rotifers, i.e. less than 120 μm, and later assorted-sized rotifers can be fed. Brine shrimp, *Artemia*, in the nauplii stage are required for feeding the larvae from day 9 to day 21, and afterwards *Artemia* biomass can be given. Rotifer (*Brachionus plicatilis*) is given as feed to the larvae from the third day. Rotifer is maintained in the larval rearing tanks at a concentration of 20 nos./ml initially. From the 4th to 15th day, the rotifer concentration is increased to 30–50 nos./ml gradually. And concentration is increased to 6–10 nos/ml from the 9th to 15th day of rearing. Every day after water exchange, the food concentration in the tank should be assessed, and fresh rotifers should be added to the required concentration. *Artemia* nauplii are given as feed along with rotifers and green water from the tenth day. By this time the larvae will be around 4 mm TL in size. Larvae can be fed exclusively with *Artemia* from the 16th day to 24th day. The density of the brine shrimp nauplii in the rearing medium is maintained at 2,000 nos./l initially and gradually increased to 6,000/l as the rearing days progress. The daily ration of *Artemia* nauplii feeding is adjusted after assessing the unfed *Artemia* in the rearing tank at the time of water exchange and the larval density.

Water Exchange

To maintain water quality in the larval rearing tanks, 30–40 % water change is done daily. The salinity should be maintained around 30 ppt. And the desirable range of temperature is 27–29 °C. The water level reduced (30–40 %) in the rearing tank is levelled up with filtered quality seawater and green water after taking cell count of the algae in the rearing tank. Algal water is added daily up to the 15th day. After bottom cleaning and water reduction, while water change is done, algal water is also added depending upon the concentration (around 20,000 cells/ml in the rearing tank).

Nursery Rearing

Nursery Rearing in Hatcheries

Seabass fry of 25–30 days old in the size of 1.0–1.5 cm can be stocked in 5–10 tonne capacity circular or rectangular (RCC or FRP) nursery tanks. Outdoor tanks are preferable. The tanks should have water inlet and outlet provision. Flow-through provision is desirable. In situ biological filter outside the rearing tanks would help in the maintenance of water quality. The water level in the rearing tanks should be 70–80 cm. Good aeration facility should be provided in the nursery tanks. Nursery tanks are prepared a week before stocking. After filling with water 30–40 cm, fertilize with ammonium sulphate, urea and superphosphate at 50, 5 and 5 g (10:1:1 ratio) per 10 tonnes of water, respectively. The natural algal growth would appear within 2–4 days. In these tanks freshly hatched *Artemia* nauplii at 500–1,000 l are stocked after levelling the water to 70–80 cm. The nauplii stocked are allowed to grow into biomass (refer to biomass production of *Artemia*) feeding with rice bran. When sufficient *Artemia* biomass is seen, seabass fry are stocked at 800–1,000 nos/m^3. The preadult *Artemia* would form good food for seabass fry. The fry would not suffer for want of food in the transitional nursery phase in the tank since the larvae are habituated to feed on *Artemia* in the larval rearing phase. Along with '*Artemia* biomass' available as feed inside the tank, supplementary feed mainly minced fish/shrimp meat is passed through a mesh net to make each particle size of around 3–5 mm, and cladocerans like *Moina* sp. can also be given. The fish/shrimp meat feeding has to be done daily three to four times. Feeding rate is 100 % of the body weight in the first week of rearing. This is gradually reduced to 80, 60, 40 and 20 % during the second, third, fourth and fifth week, respectively. Regular water change to an extent of 70 % is to be done daily. The leftover feed and the metabolites have to be removed daily, and aeration should be provided. In a rearing period of 4–5 weeks in the nursery, the seed will be in the size of 1.5–3.0 g/4–6 cm with a survival rate of 60–70 %. Adopting this technique at a stocking density at 1,000 nos/m^3 in the hatchery, a survival rate of up to 80 % has been achieved.

Nursery Rearing in Ponds

Nursery ponds can have an area of around 200–500 m^2 with a provision to retain at least 70–80 cm water level. The pond is prepared before stocking. If there are any predator/pest fishes, they have to be removed. Repeated netting, draining and drying of the pond are done. In case where complete draining is not possible, the water level is reduced to the lowest extent possible and treated with derris root powder at 20 kg/ha added or mahua oil cake (MOC) at 200–300 kg to eradicate unwanted fishes. The use of other inorganic chemicals or pesticides is avoided because these may have a residual effect. After checking the pond bottom quality water is filled. If the pond bottom is acidic, neutralization is done with lime application.

In order to make the natural food abundant, the pond is fertilized with chicken manure at 500 kg/ha keeping the pond water level at 40–50 cm. The water level is gradually increased. After a 2–3-week period when the natural algal food is more, freshly hatched *Artemia* nauplii are introduced. Normally 1 kg of cyst is used for a 1 ha pond. These stocked nauplii grow and become biomass in the pond forming food for the seabass fry. Seabass fry are stocked at 20–30 nos/m^2. Stocking should be done in the early hours of the day. Fry should be acclimatized to the pond condition.

Nursery Rearing in Cages/Hapas

Floating net cages/hapas can be in the size of 2 × 1 × 1 to 2 × 2 × 1 m depending upon necessity. Cages are made with nylon/polyethylene webbing with a mesh size of 1 mm. Fry can be stocked at 400–500/m^2. The feeding rate can be as that described to tank nursery. The net cages have to be checked daily for damages that may be caused by other animals like crabs. The net cages will be clogged by the adherence of suspended and detritus materials and siltation or due to foulers resulting in the restriction of water flow. This would create confinement in the cages and unhealthy conditions. To avoid this, cages/hapas should be cleaned every day. Regular grading should be done to avoid cannibalism and increase the survival rate. Even in higher stocking density at 500/m^2, a farmer could get a survival rate of 80 % in the farm site when the fry were reared in hapas adopting the trash fish feeding and other management strategies mentioned above.

Farming

Traditional Coastal Aquaculture in India

Seabass is cultured in the ponds traditionally as an extensive-type culture throughout the areas in the Indo-pacific region where seabass is distributed. In low-lying excavated ponds, whenever the seabass juveniles are available in the wild seed collection centres (e.g. April–June in

Hapa Nursery

West Bengal, May–August in Andhra Pradesh, Sept–Nov in Tamil Nadu, May to July in Kerala and June–July in Maharashtra), juveniles of assorted size seabass are collected and introduced into the traditional ponds which will be already with some species of fish, shrimps and prawns. Forage fishes like tilapia will also be available in these types of ponds. These ponds will have water source from adjoining brackishwater or freshwater canals or from monsoon flood. The juvenile seabass introduced in the pond will prey upon the available fish or shrimp juveniles as much as possible and grow (Ravisankar and Thirunavukkarasu 2010). Since seabass by nature is a species with different growth that are introduced into the pond at times of food scarce, the larger ones may resort to feed upon the smaller ones reducing the number.

Seabass are allowed to grow for 6–7 months of the culture period till such time when water level is available in these ponds and then harvested. At the time of harvesting, there will be large fishes of 4–5 kg as well as very small fishes. This is a common scenario in many coastal areas. In this manner production up to 2 tonnes/ha/7–8 months have been obtained depending upon the number and size of the fishes entered/introduced into the pond and the feed available in the pond. However, this practice is highly unorganized and without any guarantee on production or return for the aquaculturists. With the advances in the technology in the production of seed under captivity assuring the supply of uniform-sized seed for stocking and quality feed for feeding, the seabass culture is done in Southeast Asian Countries and Australia in a more organized manner.

Pond-Based Fish Farming

Seabass seed can be stocked in a prepared pond at 10,000/ha. The seed size of 2.0 g and above is preferable for stocking in the growout farms. Water depth should be maintained not less than 1.0 M. Seabass fishes stocked can be fed with minced meat of trash fish. Cheaper fishes like tilapia, sardines and horse mackerels which may not fetch more than Rs. 5/per kg can be bought from the commercial fish landing centres, washed and frozen in cold storages as required. The fish can be taken out an hour prior to feeding, thawed and minced as meat using meat mincer. Feed can be made as dough ball like paste and placed in trays, kept hanging in four or five places in the pond. The feeding rate is ad libitum in any case not more than 100 % body weight on wet weight basis of the biomass initially and gradually reduced to 10 % at the last phase of the culture period. Feed rations can be given in two doses in the forenoon and afternoon.

Fish Farming with Formulated Feed

Seabass is cultured with extruded floating pellets in Australia, Thailand, Malaysia and Singapore. Being a carnivorous fish seabass needs a high-protein diet. Normally, in the preparation of diet for seabass, the animal ingredients are added more than 60 % so that the required protein levels can be kept. The nutritional requirements of the seabass are as follows: protein around 55 %, lipid 15 %, fatty acids 2 % and carbohydrates 15 %. Since seabass is a fish feeding mainly on the fishes and shrimps moving in the water column (pelagic), the pellet should be slow sinking and should be in the column for a reasonable time so that the fish can ingest the food before it reaches the bottom. The extrude pellets will have reduced loss, the digestibility will be good due to precooking, the feed mixture can be with higher moisture and the flavour of feed also can be retained with the addition of excess fish oil. The pellet size should be from 2.0 to 6.0 mm as per the size of the fish.

Growout Culture of Seabass in Cages

Fish culture in cages has been identified as one of the eco-friendly at the same time intensive culture practice for increasing fish production. Cages can be installed in open sea or in coastal area. The former is yet to be developed in many countries where seabass is cultured, but coastal cage culture is an established household activity in Southeast Asian countries. There are abundant potential as in India also for cage culture in the lagoons, protected coastal areas, estuaries and creeks. Since cage culture of seabass has been proved to be a technically feasible and viable proposition, this can be taken up in a large scale in suitable areas. The cage culture system allows high stocking density and assures high survival rate. It is natural and eco-friendly and can be adopted to any scale. Feeding can be controlled and cages can be easily managed. Fishes in the cages can be harvested as per the requirement of the consumers, which will fetch high unit price. Above all, cage culture has got low capital input, and operating costs are minimal. Cages can be relocated whenever necessary to avoid any unfavourable condition.

Stocking Density

In the cages, fishes can be stocked at 25–30 nos/m^2 initially when they are in the size of 10–15 g. As they grow, after 2–3 months culture, when the fish attained a mean body weight of 150 g, stocking density has to be reduced to 10–12 nos/m^2 for space. Cage culture is normally done in two phases – till they attain 100–150 g size in 2–3 months and afterwards till they attain 600–800 in 5 months.

Feeding in Cages

Fishes in the cage can be fed with either extruded pellets or with low-cost fishes as per the availability and cost. Floating pellets have advantages of procurement, storage and feeding. Since a lot of low-cost fishes are landed in the commercial landings in the coastal areas which are fetching around Rs. 3–5/kg, they are only used as feed for seabass culture. Low-cost fishes also serve as feed for seabass in ponds and in many cage culture operations. The rate of feeding can be maintained around 20 % initially and reduced to 10 and 5 % gradually in the case of trash fish feeding, and in the pellet feeding, the feeding rate can be around 5 % initially and gradually reduced to 2–3 % at a later stage.

In the feeding of low-cost fish, the feed conversion ratio (FCR) works out around 6 or 7. In the case of pellet feeding, the FCR is to be 1–1.2 in Australia. However, the cost-effectiveness of the pellet feeding for seabass in growout culture has to be tested.

Groupers

Groupers also migrate for maturation and spawning to deeper waters in the sea. The groupers attain maturity after 2 years at their age when they are around 2–3 kg in size. Groupers are protogynous and herbivorous where many are females in the early period and reverse to male when they are larger in size. In hatchery operations, obtaining males sometimes require intervention through exogenous hormone administration. Successful breeding of some species of groupers has been reported from different R&D institutions like CMFRI, CIBA and RGCA.

The techniques for reverting female to male and retaining them as male have been developed in CIBA through oral administration (through feed) of 17 -methyl testosterone hormone in

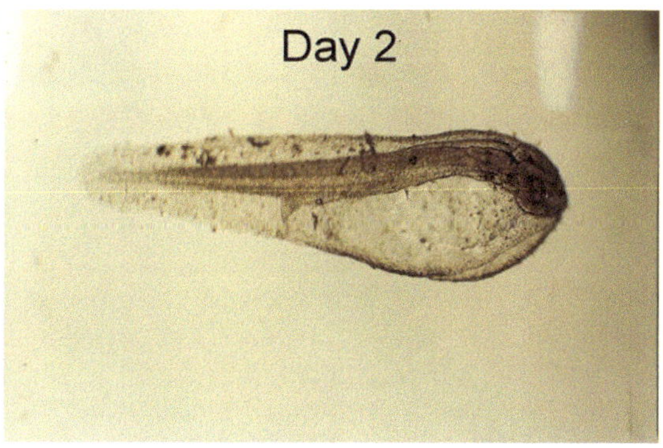

doses of 2 mg/kg body weight on alternate days. The breeding protocols include the selection of females with ova diameter of above 450 μm and administration of HCH hormone at 750–1,000 IU/kg body weight for females and LHRHa at 40 μg/kg body weight. Successful spawnings were observed after 72–144 h of hormonal administration. Hatching took place after 22–24 h of incubation. Rearing the larvae feeding with rotifers SS strain where the size is less than 80 μmm following green water technology has been successful. However, the survival rate is very low (around 5 %) in many cases for a month's rearing.

Grey Mullet (*Mugil Cephalus*) Breeding and Seed Production

Grey mullet *Mugil cephalus* is a herbivorous fish. Considering its high potentiality for farming along with other fishes and shell fishes with low-cost inputs, there is good market demand in some parts of India like Kerala and West Bengal. It is felt that it will be highly useful for sustainable farming in traditional coastal farms. However, breeding of grey mullet under controlled conditions, though being attempted for some years, is yet to take off as a standardized technology for commercial venture.

Grey mullet in the size of 300 g–1.5 kg collected from wild catch or farm-reared stock could be maintained in earthen ponds or broodstock holding tanks feeding with formulated feed at 2–3 % of body weight daily providing with quality seawater with the desirable parameters prevailing in the open sea and taking care of the regular health monitoring protocols. Matured fishes could be obtained during the spawning season, normally in the months of October-–January. Breeding protocols include selection of females with ova diameter around 0.58–0.6 mm and administration of a prime dose of HCG at 1,000 IU and a resolving dose of LHRH at 40–50 μg/kg body weight, and half the dose for the males was found to make successful spawning. The larvae also could be reared following the protocols as for other marine fish larvae. In India though success in captive broodstock development, maturation and spawning has been achieved, the technology for commercial venture is yet to be developed.

Milkfish (*Chanos Chanos)*

Milkfish breeding and seed production have become a household activity in countries like the Philippines, Indonesia and Taiwan. However, in Indian context, breeding of milkfish under

Gravid Fish ready for spawning

Hatched out larvae

captivity is yet to make a beginning. Captive broodstock of milkfish which developed after being fed with formulated feed at 2–3 % body weight and after 5 years of being held under captive conditions have shown male maturation, and the female fishes have not attained gonadal maturity.

Milkfish brooders

Pearlspot (*Etroplus suratensis*)

Pearlspot *Etroplus suratensis*, an indigenous cichlid having a high market value in some parts of India like Kerala, is considered as a highly suitable table fish which can be farmed in ponds or cages with low input in shallow/freshwater/brackishwater systems. Pearlspot breeds in confinement (Padmakumar et al. 2009). After pair formation and selecting a suitable hard substrate, the eggs are laid in a mosaic manner by the female and fertilized by the sperm released by the males by following the course of the female. The eggs are guarded and cleaned periodically for a period of 6–7 days after which they are transferred to nests (pits), at the time of hatching; the hatchlings subsist with yolks for 3–4 days after which they are guarded by the parent fishes till they attain the advanced fry or fingerling stage. To increase the survival rate in the early stages, the eggs at the time of hatching are transferred to tanks and maintained with good aeration through which the hatching rate is improved. Afterwards the juveniles are fed with live zooplankton initially and/or later with egg custards and formulated feed.

Cobia Breeding

Cobia, *Rachycentron canadum*, has also been identified as a potential candidate species for farming in cages in the open waters. Initial trials have shown encouraging results. In India, within 8 months' duration a marketable-sized fish could be grown. Considering the palatability, texture

Pearlspot (*Etroplus suratensis*)

and the market demand, interest is shown by many for the farming of Cobia. A national collaborative project involving CIBA, CMFRI and Fisheries College, Tuticorin, has been launched with the objective of developing broodstock in different ecosystems like open sea cages, earthen ponds and polythene-lined ponds and developing a viable technology for seed production by entrepreneurs (Benetti et al. 2008). Initially success has been achieved by CMFRI using a cage-reared stock wherein females with ova diameter more than 0.7 mm has been selected and induced through hormonal administration.

At CIBA matured fishes could be obtained from the broodstock maintained in earthen pond conditions wherein the routine protocols like water exchange to the extent of 30 % daily feeding with forage fishes at 5 % of body weight and the routine health management protocols for ectoparasitic infection through treatment with 100–150 ppm formalin for one hour and/or dip treatment in freshwater for 5 min providing water quality conditions of salinity ranging from 24 to 32 ppt, pH 7.5–8.2, ammonia less than 1.5 ppm and dissolved oxygen 4.8–6.2 ppm are followed. Fishes in the size range of 5–15 kg were

Pond reared Cobia broodstock

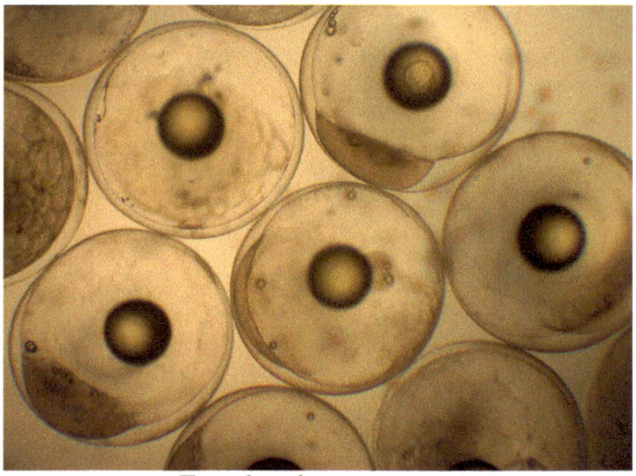
Egg development

maintained in an earthen pond at the density of 1 kg/m^3. Fishes of size 12 kg introduced in the pond during the month of March 2010, after 8 months of maintenance, following the above-mentioned protocols, showed maturity, where females with ova size of 0.5–7 mm and oozing males could be obtained. An effort made under controlled breeding with exogenous hormone of HCG administration at 300 IU intramuscularly in a fish with ova diameter of 0.7 as a priming dose and after 36 h with a resolving dose of LHRH hormone at 10 μg/kg body weight for females and a single dose of LHRH at half the rate for that of females yielded successful spawning after 20 h of hormonal administration where about a million eggs were spawned with 30 % fertilization. This success has showed that broodstock can be maintained in small ponds also which will simplify the operation of hatchery for Cobia.

The protocols developed in some of the above-mentioned species can be applied for similar species with modifications as per the requirement.

Conclusion

Coastal aquaculture has become an important source for the production of food for the alleviation of poverty and generation of employment and wealth for the people living in coastal areas many of whom are underprivileged. The assurance of comprehensive technology puts alone support to sustainable aquaculture in the future. The future areas of research and development must focus on the following to achieve sustainability of aquaculture in the long run:

- Captive broodstock development, captive maturation and induced breeding technology of different cultivable marine finfish
- Live feed culture technology development (Rotifer, *Artemia*, *Moina*, *Daphnia*, Copepod, etc.)
- Health management in broodstock development and post hatching phase
- Genetic improvement and selective breeding
- Development of transgenic fishes for better growth and health
- Feed development for maturation process of fish and different fish larvae
- Development of pen and cage culture system for finfish
- Bio-security in hatchery and farming system

References

Benetti DD, Orhun MR, Sardenberg B, O'Hanlon B, Welch A, Hoenig R, Cavalin F (2008) Advances in hatchery and grow-out technology of cobia *Rachycentron canadum* (Linnaeus). Aquac Res 39:701–711

Davis TLO (1982) Maturity and sexuality in barramundi, *Lates calcarifer* (Bloch) in the Northern Territory and south-eastern Gulf of Carpentaria. Mar Freshw Res 33:529–545

Guiguen Y, Cauty C, Fostier A, Fuchs J, Jalabert B (1994) Reproductive cycle and sex inversion of the seabass, *Lates calcarifer*, reared in sea cages in French Polynesia: histological and morphometric description. Environ Biol Fish 39:231–247

Padmakumar KG, Bindu L, Manu PS (2009) Captive breeding and seed production of *Etroplus suratensis* in controlled systems. Asian Fish Sci 22:51–60

Patnaik S, Jena S (1976) Some aspects of biology of *Lates calcarifer* (Bloch) from Chilka Lake. Indian J Fish 23:65–71

Ravisankar T, Thirunavukkarasu AR (2010) Market prospects of farmed Asian seabass *Lates calcarifer* (Bloch). Indian J Fish 57:49–53

Szentes K, Mszros E, Szab T, Csorbai B, Borbly G, Bernth G, Urbnyi B, Horvth Á (2012) Gonad development and gametogenesis in the Asian sea bass (*Lates calcarifer*) grown in an intensive aquaculture system. J Appl Ichthyol 28:883–885

Culture of Sea bass (*Lates calcarifer*) in Cages in Ponds

S. Kandan

Introduction

Asian sea bass (*Lates calcarifer*), a popular edible marine fin fish, commands consistent demand in domestic and international markets. It has widely been distributed in the Indo-Pacific region and extending up to Taiwan, the southeast Australian coast, Papua New Guinea, the Arabian Sea and the Bay of Bengal and extends to the Persian Gulf region. In India, sea bass fishery is reported from all along the coast including the Andaman and Nicobar Islands. Having the characteristics of catadromous pattern of life cycle, its population occupies a wide range of habitats starting from freshwater lakes, rivers, estuaries and inshore coastal waters. However, the adult fish migrate to deeper inshore sea areas for spawning and as such the early cycle is restricted in seawater areas. Besides, exploiting its natural resources from different environmental conditions, the sea bass become a compatible species for aquaculture in saline water as well as freshwater conditions.

Status of Sea Bass Culture in India

In India, the sea bass has been cultured in brackishwater and freshwater by stocking wild seeds in some part of West Bengal, Tamil Nadu and Kerala. The cage culture sea bass farming is still in developmental stages even though the culture of sea bass in cages is now established by MPEDA, RGCA, CMFRI and other research institutions. For the past five years, a considerable development has been made to culture the species in cages in ponds of all bio categories and hi-tech cages in open sea. But, many problems are remaining unsolved.

1. Cannibalism during fingerling production from fry (1.0–1.5 cm) to 6–7 cm fingerlings
2. Lack of availability weaning diet required for nursery rearing in India
3. Non-availability of extruded pellet feed for grow out in India
4. Non-availability of proper culture techniques in different bio categories in India

Besides these problems, the basic technique for growing sea bass in cages in the pond or in the open water is being developed according to the Indian climatic conditions for the past 3 years.

Culture Technology for Growing Fish in Cages in Ponds

The sea bass can be cultured in freshwater or brackishwater ponds; Cannibalism is one of the most serious problems in sea bass culture especially during larval rearing, fingerling production and up to Juvenile stage (100 g ABW) (Madrones and Catacutan 2012). In order to minimize the

S. Kandan (✉)
The Marine Products Export Development Authority, Regional Centre (Aquaculture), No. 1, Cooks Road, Nagapattinam 611001, Tamil Nadu, India
e-mail: shanmugakandan@yahoo.co.in

S. Perumal et al. (eds.), *Advances in Marine and Brackishwater Aquaculture*,
DOI 10.1007/978-81-322-2271-2_9, © Springer India 2015

Table 1 Optimal environmental parameters for cage culture in ponds

Parameters	Freshwater pond	Brackishwater ponds
Salinity (ppt)	Nil < 5	25–32
DO (ppm)	>4	>4
pH	7.5–8.5	7.5–8.5
Temperature (°C)	24–32	24–32
Transparency (cm)	25–35	25–35
Ammonia (ppm)	<0.1	<0.1
Soil pH	7.5–8.5	7.5–8.5

chances of cannibalism, culture is being carried out in two phases, i.e. the nursery phase and grow-out phase.

Nursery Phase

The main purpose of the nursery is to culture the fry from hatchery (1–2.5 cm in size) to juvenile size (8–10 cm). This can solve the problem of space competition in the nursery tanks. Beyond the nursing period, the juveniles can be graded into different size groups and stocked in separate grow-out ponds. It has been observed that the juveniles from the nurseries perform better in terms of growth and survival than those stocked directly into the grow-out ponds (Tookwinas 1987; Tookwinas and Charearnrid 1988).

Nursing the fry in concrete tanks is not recommended as accumulation of excess feed on the bottom of the tank cannot be avoided. Such accumulation can cause bacterial disease. In addition, constant contact with the tank wall results in wounded fish and subsequent bacterial infection

For nursery rearing nylon happa of 1 m× 1 m× 1 m can be used or knotless nylon cages also can be used for best survival of the seeds.

Grow-Out Phase

Pond Preparation and Elimination of Predators

Prior to the beginning of demonstration, the bottom of the pond should be dried under the sunlight up to cracking stage so as to eliminate the unwanted gases from the pond bottom. The black soil and organic debris have to be removed carefully from the pond and ploughing has to be done. The predators have to be eliminated by applying tea seed cake, dichlorvos and calcium hydroxide at the level of 200 ppm. Later on water should be filled after careful screening of unwanted aquatic animals entering into the pond by using filter bags in the water inlets. Algal bloom has to be cultured by using suitable organic and inorganic fertilizers and the transparency has to be maintained at the level of 25–35 cm. The water column should be maintained 1.8–2 m (Table 1).

Grow-Out Technology

In India, a technology has been developed and perfected for culturing of sea bass in cages in ponds by RGCA, an R&D, and the arm of the Marine Products Export Development Authority. In this method, the pond cages have the dimension of 2 m × 2 m × 1.3 m (approx. 5.0 cu.m.) using PVC pipe frames of 40 mm (floating frame), 32 mm (sinker) and 25 mm (top lid). The cages are fastened to the bamboo or wooden poles of the catwalks fixed in the ponds. The catwalks are provided for the purpose of day-to-day management activities, such as feeding, sampling, grading, etc.

Sea bass cages usually are made of nylon or polyethylene or HDPE netting with varying mesh size depending on the size of the fish grown (Table 2).

The stocking densities in the cages vary according to the size of the fish: as the culture progresses and the fish grow in size, the density

Table 2 Different cage mesh sizes and sizes of the fish to be stocked

Total length of fish (cm)	Cage mesh size (mm)
15–18	20
18–22	24
22–26	32
26–32	38
32 and above	44

Table 3 Suggested stocking density in cages based on number/cu. m.

	Stocking density no./cu. m.	
Size (cm)	With aeration	Without aeration
15.0–20.0	300	180
20.0–24.0	200	140
24.0–28.0	150	100
28.0–30.0	100	70
30.0–32.0	50	30
32.0–34.0	30	15

has to be adjusted suitably. The suggested stocking densities are given in Table 3.

Feed

At present, sea bass culture facing the non-availability of floating extruded pellet feed is the major constraint. However, few companies in India have come forward to manufacture feed for sea bass culture, which is highly suitable for cage culture. Even though the trash fish are given widely for the culture, at present in many places for sustainable aquaculture, the pellet feed is the highly recommended. The feed should be given twice daily in the morning hours, 6–7 A.M. and afternoon 6–7 P.M. at the overall rate of 8–10 % of total biomass in the first 2 months of culture. After 2 months, feeding is reduced to once daily and given in the late evening at the rate of 2–5 % of the total biomass. The floating pellet feed should be given only when the fish swim near the surface to eat. The suggested feeding schedule for extruded pellet feed is given in Table 4.

FCR

For any aquaculture practice, the FCR is the determining factor for the economic viability of the fish for growing domestic or export and also the cost of the production per unit. For sea bass, 1:1.2 FCR is recommended by using extruded pellet feed and 1: 5–7 is the observed FCR by using trash fish or farm-made feed (Schipp 1996).

For the present demonstration the expected production is 3,000 kg by using 7,130 kg of imported pellet feed and FCR is calculated approximately 1:2.37

Grading

The mechanical grader to be supplied by RGCA can be used for grading the fingerlings. Initially, once in 15 days and later monthly once the grading has to be done to separate the shooters and the bigger sea bass fingerling. This exercise will give more survival rate with better growth as the sea bass fingerlings are getting the suitable feed according to their mouth size. Also, the cannibalistic characteristics drastically come down due to timely grading.

Protection of Cages and Fish Stocks from Damage from Predators

As experienced in the past, in freshwater demonstration, water snakes, tortoise and water birds are the permanent predators, which cannot be eliminated by any means that often damaged the grow-out cages by cutting the web and eating away the sea bass fingerlings and juveniles. In brackishwater, mud crab menace is heavy which always damages the grow-out cages. Thus, the fish will be escaped from the cages and hamper the demonstration. So to prevent these permanent predators into grow-out cages, an outer cage made up of nylon (2.5 m × 2.5 m × 1.8 m) is found effective (Philipose et al. 2013).

Table 4 Suggested feeding schedule, as % of body wt., type of feed, etc.

Size (cm)	Feed as % of body weight	Pellet size (mm)	Type of feed
15–18	5.0	2	Floating
18–20	4.0	4	Floating
20–22	3.5	6	Floating
22–25	3.0	8	Floating
25–27	2.6	10	Floating
27–30	2.2	10	Floating
30–35	2.0	14	Floating

Background

The Marine Products Export Development Authority (MPEDA), Regional Centre (Aq), Thanjavur, has taken up demonstration of sea bass culture in cages in farmers' ponds during 2007–2008 and 2008–2009 and 2010–2011.

The demonstration carried out at M/s HITIDE Sea farm, Mahendrapally, Kattur

The demonstration carried out at M/s BISMI Aqua farm, Perunthottam

The demonstration carried out at M/s Aqua Nova farm (fresh water), Saliyamangalam

During 2007–2008, a very successful cage culture demonstration was conducted in the brackishwater pond of M/s Hitide Sea farms, Kattur, and 1.5 ton of sea bass from 26 cages with the DOC of 249 were harvested. This result has given very good encouragement to do diversified aquaculture among farmers in this region. Further, with the frequent disease outbreak and low market value prevailing for shrimp, many of the farmers are willing to do sea bass culture in the defunct farm, in order to popularize the demonstration. The third demonstration was conducted at the freshwater farm at Saliyamangalam, Thanjavur Dt. 5,040 seeds were stocked during the month of March 2010 and seeds were grown to an average weight of 700–800 g and harvest was done on February 2011. From 23 cages 1,121 kg sea bass were harvested; the escaped fish contributed to the 300 kg additionally and added to the successful production of sea bass in cages in freshwater.

Conclusion

In India, the aquaculture is centric to the shrimp/scampi production and these two species are contributing in total of 52 % towards export. The freshwater fish produced through aquaculture is mainly catering to the domestic market only. In Indian seawater many fin fish and shellfish are abundant for aquaculture, which is economically important, and the sea bass (*Lates calcarifer*) fish is occupying the main role at present as it is a candidate species for cage culture as it has completed a value chain approach from seed production, nursery rearing, grow out and marketing and export by RGCA – MPEDA.

Acknowledgement The author is gratefully acknowledging the approval of the honourable chairman and the director of MPEDA-HO, Kochi, for presenting the findings obtained through demonstrations and the support extended by the deputy director (Aqua) of MPEDA, Thanjavur, and other officials and staff for conducting successful demonstration. Thanks are also due to B. John Antony, MPEDA, Thanjavur, for preparing this text.

References

Madrones-Ladja JA, Catacutan MR (2012) Net cage rearing of the Asian seabass *Lates calcarifer* (Bloch) in brackishwater pond: the technical and economic efficiency of using high protein diets in fingerling production. Philipp Agric Sci 95:79–86

Philipose KK, Krupesha Sharma S, Loka J, Damodaran D, Narsimhulu S, Dube P (2013) Culture of Asian seabass (*Lates calcarifer*, Bloch) in open sea floating net cages off Karwar, South India. Indian J Fish 60:67–70

Schipp G (1996) Barramundi: barramundi farming in the northern territory. Aquaculture Branch, Department of Primary Industry and Fisheries, Darwin

Tookwinas S (1987) Cage culture of brackishwater fish in Satul Province. Technical paper, Brackishwater Fisheries Division, Dept. of Fisheries, Bangkok, Thailand, 30 p

Tookwinas S (1988). Some aspects of water quality management for cage culture of seabass. In: Culture of the seabass (*Lates calcarifer*) in Thailand, NACA training manual. UNDP/FAO, Bangkok, pp 59–68

Tookwinas S, Charearnrid B (1988) Cage culture of sea bass *Lates calcarifer* in Thailand. In: Culture of seabass *Lates calcarifer* in Thailand. UNDP/FAO, Bangkok, pp 50–58

Concept of Stress and Its Mitigation in Aquaculture

Prem Kumar, Thirunavukkarasu A.R., R. Subburaj, and G. Thiagarajan

Introduction

Fishes are cold-blooded animals (ectothermic/poikilothermic) that cannot regulate their body temperature; hence, their physiological mechanisms are directly or indirectly temperature dependent. Some degree of thermoregulation is present in some fishes like tunas and *Lamnidae* sharks, in which the brain, eye and muscle temperature have been reported to be 2–12 °C higher than the ambient water temperature (Steven and Neil 1978). Hence, the temperature is one of the most important factors affecting the poikilothermic organisms.

Different stressors such as physical, chemical, biological and procedural exist in different stages of aquaculture practices. The physiological responses (stress response) of the animals exposed to stressor are of three types, i.e. primary response, secondary response and tertiary response (Wedemeyer and Mcleay 1981).

It is generally accepted that the animal responds to a variety of stressors, which is an adaptive mechanism and called as general adaptation syndrome (GAS) (Barton and Iwama 1991).

There are two approaches to mitigate the stress: the first one is non-chemical (biological method) and the other one is the chemical method. The non-chemical (biological) method includes the entire environment management which includes water quality management such as temperature, dissolved oxygen, ammonia, nitrogen, nitrite and salinity and stocking density, uniform-size stocking, stocking ratio in polyculture, etc. The chemical method includes dietary supplementation of vitamin C, vitamin E, tryptophan, immunostimulants, etc.

P. Kumar (✉) • Thirunavukkarasu A.R. • R. Subburaj
G. Thiagarajan
Fish Culture Division, Central Institute of Brackishwater Aquaculture, No. 75, Santhome High Road, R.A. Puram, Chennai 600028, Tamil Nadu, India
e-mail: prem@ciba.res.in

Concept of Stress and Stressors

Stress

Stress was defined by Seyle (1950): Stress means the sum of all the physiological response by which an animal tries to maintain or re-establish a normal metabolism on the face of physical or chemical force.

Stressors

The factor that causes stress is called stressors. Rearing of aquatic organisms in a man-made environment has resulted in exposure to a number of stressors, which may not be experienced to the same degree in natural environments. Stressors in aquaculture systems may be categorised as shown in Table 1 (Wedemeyer et al. 1999).

S. Perumal et al. (eds.), *Advances in Marine and Brackishwater Aquaculture*,
DOI 10.1007/978-81-322-2271-2_10, © Springer India 2015

Table 1 Type of stressors and its examples

Sr. no.	Type of stress	Examples
1.	Physical	Temperature, light, dissolved oxygen, sound
2.	Chemical	Water quality, pollution, diet, metabolic waste
3.	Biological	Stocking density, microorganisms (pathogenic and nonpathogenic), macro-organisms (parasites), lateral swimming space requirements
4.	Procedural	Handling, hauling, stocking, disease treatment, feeding methods (manual and automated)

Stress Responses

Fishes are ectothermic organisms and therefore are unable to control their body temperature within narrow temperature limits (Hazel 1993). As a result, their physiology is dependent on ambient temperature. The environmental temperature determines the metabolic rate, physiology and growth of fishes (Morgan et al. 1999). Water temperature influences O_2 concentration, metabolism and growth of fishes (Langston et al. 2002).

Stressor is the causative factor and stress is a response. The stress response of fishes follows the general vertebrate pattern. A key element in the stress response is a switch from anabolism to catabolism. The quantum of response may vary with the nature of stress and its duration, and it also depends on factors like age, sex, maturation stage, species and strain of the fish.

When fishes and other aquatic organisms experience environmental disturbances that lie outside the normal range, the effect may be dramatic. At the organism level, a series of physiological changes occur following stressful challenges, which are adaptive in nature. These physiological responses are termed as "general adaptation syndrome" (GAS), Seyle 1950. It consists of:

1. An *alarm reaction* in which "stress hormones" (catecholamine and corticosteroids) are released
2. A *stage of resistance* during which adaptation occurs
3. A *stage of exhaustion* in which adaptation is lost because the stress was too severe or long lasting

Stress response has been classified into primary (neural and neuroendocrine responses), secondary (physiological consequence of such primary response) and tertiary response (changes in behaviour, growth rate, increased susceptibility to diseases and change in population) (Wedemeyer and Mcleay (1981)). Primary stress response results in the activation of the neuroendocrine system, which brings about changes in metabolism, osmoregulation and haematology (Barton 2002). The primary stress response is the physiological alarm, during which there is an increase in the levels of "stress hormones", i.e. corticosteroids and catecholamines. These primary responses result in a suite of biochemical and physiological changes leading to secondary stress response, characterised by elevated glucose levels, haemato-immunological changes and change in other metabolic enzymes. Secondary stress responses are believed to be adaptive mechanisms and are particularly important for fishes to recover from stress. However, during chronic conditions the fish loses its capacity to adapt. This results in pathological changes, reduction in reproductive success, decrease in growth rate and decreased disease resistance that leads to death of the organism thereby causing change in the population and biodiversity of that species, which is termed as tertiary stress response.

Primary Stress Responses

1. Release of adrenocorticotropic hormones (ACTH) from the adenohypophysis
2. Release of "stress hormones" (catecholamines, i.e. adrenalin, noradrenalin and dopamine and corticosteroids especially cortisol) from the head kidney

Sensory perception of stress is a prerequisite for stress response in animals. In fishes, an adverse condition stimulates the afferent neural pathway that runs in the sympathetic nervous system from the hypothalamus to the chromaffin tissue of the head kidney and stimulates the chromaffin tissue, which leads to the release of catecholamines. Catecholamine (adrenalin/epinephrine) is released from the chromaffin tissue in the head kidney of teleosts and also from the ending of adrenergic nerves. Because catecholamines, predominantly epinephrine in teleostean fishes, are stored in chromaffin cells, their release is rapid and the circulating levels of these hormones increase immediately with stress. The release of catecholamines is extremely rapid compared to the release of cortisol.

Corticotrophin-releasing hormone (CRH) or factor (CRF) is released from the hypothalamus of the brain, which stimulates corticotrophic cells of anterior pituitary (adenohypophysis) to secrete adrenocorticotropic hormone (ACTH), which stimulates interrenal cells (adrenal cortex homologue) to synthesise and release corticosteroids particularly cortisol which is the principal corticosteroid in fish. The release of cortisol in teleost is delayed relative to catecholamine release. Resting and unstressed levels of circulating corticosteroids in fishes are less than 30–40 ng/ml (Wedemeyer et al. 1990). Characteristic cortisol elevation of fishes in response to acute stressors tends to range within about 30–300 ng/ml (Wedemeyer et al. 1990; Barton and Iwama 1991). Elevation of plasma catecholamines and cortisol due to primary stress leads to secondary stress responses.

Secondary Stress Responses

In fishes, cortisol enters the liver cells where it binds to nuclear receptor, resulting in activation of genes that produce a series of enzymes that have a range of metabolic effects. This results in a suite of biochemical and physiological changes which may include hyperglycaemia, hyperlacticaemia, depletion of tissue glycogen reserves, lipolysis and inhibition of protein synthesis. Other changes may include the osmotic and ionic disturbances due to diuresis and loss of electrolyte from the blood and change in haematology (reduction of white blood cells, leucopenia) (Barton and Iwama 1991). Stress is an energy demanding process and the animal mobilises energy substrate to cope with stress metabolically. The production of glucose under stress assists animals to cope with the increased energy demand. The stress hormones adrenalin and cortisol have been shown to increase plasma glucose production in fishes by both gluconeogenesis and glycogenolysis.

Catecholamines, in particular, have marked influence on cardiovascular functions leading to change of blood circulation, gill perfusion and oxygen carrying capacity of blood. Corticosteroids on the other hand are known to stimulate the ion-transport mechanism in the gill and kidney.

These secondary stress responses are believed to be adaptive mechanisms and are particularly important for fishes to recover from stress by maintaining oxygen supply to the tissues, to regain osmotic and ionic equilibrium and to meet the increased energy demands imposed by exposure to environmental stressor. Typically these changes persists only for few hours or days following acute exposure to the stressor and therefore do not result in any deleterious effect on the animal.

Intracellular stress response is characterised by the production of a family of proteins known as heat shock proteins (HSP). Exposure of cells or whole organisms to heat shock results in a reversible increase in the synthesis of some acute phase proteins against subsequent shock known as HSP (Palmisano et al. 2000; Ming et al. 2003), which play an important role in homeostasis. HSPs are a family of highly conserved cellular proteins that have been observed in all organisms (Feder and Hofmann 1999) including fishes (Iwama et al. 1998). They were first discovered in the chromosomal puffs of drosophila salivary glands after thermal shock (Ritossa 1962). In the normal unstressed cells, heat shock proteins are essential for folding and translocation of newly formed proteins and renaturation of denatured proteins. The expression of these proteins increases manifolds in the cells during stress. HSP has an ability to mediate misfolded or denatured functional proteins caused by various stressors in the cell; hence, this protein is also known as molecular chaperone.

Tertiary Stress Responses

Chronic exposure to stressors provokes tertiary stress responses that result in a number of pathological changes and reduction in reproductive success, depression of growth rate and decreased disease resistance. Tertiary stress response represents whole animal and population level changes associated with stress.

1. Whole animal:
   (a) Impaired growth, parr-smolt transformation (smelting), spawning success and migration behaviour and spawning
   (b) Increased disease incidence (infectious and noninfectious)
2. Population parameters:
   (a) Reduced intrinsic growth rate, recruitment, compensatory reserve and productivity
   (b) Altered community species abundance and diversity

Thus, when fishes are exposed to environmental stressor, a hierarchy of responses is initiated and if the stress is severe or long lasting, successively higher levels of biological organisation get affected. This signifies that the primary responses are the changes at the endocrine level, whereas the tertiary responses refer to those changes that can be easily seen by observing the animal.

Stress Mitigation Methods

One of the most promising areas of research is the development of strategies to reduce the stress during various aquaculture practices. There are two approaches to mitigate stress, the first one is non-chemical (biological method) and the other is chemical method. The non-chemical (biological) method includes the entire environmental management, which includes water quality management such as temperature, dissolved oxygen, ammonia, nitrogen, nitrite, salinity, etc., and optimum stocking density, uniform-size stocking, stocking ratio in polyculture, etc. The chemical method includes dietary supplementation of vitamin C, vitamin E and tryptophan and immunostimulants, etc.

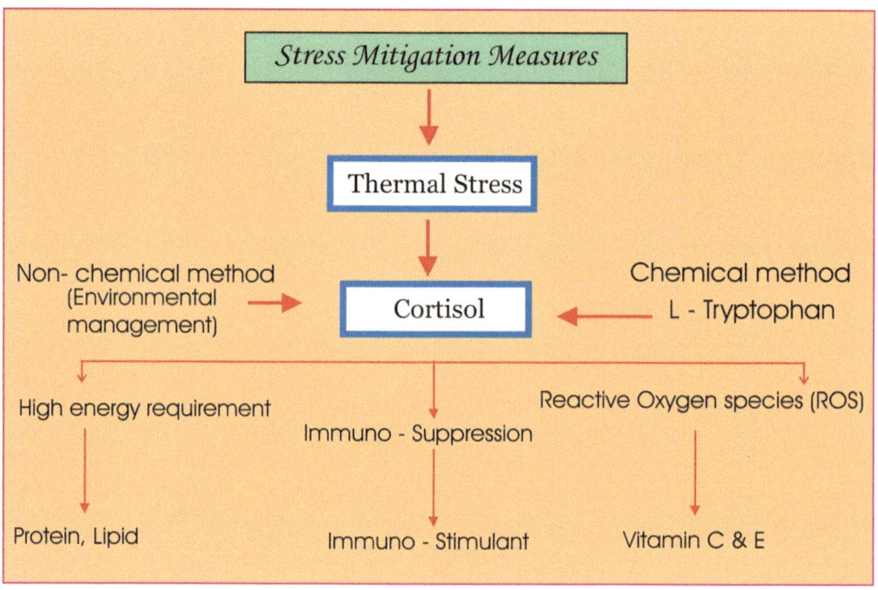

Flow diagram: stress mitigation measures

Table 2 Showing some of the important water quality parameters and their optimum range for freshwater fish culture

Sl. no.	Water quality parameters	Optimum range
1	Dissolved oxygen	>5 ppm
2	Temperature	28–31 °C
3	Turbidity	20–60 cm
4	Total hardness	40–100 ppm
5	Total alkalinity	60–300 ppm
6	Ph	7.0–7.5
7.	NH_3–N	<0.05

Non-chemical Methods (Environmental Management)

Water quality is defined as the suitability of water for the survival and growth of fishes, whereas water quality management is a technique to bring the water quality in desirable level for the growth and survival of fishes. Environment management in aquaculture means water quality management such as dissolved oxygen (DO), water temperature, alkalinity, water Ph, turbidity, hardness, etc. (Table 2).

Chemical Methods

Dietary Supplementation of Protein, Vitamin C and Vitamin E

Oxidative stress is caused by the production of reactive oxygen species and nitrogen species during stress (Gordon 2001). Oxygen radicals is generated due to respiratory burst activity of phagocytes, present in cells under normal conditions but its production increases during pathophysiological conditions and stress. The use of immunostimulants and antioxidants to ameliorate the damage to immune system by stress has been studied by many workers. Fishes fed with immunostimulant glucan prior to transportation have increased specific and nonspecific immune response as is evident by higher number of lymphocytes and enhanced phagocytosis (Jeney et al. 1997).

It is reported that supplementation of high protein and vitamin C reduced the bioaccumulation and stress responses in *Channa punctatus* (Sarma 2004). Dietary high protein and vitamin C were supplemented for ameliorating stress (Manush et al. 2005) in *Macrobrachium rosenbergii*. Vitamin C is considered to play an important role in animal health as antioxidants inactivate damage of free radicals produced during normal cellular activity from various stressors that have been reported confirming the protective role of vitamin C. Vitamin C can interact with other antioxidants such as carotenoids and vitamin E. A high concentration of vitamin C at the cell membrane regenerates the reduced vitamin E created during oxidation and reduction process. Vitamin C is highly interactive and may fortify antioxidant defences and enhance immune response indirectly by maintaining optimal vitamin E levels. Vitamin C and vitamin E act as membranes protecting agents and give stability to lipid bilayer.

Dietary Supplementation of L-Tryptophan

Tryptophan is one of the eight essential amino acids, necessary for the development of the vitamin niacin/nicotinic acid and serotonin. It cannot be synthesised in the body and thus must be obtained from food or supplements. L-tryptophan acts as cortisol blocker hence reduces stress-induced production of plasma cortisol.

Dietary Supplementation of Lactoferrin

Lactoferrin (LF) is a family of iron-binding glycoproteins having molecular weight of 80 KDa that originated from some secretion of mammals. It is a kind of immunostimulant and has a lot of biological functions, e.g. iron absorption and transportation, bacteriostatic effects and enhancement of mucosal immunity system in mammals. Orally administration of LF enhances nonspecific defence system and phagocytic activity in rainbow trout and decreases plasma cortisol level in goldfish by LF administration. Dietary LF also enhances tolerance to physiological stressors such as air-exposed stress in juvenile Japanese flounder against high stocking density stress in rainbow trout and common carp and low salinity stress in shrimp (Koshio et al. 2000).

Conclusion

During aquaculture practices animals come across many different types of stressors such as physical, chemical and biological. To cope up with these stressors physico-biochemical process of biomolecules, cells, organelles and organisms vary from species to species. Exposure of fishes to extreme environmental conditions elicits a cascade of physiological and biochemical changes characterised as primary, secondary and tertiary stress responses. Stress can be mitigated by biological and chemical methods.

References

Barton BA (2002) Stress in fishes: a diversity of responses with particular reference to changes in circulation corticosteroids. Integr Comp Biol 42:517–525

Barton BA, Iwama GK (1991) Physiological changes in fish from stress in aquaculture with emphasis on the response and effects of corticosteroids. Annu Rev Fish Dis 1:3–26

Feder ME, Hofmann GE (1999) Heat shock proteins, molecular chaperons, and the stress response: evolutionary and ecological physiology. Annu Rev Phys 61:243–282

Gordon S (2001) Mononuclear phagocytes in immune defense. In: Roitt I, Brostoff J, Male D (eds) Immunology. Mosby, Edinburgh, pp 148–160

Hazel JR (1993) Thermal biology. In: Evans DH (ed) The physiology of fishes. CRC Press, Boca Raton, pp 427–468

Iwama GK, Thomas PT, Forsyth RB, Vijayan MM (1998) Heat shock protein expression in fish. Rev Fish Biol Fish 8:35–56

Jeney G, Galeotti M, Volpattil D, Jeney Z, Anderson DP (1997) Prevention of stress in rainbow trout *(Oncorhynchus mykiss)* fed diets containing different doses of glucan. Aquaculture 154:1–15

Koshio S, Yokoyama S, Ishikawa M, Hayasawa H, Oshida K (2000) Effect of bovine lactoferrin on growth, survival and tolerance to stress of Kuruma prawn *Penaeus japonicus*. In: Shimazaki K (ed) Lactoferrin: structure, function and applications. Elsevier, Amsterdam, pp 443–449

Langston AL, Hoare R, Stefansson M, Fitzgerald R, Wergeland H, Mulcahy M (2002) The effect of temperature on non-specific defence parameters of three strains of juvenile Atlantic halibut (*Hippoglossus hippoglossus* L.) Fish Shellfish Immunol 12:61–76

Manush SM, Pal AK, Das T, Mukherjee SC (2005) Dietary high protein and vitamin C mitigate stress due to chelate claw ablation in *Macrobrachium rosenbergii* males. Comp Biochem Physiol A 142:10–18

Ming YC, Hung SY, Lin EC, Hseu TH, Lee WC (2003) Association of a single nucleotide polymorphism in the 5′-flanking region of porcine HSP70.2 with backfat thickness in duroc breed. Asian-Aust J Anim Sci 16:100–103

Morgan IJ, D'Cruz LM, Dockray JJ, Linton TK, Wood CM (1999) The effect of elevated summer temperature and sub-lethal pollutants (ammonia, Low pH) on protein turn over in the gill and liver of rainbow trout (*Oncorhynchus mykiss*) on a limited food ration. Comp Biochem Physiol A 123:653–654

Palmisano AN, Winton JR, Dickhoff WW (2000) Tissue specific induction of Hsp90 mRNA and plasma Cortisol response in chinook salmon following heat shock seawater challenge and handling challenge. Mar Biotechnol 2:329–338

Pickering AD (1981) Introduction: the concept of biological stress. In: Pickering AD (ed) Stress and fish. Academic, London, pp 1–9

Ritossa FM (1962) A new puffing pattern induced by a temperature shock and DNP in *Drosophila*. Experientia 18:571–573

Seyle H (1950) Stress and the general adaptation syndrome. J Brit Med 1:1383–1392

Steven ED, Neil WH (1978) Body temperature relations of tunas, especially Skipjack. In: Hoar WS, Randal DJ (eds) Fish physiology, vol VII. Academic, London, pp 316–359

Wedemeyer GA, Mcleay DJ (1981) Methods for determining the tolerance of fishes to environmental stressors. In: Pickering AD (eds) Stress and fish. Academic Press, New York, pp 247–268

Wedemeyer GG, Barton BA, McLeay DJ (1990) Stress and acclimation. In: Schreck CB, Moyel PB (eds) Methods for fish biology. American Fisheries Society, Bethesda, pp 451–489

Wedemeyer GA, Meyer FP, Smith L (1999) Environmental stress and fish diseases. Narendra Publishing House, New Delhi, 192 p

Breeding and Rearing of Marine Ornamentals

T.T. Ajith Kumar, V. Gunasundari, and S. Prakash

Introduction

The marine aquarium trade is a billion dollar business that may sustain continued growth in the coming years. In the past decade, there has been a worldwide increase in the popularity of reef tanks, which has led to an increased demand for marine ornamental organisms (Wood 2001; Green 2003). In contrast to the freshwater ornamental species, most marine organisms being marketed in the aquarium trade are collected from the wild, particularly from coral reef areas. The prevalence of destructive low-cost harvesting techniques, such as the use of cyanide and explosives, has caused dramatic and drastic impacts on the health and biodiversity of the reef ecosystems. Developing hatchery technology for marine ornamental species is therefore urgently needed to guarantee the sustainable supply for the industry while minimising the negative impacts on the natural environment (Lin et al. 2002). In the recent years, researchers, traders, collectors and hobbyists have begun a worldwide effort to minimise the growing pressure on the natural populations of marine ornamental species and to promote the sustainable use of these high-valued resources (Corbin 2001).

Although fishes and corals are still the most heavily traded ornamental marine species for the aquarium, many other organisms are also highly popular among hobbyists. The organisms receive the 'ornamental status' mainly because of their dazzling coloration and delicacy, hardiness in captivity and being 'reef safe' (do not harm other aquarium organisms) (Sprung 2001). Nevertheless, if a species presents mimetic adaptations, displays associative behaviour (particularly symbiotic associations) or performs a specific function on the reef aquarium (eating nuisance organisms), it may also be targeted by the marine ornamental hobbyists.

Presently, a percentage of commercially cultured ornamentals is limited with few fish species, mainly the clowns of the genus *Amphiprion* (Wabnitz et al. 2003; Ajith Kumar and Balasubramanian 2009; Ajith Kumar et al. 2010) and damselfishes (Gopakumar et al. 2002; Subodh Kant Setu and Ajith Kumar 2010).

T.T.A. Kumar (✉)
Centre of Advanced Study in Marine Biology, Faculty of Marine Sciences, Annamalai University, Porto Novo 608502, Tamil Nadu, India

National Bureau of Fish Genetic Resources (ICAR), Canal Ring Road, Dilkusha Post, Lucknow 226002, Uttar Pradesh, India
e-mail: ttajith87@gmail.com

V. Gunasundari
Centre of Advanced Study in Marine Biology, Faculty of Marine Sciences, Annamalai University, Porto Novo 608502, Tamil Nadu, India

S. Prakash
Centre of Advanced Study in Marine Biology, Faculty of Marine Sciences, Annamalai University, Porto Novo 608502, Tamil Nadu, India

Centre for Climate Change Studies, Sathyabama University, Jeppiaar Nagar, Rajiv Gandhi Salai, Chennai 600119, Tamil Nadu, India

S. Perumal et al. (eds.), *Advances in Marine and Brackishwater Aquaculture*,
DOI 10.1007/978-81-322-2271-2_11, © Springer India 2015

Global Status

Marine ornamental aquaculture provides employment opportunities to the rural poor and also is an earner of foreign exchange to many developing countries. It has been estimated that over 1.5 million people are engaged in this industry and over 3.5 million hobbyists constitute the trade (Dey 2010). A total of 1,471 species of fishes are traded worldwide with the best estimate of annual global trade ranging between 20 and 24 million individuals. Damselfish (Pomacentridae) make up almost half of the trade and other species such as angels (Pomacanthidae), surgeons (Acanthuridae), wrasses (Labridae), gobies (Gobiidae) and butterflyfishes (Chaetodontidae) accounting another 25–30 %. The most traded species are blue-green damselfish (*Chromis viridis*), anemonefish (*Amphiprion ocellaris*), whitetail dascyllus (*Dascyllus aruanus*), sapphire devil (*Chrysiptera cyanea*) and three-spot dascyllus (*D. trimaculatus*) (Wabnitz et al. 2003).

The international ornamental fish trade at the retail level is estimated to be more than US$ 8 billion, while the entire industry including aquarium tanks, plants, accessories, feed, medications, etc. is at US$ 20 billion. Up to 2008, more than 125 countries were involved in the ornamental fish trade, of which 15 countries exported fishes, worth more than US$ 5 billion. In Asian countries, Singapore stands first with an estimated share of 39.3 % with total exports, followed by Malaysia. The value of a fish from marine origin in the trade has increased from US$ 9 million in 2003 to reach almost US$ 29 million in 2007. The impact trend during a 10-year period (1999–2008) has shown that the imports rise from US$ 245.6 million in 1999 to US$ 349.4 million in 2008 and exports grow from US$ 167.6 million in 1999 to US$ 343.9 million in 2008 (Dey 2010).

Indian Scenario

Ornamental aquaculture is an excellent business opportunity in India, since there is a strong demand from domestic and export markets. In India, there is a good potential due to enormous geographical distribution and extensive species diversity. Our country is blessed with a wide array of marine ornamental varieties, while our contribution was only 2.5 % (US$ 3.8 million) to the total Asian ornamental fish exports. Compared to other Asian countries, the Indian ornamental fish sector is small, but exciting, with tremendous growth and large-scale gainful employment generation. At present, the ornamental fish export from India is dominated by the wild-caught species, which cater to a small portion of the global market, and we contribute the least (0.9 %) to the global market. In 2007, the marine component of the trade in India reached 48 % and the freshwater component 52 %, and the value of brackishwater fish was trifling (FAO 2009). Currently, India has also started the breeding and rearing of marine ornamental organisms, and the possibilities are more to reach greater heights, as our natural wealth of ornamentals in the Andaman and Lakshadweep islands are very high. Despite having cheap labour, quality water and sufficient manpower, India lacks the appropriate infrastructure, technology and trainings in this sector.

Captive Breeding

In the recent years, there has been an increased focus on supplying aquarium fishes through closed system culturing. Ornamental aquaculture can be an environmentally friendly way to increase the supply of such organisms, by helping to reduce the pressure on wild fish populations and producing a wide variety of species throughout the year. Furthermore, rearing aquarium fish in captivity is likely to lead the production of hardier species, which fare better in captivity and survive longer (Olivier 2003; Ogawa and Brown 2001).

To date, India has proved successful breeding only in a few species. It is hoped that much of the market demand for the more popular ornamentals such as clown, damsel and angelfish may eventually be satisfied by cultured fish, once culture technologies have been established successfully. Closing life cycles in captivity and suitable live feeds are the challenges for most of the marine species.

Hatchery Technology in Hand

Taking the facts discussed above and also in view of the growing demand on marine aquarium organisms, an attempt was made to establish the Marine Ornamental Fish Hatchery at the Centre of Advanced Study in Marine Biology, Annamalai University, Parangipettai and Agatti Island, Lakshadweep, with the financial support of the Govt. of India to develop a hatchery pro duction technology for marine ornamental fishes and transfer the technology to the coastal and island communities of the country for their sustainable development, and this task would help in conserving the marine biodiversity. In India, the Central Marine Fisheries Research Institute is the pioneer concentrating on the ornamental aquaculture from the last two decades and succeeded with its technologies.

In recent years, many worldwide fisheries organisations are showing keen interest in breeding and rearing of marine ornamental fishes. In this context, Indian fishery organisations have successfully developed broodstocks for 13 species of clowns such as *Amphiprion ocellaris*, *A. percula*, *A. sebae*, *A. clarkii*, *A. nigripes*, *A. frenatus*, *A. akallopisos*, *A. sandaracinos*, *A. perideraion*, *A. polymnus*, *A. ephippium*, *A. thiellei* and *Premnas biaculeatus*, and the hatchery technology has been developed for ten species (Fig. 1). Annamalai University has succeeded with this, using estuarine water, a milestone work done in the aquaculture history of the country. This is an excellent effort in the ornamental aquaculture in general and clownfishes in particular, since these fishes belong to coral reefs. The university has also developed broodstock for ten different species of damselfishes, *viz.* *Pomacentrus caeruleus*, *P. pavo*, *Neopomacentrus cyanomos*, *Dascyllus trimaculatus*, *D. aruanus*, *D. carneus*, *D. reticulatus*, *Chromis viridis*, *Chrysiptera cyanea* and *Neoglyphidodon oxyodon*, and experimental successes in hatchery production were obtained for four species (Fig. 1).

The development of artificial culture methodologies for marine ornamental crustaceans has been focused on a limited number of shrimp species. Among them, the tropical species of the genera *Lysmata*, *Periclimenes*, *Hymenocera* and *Stenopus* have received special attention, mainly due to their growing demand in the aquarium trade (Lin et al. 2001). Recently, in India, initiatives have been taken in rearing certain species of these marine ornamental shrimps such as *Stenopus hispidus*, *Rhynchocinetes durbanensis*, *Hymenocera picta*, *Lysmata debelius* and *L. amboinensis* (Fig. 2) successful captive spawning was achieved, and larval rearing practices are under progress.

The cardinal fishes and dottyback (Fig. 3), which are extremely attractive in appearance, are very hardy in captivity. The cardinals exhibit an unusual mode of reproduction in that the males incubate their female partner's eggs in their mouth (Tullock 1999). They make outstanding tank companions with all fishes, coral and other marine ornamentals; hence, they become very popular in the marine ornamental trade (Michael 1996). Because of the remarkable rise in the popularity of this species, various stakeholders in the aquarium industry are keen in producing them in captivity.

Live Feeds

Phytoplankton: Microalgae

Microalgae are the floating microscopic plants which constitute the base of the food chain in an aquatic ecosystem. They are being used to produce mass quantity of rotifers and brine shrimps which serve as food for fish larval stages. The mostly used marine microalgal species *Nannochloropsis salina*, *Chlorella marina* and *Isochrysis galbana* were stocked using Conway - Walnes, F/2 medium and agricultural fertilisers for mass culture. The 'green water technique' is used for rearing fish

Fig. 1 Hatchery-bred clowns and damsels

Stenopus hispidus

Lysmata debelius

Rhynchocinetes durbanensis

Fig. 2 Broodstock developed for marine ornamental shrimps

Pterapogon kauderni

Sphaeramia nematoptera

Pseudochromis dilectus

Fig. 3 Broodstock developed for other marine ornamental fishes [*Cardinals & Dotty backs*]

larvae; microalgae are directly pumped into the larval rearing tanks, where they serve as water conditioner by stabilising the water quality, nutrition of the larvae and microbial content.

Zooplankton: Rotifer

They form an excellent feed for fish larvae due to their size, active movement, ability to be cultured in high densities and high reproductive ratio. The mostly used rotifers (*Brachionus rotundiformis* and *B. plicatilis*) were stocked and mass cultured using algal mass culture, and within 5–7 days, the culture contained a maximum concentration of 50–60 nos/ml.

Artemia

Artemia cysts were purchased commercially and allowed to hatch in a cylindrical tank with transparent bottom. Vigorous aeration and an artificial light for 24 h were essential for good hatching rate, and it will be hatched out after 18–24 h.

First Feeding of Larvae

The initial nourishment to the developing larvae will be obtained from the yolk. When the yolk reserves have been completely utilised, the larval feeding capabilities are developed, and supplement food in sufficient quantities is needed. However, they have limited yolk reserves and have to resort to exogenous feeding even though they have small mouths and primitive digestive system.

Lab to Land

The hatchery technology developed by the Annamalai University and CMFRI has been transferred besides training imparted to the coastal people of the country which is considered as a significant step towards the conservation of marine biodiversity and yet another way to enhance the marine products export through hatchery-bred fishes. During the training period, various aspects such as fish handling, feeding, water quality, larval rearing, live feed culture, disease diagnosis, packing, transportation, marketing and establishment of backyard hatchery were taught. These institutes have developed different packages to suit varied interests and also published different handbooks on the breeding and rearing of clowns and damsels. In addition, the researchers are concentrating on value addition to the hatchery-bred organisms such as colour and growth enhancement, stress tolerance and disease resistance, etc.

Conclusion

In recent years, marine aquarium keeping has become a popular activity worldwide, with reef tanks being widely considered as the most challenging and spectacular display. This highly profitable industry relies almost exclusively on the wild organisms, mainly caught in the reef areas, causing destruction of the coral reefs. This current state of crisis on the coral reefs has put the marine aquarium industry 'in the line of fire'. The devastating and long-lasting effects of dynamite fishing, water quality degradation because of anthropogenic pollution and global warming are the threats that are certainly more relevant. Nevertheless, there is a current effort to advocate and enforce sustainable collection and trade of the marine ornamental species. In an attempt to minimise the industry's dependence on wild collections, research institutes and private entrepreneurs have started to address the culture of marine ornamental species. However, the lack of basic scientific knowledge on many of the targeted species by culture efforts has caused serious bottlenecks that impair commercial culture of the most highly demanded species. Despite such difficulties, the strong belief of some researchers and traders is that the captive culture is more potentially profitable commercial venture and there is a need for the sustainable development of this industry.

However, further research studies are still needed to allow the regular supply of a broader number of cultured marine ornamentals for the marine aquarium trade. The major goal of the marine ornamental aquaculture is not only to promote the coral reef conservation but to develop a sustainable alternative to all those involved in the collection and supply of these remarkable organisms to the aquarium trade.

References

Ajith Kumar TT, Balasubramanian T (2009) Broodstock development, spawning and larval rearing of the false clown fish, *Amphiprion ocellaris* in captivity using estuarine water. Curr Sci 97(10):1483–1486

Ajith Kumar TT, Sethu SK, Murugesan P, Balasubramanian T (2010) Studies on captive breeding and larval rearing of clownfish *Amphiprion sebae* (Bleeker, 1835) using estuarine water. Indian J Mar Sci 39(1):114–119

Corbin JS (2001) Marine Ornamentals'99, conference highlights and priority recommendations. In: Burgess P (ed) Marine ornamental conference, Kluwer Academic Publishers, Netherlands, Hawaii. Aqua Sci Conserv 3:3–11

Dey VK (2010) Ornamental fish trade – recent trends in Asia. In: Souvenir, ornamentals Kerala 2010. Department of Fisheries, Government of Kerala. pp 39–45

FAO (2009) Fish Stat Plus. Universal software for fishery statistical time series. Version 2.3. www.fao.org

Gopakumar G, Sreeraj G, Ajith Kumar TT, Sukumaran TN, Raju B, Unnikrishnan C, Hillary P, Benziger VP (2002) Breeding and larval rearing of three species of damsel fishes (Family: Pomacentridae). Mar Fish Inform Serv T & E Ser 171:3–5

Green E (2003) International trade in marine aquarium species: using the global marine aquarium database. In: Cato JC, Brown CL (eds) Marine ornamental species: collection, culture and conservation. Iowa State Press, Ames, pp 31–47

Lin J, Zhang D, Creswell R (2001) Aquaculture of marine ornamental shrimps: an overview. Aquaculture 2001 book of abstracts. World Aquaculture Society, Orlando, p 378

Lin J, Zhang D, Rhyne A (2002) Broodstock and larval nutrition of marine ornamental shrimp. In: Cruz-Suárez LE, Ricque-Marie D, Tapia-Salazar M, Gaxiola-Cortés MG, Simoes N (eds) Avances en Nutrición Acuícola VI. Memorias del VI Simposium Internacional de Nutrición Acuícola. 3 al 6 de Septiembre del 2002. Cancún, Quintana Roo, México, pp 589

Michael S (1996) The Banggai Cardinalfish: a newly available species that may become too popular for its own good. Aquar Fish Mag 8(8):86 87

Ogawa T, Brown C (2001) Ornamental fish aquaculture and collection in Hawaii. Aquar Sci Conserv 3 (1–3):151–169

Olivier K (2003) World trade in ornamental species. In: Cato J, Brown C (eds) Marine ornamental species: collection, culture and conservation. Iowa State Press, Ames, pp 49–63

Setu SK, Ajith Kumar TT (2010) Spawning behaviour and Embryonic development of Regal damsel fish, *Neopomacentrus cyanomus* (Bleeker, 1856). World J Fish Mar Sci 2(5):410–415

Sprung J (2001) Oceanographic series invertebrates:, a quick reference guide. Ricordea Publishing, Miami, 240 pp

Tullock J (1999) Banggai cardinalfish alert. Aquar Front. http://www.aquariumfrontiers.net/EnvironmentalAquarist/html

Wabnitz C, Taylor M, Green E, Razak T (2003) From ocean to aquarium. UNEP-WCMC, Cambridge, p 64

Wood EM (2001) Collection of coral reef fish for aquaria: global trade, conservation issues and management strategies. Marine Conservation Society, UK, 80 pp

Breeding and Seed Production of Tiger Shrimp *Penaeus monodon*

P. Soundarapandian

Introduction

In the past, shrimp remained as a luxury commodity because of its high popularity and cost. Hence, it was considered as favourite food stuff for the high society people. However, in recent years, due to increase of population and health consciousness, the modern world made people to consume more of shrimp and shrimp product because of its delicious and safeness among animal origin. In recent years, shrimp has become a more popular source of protein food, which commands high prices in the international market. This great importance increased the exploitation of shrimp, which lead to sudden decline in natural stock. Thus, it has become imperative to culture commercially important shrimp species.

The aquaculture operation of the shrimp basically requires an abundant, reliable and inexpensive source of seed supply: without such source, the productions will not be possible. Hence, the captive shrimp seed production has received much attention.

A variety of techniques for hatchery production of shrimp seed have been developed in different parts of the world, which involved generally capital intensive and high technology. A developing country such as India with limited resources needs a low-cost technology, which should be simple enough to be used by even semi-skilled labourer. The technology presented here is based on a multidisciplinary research experience from experimental to commercial level (Paul Raj 1999).

P. Soundarapandian (✉)
Centre of Advanced Study in Marine Biology, Faculty of Marine Sciences, Annamalai University, Parangipettai 608502, Tamil Nadu, India
e-mail: soundsuma@gmail.com

Biology of Shrimp

Systematic Position

Phylum	:	Arthropoda
Class	:	Crustacea
Subclass	:	Malacostraca
Order	:	Decapoda
Suborder	:	Natantia
Family	:	Penaeidae
Genus	:	*Penaeus*
Species	:	*monodon*

Distribution

Penaeus monodon occurs in the brackishwater areas, in estuaries and in the inshore waters of the east and west coasts, but its availability in the world is much lesser when compared to other marine shrimps like *P. indicus*; *P. monodon* completes its life cycle in two environments, namely, marine and estuarine environment. It is a euryhaline species with high tolerance to fluctuations of salinity.

S. Perumal et al. (eds.), *Advances in Marine and Brackishwater Aquaculture*,
DOI 10.1007/978-81-322-2271-2_12, © Springer India 2015

Its general distribution is along the entire east coast of Africa, Madagascar, Mauritius, Pakistan, east and west coasts of India, Sri Lanka, Andaman and Nicobar Islands, Singapore, Malaysia, the Philippines, Southern Japan, Celebes and Nearby Islands, Taiwan and Queensland and New South Wales in Australia.

Identification (Plate 1)

Alternating dark and light bands on its abdominal segments and also uniform coloured internal flagella

- Swimming legs fringed with bright red setae
- Yellow colour blotch at the base of the walking and swimming legs
- Rostrum long, sigmoid with 7–8 dorsal teeth and 2–3 ventral teeth

Food and Feeding Habits

Crustaceans form the main food item comprising the harpacticoid copepods and parts of brachyurans. Other item of food is bivalve molluscs, parts of fish, polychaete worms and vegetable matters.

The food of the fry in the cultivating ponds is 'lab lab', blue green algae and mixture of minute plant and animals.

The larvae during zoeal stages feeds on algae, mainly diatoms like *Chaetoceros* sp. Mysis stage onwards feeds on minute zooplankton like nauplii of *Artemia* sp. or phytoplankton. Postlarvae are carnivorous and prefer to feed on zooplankton.

Growth

This is the fast-growing and largest shrimp in the world. The female grows to a length of 363 mm (440 g), while the maximum size attained by males is 270 mm (180 g). In shrimp culture farms, they are usually harvested about 160–165 mm (30–35 g) in size. The stocked seed (15–20 mm) attain this harvest size in about 4 months in 15–25 ppt salinity.

Male and Female Identification (Plate 2)

The males are usually smaller in size and have external reproductive organ called petasma, which is found in the first pair of pleopods.

Plate 1 *Penaeus monodon*

Thelycum of Female

Petasma of Male

Plate 2 Male and female

The females are larger in size and can be identified by the presence of the external reproductive organ called thelycum, which is situated at the base of 4th and 5th pair of walking legs.

Maturity

The species grows to maturity, when their sizes are around 200 mm (weighing about 65 g) in male and 240 mm (90 g) in female.

Fecundity

Fecundity ranges from 2 to 19 lakh eggs depending upon the size of the female. Fecundity is more in the wild than in pond-reared broodstock on account of differential growth rate. In nature, 3 or 4 spawning may occur in a year.

Life Cycle

Though it dwells in the littoral waters, it has an estuarine phase in its life cycle. The adults mature, mate and release eggs in the deep part of littoral waters. The hatched out larvae dwell in the surface waters of the ocean till they reach postlarval stage. The postlarvae will drift towards coastal estuarine/brackishwater areas, which serve as nursery ground for them. As these grow to subadult stage, they start migrating towards inshore sea, where they mature and reproduce.

BRACKISHWATER (5 TO 25 PPT)

COASTAL/ESTUARINE OR OCEANIC WATERS (28 TO 35 PPT)

POSTLARVAE ← MIGRATION ← POSTLARVAE ← MYSIS ← ZOEA

↓ ↑

JUVENILES SHRIMP LIFE CYCLE NAUPLII

↓ ↑

ADOLESCENT → MIGRATION → SUB ADULTS → ADULTS → EGGS

Larval Stages (Plate 3)

The larval phase of the shrimp life cycle is spent in inshore sea in nature, and hence they need an environment with a salinity of 28–35 ppt. Nauplii subsist on yolk and mysis feed on phytoplankton (unicellular algae) and zooplankton (microscopic animals). Postlarvae are carnivorous and prefer to feed on the zooplankton. The identification features of larval stages are given in Table 1.

Nauplius

Nauplii are pyriform in shape, light brown in colour, opaque and swim in short jerks by means of three pairs of appendages, namely, first antennae, second antennae and mandibles. This stage has six substages.

Zoea

This is the first feeding stage. Its body is elongated and carapace, thorax and abdomen are distinct. Antennae, mandibles, maxillae and maxillipeds are also distinct and are useful in filter feeding and swimming (7–8 pairs of appendages). It has three substages.

Mysis

Mysis has well-developed carapace and abdomen. The five pereopods have undergone considerable enlargements. The exopods serve as swimming organs (13 pairs of appendages). The Mysis keeps its head down and swims. It has three substages.

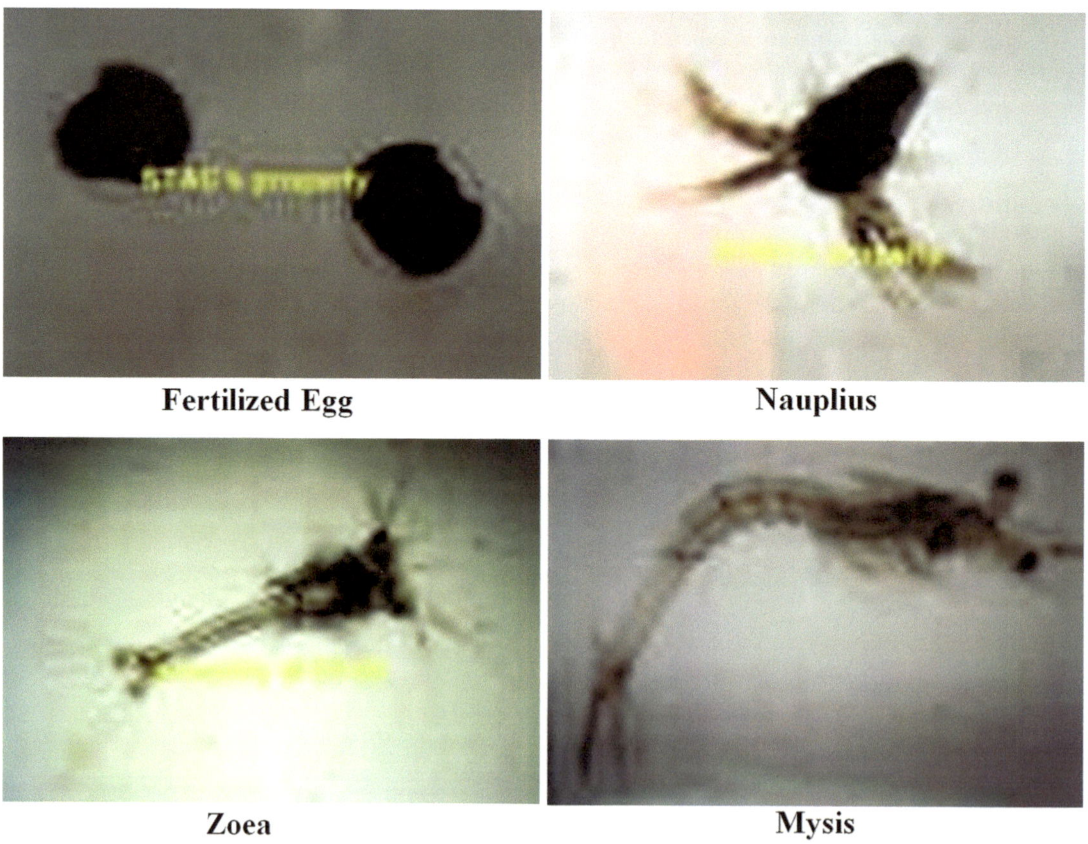

Plate 3 Larval stages

Table 1 Identification of larval stages of *P. monodon*

Stage	Duration	Feed	Identification features
Egg	12–15 h after spawning	–	–
N 1	5–6 h	Subsists on yolk	Furcal spines 1 + 1. The setae on appendages are not plumose
N 2	5–6 h	Subsists on yolk	Furcal spines 1 + 1. The setae on appendages are not plumose (well developed)
N 3	5–6 h	Subsists on yolk	Furcal spines 3 + 3
N 4	5–6 h	Subsists on yolk	Furcal spines 4 + 4
N 5	5–6 h	Subsists on yolk	Furcal spines 5 + 5
N 6	15–20 h	Subsists on yolk	Furcal spines 6 + 6, 6 + 7
Z 1	1½–2 days	Unicellular algae	No eye stalks
Z 2	1 day	Unicellular algae	Stalked eyes
Z 3	1½–2 days	Unicellular algae	A pair of biramous uropods and dorsal median spines on abdomen
M 1	1½–2 days	Unicellular algae and minute animals (zooplankton)	Rudiments of pleopods. (Mysis keeps its head down and glides)
M 2	1 day	Unicellular algae and minute animals (zooplankton)	Unsegmented pleopods
M 3	1½–2 days	Unicellular algae and minute animals (zooplankton)	Segmented pleopods with terminal setae
PL 1	1 day	Zooplankton	Segmented pleopods with setae will over (postlarvae swim horizontally)

Postlarvae

Fully developed postlarva resembles the adult in its shape. It has well-developed pleopods, which serve as main swimming organs (19 pairs of appendages). It swims horizontally like an adult. The postlarvae are aged in days, after its first appearance, calling it PL 1 on its first day and PL 2 on its second day and so on. Usually in hatchery, they are reared up to PL 3–5 in larval section and then transferred to postlarval section for further rearing up to PL 20. These stages can easily be acclimatised to lower salinities.

Site Selection

Major Criteria for Site Selection

Proper site selection is the key to a successful hatchery. The major criteria for site selection are as follows:

Seawater

- The hatchery site should be located close to the sea. The seashore should have gradual gradient and surf action should be minimal for easy drawing of seawater.
- The seawater salinity should be within the range of 28–35 ppt.
- The seawater pH should be within the range of 7.7–8.2.
- The seawater should have very low suspended solids preferably below 20 ppm at any given time.
- The seawater should be free from agricultural, industrial and sewage pollution.

Freshwater

Freshwater should be available in sufficient quantity for daily hatchery operations such as salinity adjustment, cleaning and domestic use. Ground water is preferred because of its better quality than surface water.

Climate

Rainfall: Total annual and monthly average rainfall should be carefully studied, as it influences on seawater quality and hatchery operations like outdoor algal culture, drying, etc.

Temperature: Atmospheric temperature variations are also an important climatic factor to be taken into consideration as it influences water temperature in the hatchery areas.

Wind: Wind direction and speed is important as it creates swells in the sea, resulting in turbidity of seawater. It should also be considered for design of structures and buildings.

Humidity and sunshine: These climatic factors are important for various aspects of hatchery operations like drying of tanks, rearing areas and heating of water during low-temperature period.

Topography

Topography means the 'lag of land'. The site for hatchery should be elevated, flat and easily drainable. The seabed should have gradient for installation of low-cost seawater intake system.

Soil

Physical composition, i.e. sand, silt and clay content of the soil, should be checked for determining the soil type. Load-bearing capacity of the soil should also be determined for the design of foundations for buildings.

Geographical Location

Availability of broodstocks: Broodstock should be regularly available year round for successful hatchery operation. The landing centre should not be beyond 2–4 h drive; otherwise transportation loss and high transportation cost will increase the cost of production.

Proximity to market: The farming areas, where the fry is expected to be marketed, should be within 300–400 km.

Availability of electricity: Electricity is essential for shrimp hatchery. High-tension power supply should be available close to the site.

Accessibility: The site should be easily accessible by an all-weather motorable road. Accessibility is important for supply of building construction materials to site, procurement of inputs for seed production and transportation of broodstock and fry.

Availability of labour: Labour should be available locally for daily hatchery operations.

Seawater Intake System

Seawater intake system used here to draw seawater for hatchery operations is called sub-sand filter intake system (Plate 4). In this system, a PVC pipe of 250 mm diameter having perforation on its surface and wound with two layers of plankton mesh and one layer of coconut coir rope is buried vertically under the sand bed to about 3 m below the existing bed level.

The water, which enters into the filter by in flow suction, is pumped through a 90 mm PVC suction pipe. A non-return flap-type PVC check valve is fitted with suction line either inside or outside the filter. Pumping can be done by using 5 HP monobloc pump installed in the beach at about 100 m away from the intake point. The water from the surface of the sea enters into the filter through different layers of sand, gets filtered and comes to the suction point. In the initial stage of pumping, all fine sand present around the filter is pumped along with the inflow water. On continuous pumping, coarser sand, which cannot enter into the filter, forms a layer around the filter and prevents the entry of fine sand thereafter. The water pumped through this filter is very clean and totally free from debris, organic matter and plankton and to some extend pathogen also

Precautions

- Pumping has to be started immediately after installation of filter; otherwise, the filter will get filled with fine sand particles which allow less or no flow of water into the filter and the system will become nonfunctional.
- Pumping has to be done at least 2 h/day from each filter, even if there is no requirement of water, to avoid chocking of filter.
- Filter has to be removed at least once in 3 months, cleaned and reinstalled to avoid the problem of biological degradation.

Plate 4 Seawater intake system

- Filter has to be installed about 1 m below the existing bed level and make sure that a water depth of 1 m must be available above sand bed even at low tide level.
- Avoid exposing filters to direct wave action.

Advantages of the System

- Cheapest and can be fabricated available at the site itself.
- Supplied good quality of water, which may sometimes pass through a series of filtration units.
- Can be installed instantly.
- Can be removed cleaned and reinstalled. Therefore, no wastage of material.
- No massive seawater intake structure is required.
- Can be installed by trained and skilled local labours.

In order to ensure good quality of seawater, the following steps have to be adopted.

Seawater Treatment Plant

Treatment plant is a place where micro-level filtration of seawater is done. Basically treatment plant has 3 major set-ups, namely, sand filters, cartridge filters and UV steriliser (Plates 5 and 6). The capacity of the pump required to run whole set-up is decided based on the rate of flow of water and total head loss due to pipes, valves, sand filters and cartridge filters. The capacity of sand filters and cartridge filters is decided based on every day water requirement for total hatchery operation and period in which the filters are to be operated.

Design of Pump

Total water requirement for hatchery operation is 370 t/day, of which, the PL rearing section requires about 200 t and the remaining water is used for the other sections. Except postlarval section, all other sections required either sand-filtered or cartridge-filtered water. The postlarval section required water only when stocking or water exchange is done.

Therefore, a sand filter supplying approximately 20–25 t of water per hour will be sufficient to meet the requirement.

Head Losses (Approximately)

1.	Suction pipe	1 M
2.	Deliver by pipe	3 M
3.	Valves and bend	3 M

(continued)

4.	Due to sand filter	5 M
5.	Due to cartridge filter	5 M
	Total	17 M

$$\text{Discharge required} = 0.0069\ \text{M}^2/\text{s}$$

$$\text{Capacity pump required} = \frac{1{,}000 \times 0.0069 \times 17}{75 \times 0.75 \times 0.75} = 2.78\ \text{HP} = 3\ \text{HP}$$

In treatment plant, two sets of filtration system are to be installed to assure supply of water at any time since cartridge filters require periodic cleaning and maintenance.

It Is Recommended to Buy:
2 numbers of 3 HP pump
2 numbers of sand filter
2 numbers of cartridge filters (5 μm)
2 numbers of cartridge filters (1 μm)

Filtration in intake system: The seawater intake system should be designed in such way that the water passes through a filter bed before it is pumped into the reservoir. In the process, most of the particulate matter will be filtered in the intake system itself.

Chlorination: While pumping, seawater should be chlorinated (5–10 ppm) before it reaches the reservoir. In the reservoir, the chlorine should be allowed to react for a period of 1–2 h. This chlorination kills all pathogenic

Plate 5 Treatment tank

Sand Filter **UV Sterilizer**

Plate 6 Filtering system

microbes and also chemically removes iron by forming a red precipitate with it.

Filtration: The seawater should be recirculated through a rapid sand filter (50 μ) till all the particulate matters and other precipitates are completely filtered.

Dechlorination: The residual chlorine available in the seawater should be estimated with chlorine test kits by using O-tolidine. Adding sodium thiosulphate pentahydrate can chemically neutralise the residual chlorine. This chemical reaction is as follows:

$$2\ Cl_2 + 2\ Na_2S_2O_35\ H_2O \rightarrow 2\ NaCl\ + NaO_2S_4O_6 + 10H_2$$

It is observed that sodium thiosulphate (hypo) required to neutralise 1 ppm of residual chlorine gas will vary between 0.5 and 2 ppm due to the influence of various factors specific to site and chemicals.

Treatment with EDTA: EDTA (ethylenediaminetetraacetic acid) is a chelating agent, which can remove chemical pollutants like trace metals and other fine debris. It is safe to add 10 ppm EDTA for ensuring clear seawater. Allow at least 1 h settlement time before using the seawater in order to ensure completion of EDTA action.

Broodstock Collection and Transport

Introduction

A large-scale hatchery has to necessarily depend on wild caught breeders (Plate 7) in addition to its own reared stock. To programme the hatchery operations and to meet the production requirements, a thorough knowledge about the availability of breeders, their collection and transportation technique is a prerequisite (Chakraborty and Sadhu 2001; Handbook on shrimp hatchery operation and management by the marine products export development authority; Operation manual for tiger shrimp hatchery by Sampath Associates 1994).

Source

Live breeders can be collected either from commercial trawler operators at fishing harbours or from country boat operators at fish landing centres.

From Trawlers

To collect live spawners, the trawl operation should not exceed 30 min. Practically this may not be possible since no trawler operator will be willing to do this. What the trawler operator usually does is to collect whatever live specimen available after the hauls in the trawl net. The operator has to be provided with large fibreglass or syntax tanks and a portable battery-operated aerator.

From Country Boats

They normally operate gill nets, which are usually set at sunset and the catch is checked at midnight or at sunrise. The traditional fishermen are adopting at bringing the spawner live by their own ingenious methods.

Temporary Storage at Collection Centre

The chances of getting live animal are totally unpredictable since the breeder availability varies seasonally as well as daily. Hence, arrangements for temporary holding of live spawners have to be made at or near the site of collection. A collapsible PVC sheet tank or a fibreglass tank of 1–2 t capacity will serve the purpose. A portable seawater pump and battery-operated aerators and oxygen cylinders also have to be provided.

Packing and Transport

Usually the spawners are packed and transported by the following methods:

Live animal holding tanks in trucks with provision for continuous aeration/oxygenation. Different types of tanks can be used for this purpose. In doublewalled plastic bags with cool seawater (22–24 °C) and oxygen which is then packed in Styrofoam boxes. The rostrum of the shrimps is covered with small rubber tube to prevent

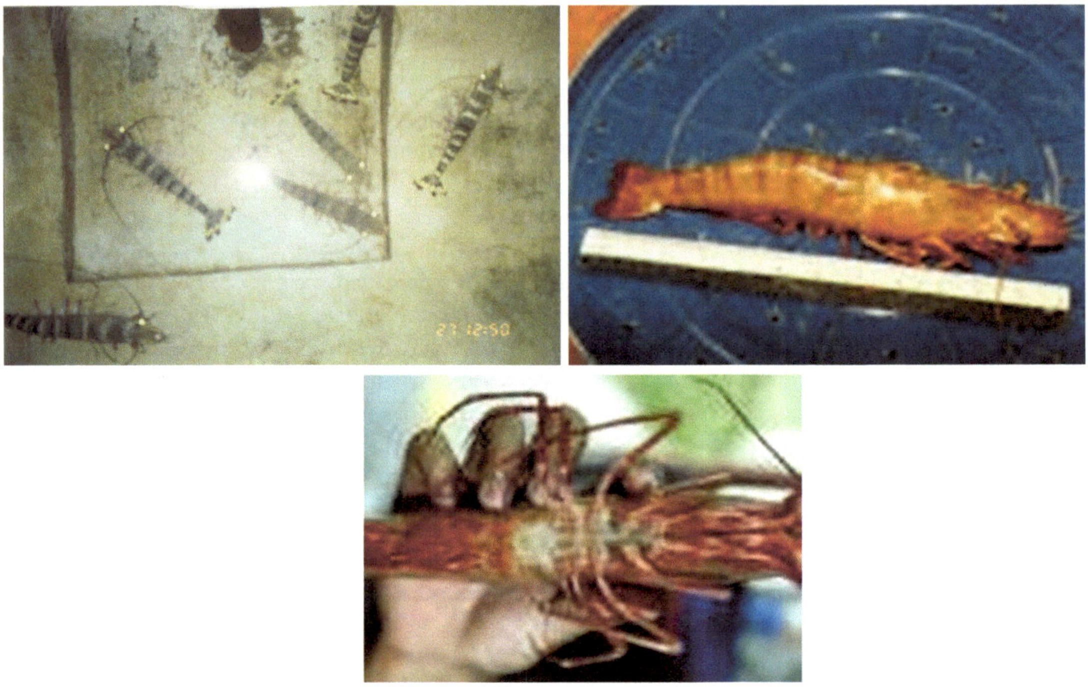

Plate 7 Brooders

puncturing of plastic bags, or the spawners are immobilised in separate PVC tubes.

Quarantine

Upon arrival in hatchery, the spawners are kept in a quarantine tank containing clear seawater with good aeration. The animals are then examined for damage, disease or parasitism. Precautionary treatment is given as follows, before they are transferred to the maturation tanks:

Formalin dip	25 ppm	15–20 min
Malachite green dip	0.5 ppm	02–03 min
Furazolidone	0.4 ppm	05–10 min

Materials Required for Breeder Collection and Transport

1.	Transport vehicle (mini truck)	1 no
2.	Fibreglass tank (small) to fit in the truck	2 nos
3.	Plastic pools/fibreglass tanks for maintaining breeders (1 t capacity)	2 nos
4.	Battery-operated aerator/air pump	8 nos
5.	Portable pump (0.5 HP)	1 no
6.	Salinity refractometer	1 no
7.	Oxygen cylinder (small)	4 nos
8.	Oxygen regulator with key	4 nos
9.	Syntax tank (500 L)	1 no
10.	Heavy duty plastic bucket	6 nos
11.	PVC pipe (1 ft size 63 mm dia. with holes)	100 nos
12.	Mosquito net	10 m
13.	Small rubber tube (3 mm dia.)	2 rolls
14.	Thermocol-lined cardboard cartons	400 nos
15.	Polyethylene tube (for bags)	100 nos
16.	Small polyethylene bags for ice packing	10 nos
17.	Rubber band	2 kg
18.	Scissor	2 nos
19.	Plastic tape	10 rolls
20.	Strapping clip	4 kg
21.	Strapping machine	2 nos
22.	Flexible pipe (20 mm dia.)	40 m

Maturation and Spawning

Introduction

This is the first major unit of any commercial shrimp hatchery, which involves in the production of the first stage of shrimp larvae, the nauplii. Operation of maturation section involves procurements of spawners, acclimatising them to local conditions, breeding techniques and spawning.

Maturation Techniques

General

Stocking density	4–6 Nos./m^2
Sex ratio (female/male)	3:1
Eye stalk ablation	All females are unilaterally eyestalk ablated
Water temperature	28–32 °C
Salinity	28–36 ppt
Filtration	Sand filter water is used
Light	Artificial low blue colour
Feeding	Fresh feeds-squids, fresh beef liver, cattle fish, clams, oysters and mussels. 10 % of the body weight. 3 times per day
Water exchange	30 % per day
Prophylactic treatment	Chloramphenicol or prefuran or erythromycin or oxytetracycline is applied at a dose of 4 ppm twice per month
Sanitation	Disinfect and wash all glasswares, PVC support columns, etc. regularly. The floor of the section as well as spawning room should be disinfected twice a week using 100 ppm chlorine water. Using long handled dip net, remove all moults and any dead animals from each tank
Siphoning	Systematically siphon the entire tank bottom making sure to remove any residual food left from previous feedings. Observe and record the residual food in each tank, and if none is left, a slightly higher dose may be given in the next feed. If any balance is left, a few grams may be reduced. Siphoning should be done twice a day, once in the morning and other in the evening

General Information About Maturation Unit

Ovarian maturation stages of tiger prawn categorised in five steps:

1. Undeveloped
2. Developing
3. Early ripe
4. Ripe
5. Spent

Mating required a minimum water volume and depth. Fertilisation of eggs from initial spawnings of captive broodstock is dependent on sperm from mating in the wild or pond environment. Stocking density of *P. monodon* in tanks will be 4–6 m^2, depending on the water quality and exchange rate. Sex ratio maintained at 1 male: 1 female to ensure mating success. Higher female ratio 3:1 would be more economical because egg and larval production per tank are maximised.

Stocking of disease-free quality broodstock of both female and male in the maturation tanks is preferred in the ratio of 2:1 or 3:1. Prior to stocking, ablate the females with one eyestalk, which induces the ovary development.

Eyestalk Ablation

In selecting females to undergo ablation, avoid soft (moulted) animals or they will not survive ablation. Determine the mean weights of the males and females based on five samples of each set and then calculate the total biomass of each tank. Before ablation, segregate the males and females in two separate tanks containing 100 L of seawater. Transfer the males to the tanks with antibiotics (20 ppm oxytetracycline or chloramphenicol or Erythromycin). Hold the animal, exactly for half an hour, and transfer them to the maturation tank.

Eyestalk ablation is the common technique which involves destroying the X-sinus gland in one of the eyestalks that produce and store gonad-inhibiting hormone (GIH), thus accelerating gonad maturation. An ablated female spawner takes 3–7 days to mature. Eyestalk ablation enables the technician to have a better control over the time of maturation for proper planning of fry production.

Eyestalk ablation may be performed in several ways. Some of the common methods are:

1. Pinching the eyestalk and eyeball with or without prior incision on the eyeball.
2. Cauterisation of the eyestalk with heated surgical clamps, forceps or soldering iron.
3. Direct cutting of eyestalk with a pair of scissors.
4. Lighting the base of eyestalk.

The ablated females are kept in the maturation tank together with unablated males. The breeders are fed daily with chopped fresh clams, mussels or oysters, and their quantity is depending on the consumption. Daily cleaning of maturation tanks and water exchange depends on the water quality and tank condition.

Spawning and Production of Nauplii

The maturity of females is determined by observing the condition of the ovary, which is located on the dorsal side of the body. It can be roughly classified according to the appearance of the ovaries into five stages.

1. *Immature or resting stage* – Ovaries extremely thin.
2. *Developing stage* – Ovaries appear as a light green in colour and straight band is visible through the shell.
3. *Early ripe stage* – Ovaries become broader and clearly visible as an olive green band.
4. *Ripe stages* – Ovaries are dark green in colour.
5. *Spent* – Complete spent ovaries are thin and appear like first stage of ovaries.

Dip the gravid spawner in 100 ppm formalin. After exactly 5 min, remove the females from formalin bath and rinse in the holding bath, and then place each female in an individual spawner tank. Record the maturation tank number from which each female was removed and the spawning tank number into which it was transferred in a data sheet.

When the ovaries have developed to stage 4, the spawners are transferred to spawning tanks. The eggs produced from early spawning are usually more and of better quality than those from subsequent spawning.

The fertilised eggs are spherical and hatch into first stage, the nauplii within 12–15 h after spawning. These nauplii can be harvested with a help of light source, as they are phototactic in nature. Harvested nauplii can be stocked in larval rearing tanks for further rearing.

Daily Work in Maturation Unit

Daily work	Fresh feed distribution
07.00 A.M.	Sanitary management
	Checking the aeration, light and water exchange of the tank
	Checking and recording temperature, salinity, pH, ammonia and nitrite in each tank and in coming water
	Removal of dead animals, if any, and uneaten feed
	Fresh feed will be cut – frozen into slices of about 0.5 × 0.5 cm^2 and then defrosted before feeding
08.00 A.M.	Distribution of feed in each tank
	Washing the floor of the maturation tank
09.00 A.M.	Maintenance works, equipment repairs, etc.
02.00 P.M.	Distribute the feed
02.30 P.M.	Arranging the water exchange
	Arranging the aeration, light, etc.
04.00 P.M.	Checking the maturity conditions in each tank
04.30 P.M.	Selecting the gravid female to spawning tank
08.00 P.M.	Distribution of feed
12.00 P.M.	Distribution of feed

Spawning Tank

07.00 P.M.	Collection of females from maturation tank
09.30 A.M.	Spawning tank one by one, observation and recording about the success of spawning
	Checking the rate of fertilisation under microscope
09.30–01.30 P.M.	Cleaning the equipments with chlorine and soap
	Eggs are allowed to hatch in the spawning tank; enough time should be given for complete hatching
	The next day, morning hatched nauplii will be harvested

Larval Rearing

Introduction

Shrimp nauplii received from maturation section are reared in larval rearing tanks for 13–15 days till they reach PL3 or PL5 stages. These pass through 6 nauplius, three zoeal and three mysis stages before they reach postlarval stage. Larval stages are very critical and sensitive phases of shrimp life cycle. They need clean and healthy environment, timely feeding, gentle and careful handling, etc. The technical personnel should have sound scientific knowledge of biology, ecology, behaviour and nutritional requirements of the larvae. They should be critical in observation, quick in decisions making and dedicated in discharging the duties. Negligence and complacency can very easily yield poor results in larval section.

Facilities Required for Larval Rearing

Larval Rearing Room

Larval rearing room, being one of the most sensitive areas of the hatchery, needs complete isolation from other sections to avoid contamination. If it is provided with natural lights in the roofing for checking low temperatures, shade clothes (preferably sky blue) should be arranged over the tanks to prevent direct sunlight on the tank, which affects the survival of larvae.

Larval Rearing Tanks

Shrimp larvae are being reared in tanks of different types of shapes (Plate 8) and capacities in different hatcheries. The recent studies and experiences have proved that the parabolic-shaped tanks with set aeration line are yielding best results owing to its ability to provide uniform aeration, uniform distribution of food materials, giving least scope for settlement of organic detritus on the tanks surface and then minimising the growth of microbes in the tank. In these tanks, it is easy to siphon out the left overfeed, organic detritus faecal strands, etc.

Disinfection Tanks

One FRP tub of suitable dimensions is required in the section for disinfecting the strainers, screens, hoses, filter bags and plastic ware after use.

Laboratory

A laboratory equipped with a microscope for evaluating the health of the larvae, seawater testing kits and equipment for conducting microbiological studies, etc. should be attached to the larval section (Plate 9).

Equipment

Strainers (Plate 10) with 100, 250, 350 and 500 nylon mesh; harvesting buckets; plastic buckets and tubs, 100 L; plastic cans for transferring larvae; glass and plastic beakers; etc. are essential equipment for water exchange, treatments, feeding and other management activities.

Chemicals and Drugs

Bleaching powder for disinfection; detergents for washing; aquatic grade antibiotics like chloramphenicol, oxytetracycline, erythromycin, furazolidone and perfuran; and fungicides like Treflan, malachite green and formalin are essential for larval rearing.

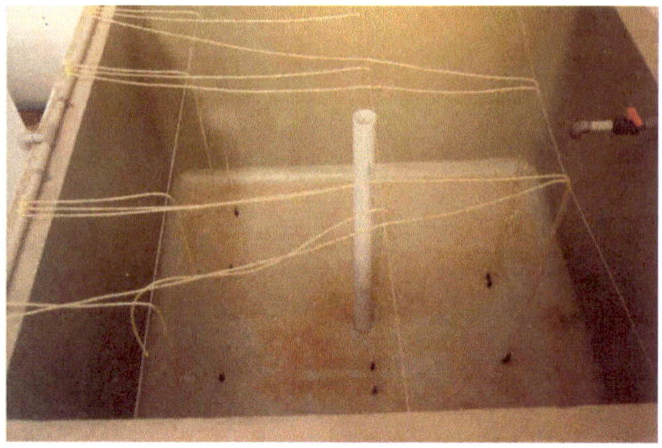

Plate 8 Larval rearing tanks

Plate 9 Laboratory

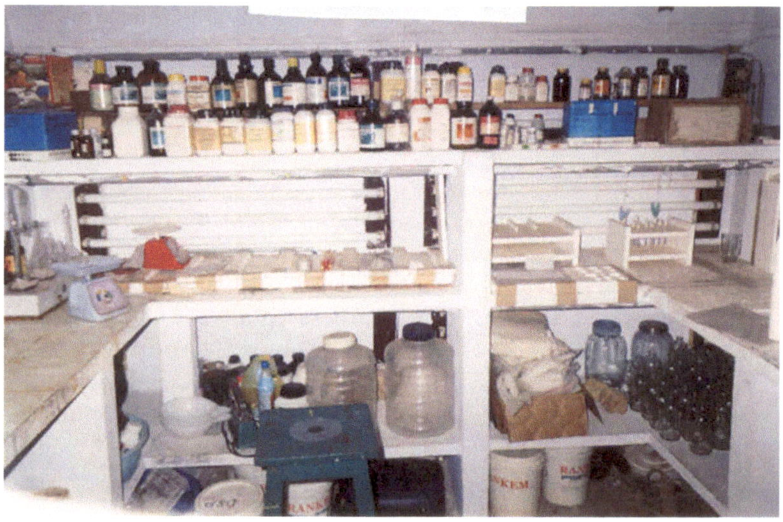

Seawater

The seawater should be made free from particulate material and turbidity, by drawing it from a subsoil filter arranged below beach sands in the intertidal zone. This water will be chlorinated (8–10 ppm) to kill all the microflora and fauna. Then it should be dechlorinated with sodium thiosulphate. EDTA (10 ppm) is added to remove the dissolved heavy metal pollutants if any. This seawater will be passed through a two μm filter bags before use in larval section.

Aeration

Filtered, dust and oil-free air is supplied continuously, by using an air blower (Plate 11).

Larval Feeds

The larvae during zoeal stages feed on algae. Diatoms like *Chaetoceros* sp. and *Skeletonema* sp. are suitable feed for shrimp larvae. A separate facility for culturing algae under aseptic conditions is an essential component of a shrimp hatchery.

Mysis stage onwards, a minute zooplankton like *Artemia* nauplii, is to be introduced as a feed along with algae. A suitable facility for hatching the *Artemia* cysts is also essential for a shrimp hatchery.

In addition, supplementary feed, microencapsulated diets of different particle sizes are essential to get better results as well as to use as substitutes during scarcity of live feeds.

Plate 10 Equipments

Air Blower

Aeration System

Plate 11 Aeration

Environmental Conditions Required for Larval Rearing

The natural environment in which the shrimp larvae live is marine. The characteristic environmental parameters of seawater which account for its quality are:

1. Salinity
2. Temperature
3. pH
4. Turbidity
5. Dissolved oxygen content
6. Microflora and fauna
7. Nutrients
8. Pollutants

Seawater Salinity

This is a natural paramcter, for which greater importance is given during site selection, since artificial manipulations to increase the salinity are laborious and uneconomical. However, adding freshwater can bring down salinities. For a shrimp hatchery, the recommended salinity range is 28–35 ppt. This should be monitored every day.

Seawater Temperature

The recommended range of seawater temperature is 28–32 °C. Temperature should be recorded twice a day at 7 A.M. and 3 P.M. - Low-temperature problems can be overcome to some extent by using the following methods.

- By providing natural light with translucent fibreglass sheet roofing
- By covering the culture tanks with PVC/plastic sheets
- By using thermostatically controlled insulated immersion heaters in the culture tanks
- By using room heaters
- By using solar water heating systems
- By passing the seawater through thermostatically controlled electrical heating system before filling the tanks

High temperatures can be checked by (1) providing shade clothes over the tanks and (2) providing proper ventilation to the culture rooms.

Seawater pH

Seawater pH is good indicator for chemical quality of the seawater. The seawater pH should be within the range of 8.2–8.5 for larval rearing operations. pH should be recorded along with salinity once in a day. Seawater pH also should be given due importance during site selection.

Larval Rearing Operations

Preparation of Larval Tanks for Stocking Nauplii

The tanks should be disinfected with 200 ppm chlorine water for 8–10 h and then thoroughly scrubbed with a mixed solution of 200 ppm chlorine and 5 % detergent by using sponge pads. Then the tank is thoroughly rinsed with freshwater and dried for at least 24 h. Just before filling the tank, it should be scrubbed with detergent and rinsed with freshwater thoroughly. Then the tank is filled to 60 % of treated and filtered seawater. Prophylactic treatment with fungicide (Treflan 0.05 ppm) and an antibiotic (Table 2) should be given to the tank water 1 h before stocking. Just before stocking, algae (*Chaetoceros* sp.) should be added at a density of 100,000 cells/ml. The tank water should be sufficiently aerated throughout the culture operation. Recommended stocking density is 100,000 nauplii/t. If the tank is newly constructed, it should be thoroughly leached with freshwater and seawater, for 2–3 days before the preparations are carried out.

Microencapsulated feeds (optional)	FRIPPAK FEEDS like CAR, CD-2 or other larval feeds may be fed at least 4 times a day as per the directions of the manufacturer
Prophylactic drugs: antifungal	Treflan (0.05 ppm); antibiotics: chloramphenicol (4 ppm) oxytetracycline (4 ppm), prefuran (1 ppm), furazolidone (1–3 ppm), neomycin sulphate (3–10 ppm)

Acclimatisation and Stocking of Nauplii

The nauplii (N5/N6) received from the maturation section should be acclimated by adding the larval tank water to the acclimation buckets slowly by using a flexible tube (1 cm dia.) for about 10–20 min. Aeration should be given during acclimation. The nauplii can be released into the tanks slowly in small quantities at different points of the tank. One hour after stocking, the population should be estimated and recorded.

Schedule of Larval Rearing Operations

The larval rearing from N6 to PL-3 takes 13 days. The schedule of different operations like water exchange, feeding regime and drug treatment is summarised chronologically in Table 2. The procedures and methods followed for each operation are explained in Table 2. Under normal conditions, this schedule provides clean and healthy environment to the larvae. But the technical personnel have to constantly observe the tank conditions and animal health and modify the water management, feeding and treatment

Table 2 Schedule of larval rearing operations

Day	Substage	Water management: Water level (in tons)	Water exchange	Mesh size of the screen (in microns)	Algal feeding (cells/ml)	*Artemia* (naups/ml)
1	N6/Z 1	6	–	100	100,000	
2	Z 1	8	Make up to 8 t	100	100,000	–
3	Z 2	10	Make up to 10 t	100	100,000	–
4	Z 3	10	30 %	250	100,000	–
5	Z 3	10	50 %	250	100,000	–
6	M 1	10	70 %	350	100,000	0.25
7	M 1	10	70 %	350	100,000	0.25
8	M 2	10	70 %	350	100,000	0.25
9	M 3	10	70 %	350	100,000	0.50
10	M 3	10	70 %	350	100,000	0.50
11	PL 1	10	100 %	350	60,000	1
12	PL 2	10	100 %	350	60,000	1
13	PL 3	10	100 %	350	60,000	1

schedule according to the need as the situation demands.

Larval Diseases and Treatment

Disease is a major threat for any biological system and hatcheries are no exception to this. In an aquatic system, prevention is the best remedy. If a hatchery system is infected with a disease, it is better to discontinue the operation till the pathogen is eradicated by disinfection and drying.

Fighting the disease with drugs leads to many complications like pathogens developing resistance to available drugs and permanent settlement of pathogens in the hatchery. Usually in all shrimp hatcheries, preventive measures like using pathogen-free treated water, administering prophylactic treatments and observing strict sanitary and hygienic principles are followed. A few common larval diseases have been listed in Table 3. The standard treatments given in the table might help for some time, but subsequently, new drugs have to be tried, after conducting sensitivity tests in the microbiological laboratory.

Usually the pathogens and parasites take advantages of the damages caused by some primary stress factors like pollution effects, undernourishment, overcrowding and bad water quality. Hence, care should be taken to prevent the primary stress factors and thus eliminate the chances for disease outbreaks.

Daily Routine of Larval Section

The daily routines of larval rearing section in chronological order are given in Table 4. Larval section needs at least two technical personnel and three aides to take care of all the activities of a day. Since all the activities are time-bound, this schedule has to be strictly followed for better results.

Walk Through

This is a physical examination of the facility, tank condition, aeration, condition of the tank water, algal density, *Artemia* density, animal behaviour, activity, health and feeding, etc. This examination will help us to know the general condition of each tank and to identify the problematic tanks. One litre glass beaker may be used for examining the animals. But it should be disinfected in chlorine bath before and after examining a tank. Walk through should be done three times a day at 7 A.M., 2 P.M. and 10 P.M.

Table 3 Diseases found in the developmental stages of penaeid prawn larvae and their control methods

Disease	Affected parts	Symptoms	Treatment	Life stages affected
Bacteria Bacterial necrosis	Appendages	Appearing as localised necrosis or discoloration on any appendage, causing high mortality of zoea and mysis stages affects postlarvae to lesser extent	Furnace, 1.1 ppm Erythromycin, 1.5 ppm Achromycin, 1.2 ppm	Z, M, PL
Vibrio infection	Haemolymph midgut gland	In initial stages of one form, some larvae will show yellow-vermilion and red colour permeating entire nervous system. Another form exhibits 'white-turbid liver' where the midgut gland of the larvae becomes generally white-turbid. Turbidity becomes more apparent and well defined as the disease progresses	Furazolidone, 2.0 ppm Terramycin, 45 mg/kg biomass Furnace, 1.3 ppm	PL

Disease	Affected parts	Symptoms	Treatment	Life stages affected
Filamentous bacteria	Gills, pleopods	Commonly found attached to the gill filaments and the pleopods turning blackish when bacteria mix with dirt. If severely affected, the respiratory function of the gill suffers damage	Cutrine plus, 0.5 ppm Malachite green, 10 ppm Potassium permanganate, 8.5 ppm Cuprous chloride, 1.0 ppm	PL
Shell disease	Exoskeleton muscles	If infected by chitinovorous bacteria, the exoskeleton will display eroded, blackened areas. The edges or tips of the exoskeleton parts are typically attacked Also bacteria can rapidly enter the body through surface breaks to cause internal damage	Malachite green, and 0.9 ppm Formalin combined, 2 2 ppm	PL
Black gill disease	Gills	In initial stages, gill colour turns dull orange-yellow or light brown. When advanced, the area darkens until it is finally black	Malachite green, 3.0 ppm Methylene blue, 3–10 ppm	PL

Disease	Affected parts	Symptoms	Treatment	Life stages affected
Fungi Lagenidium infection	Body cavity, appendages	Only thin-cuticled prawns can be infected; thus, larval prawns are highly sensitive. The hyphae appear inside the body of zoea and continue into mysis stage, resulting in massive muscle destruction and heavy mortality of zoea and mysis	Treflan, 0.1 ppm Malachite green, 0.01 ppm	Z M

(continued)

Table 3 (continued)

Disease	Affected parts	Symptoms	Treatment	Life stages affected
Ectocommensal protozoa	Gills, eyes, exoskeleton	Heavy infestation by *Zoothamnium* sp. of gills and eyes of larval prawn results in high mortality. *Epistylis* sp. seems to prefer exoskeleton as attachment site and is less harmful. When abundant on gill surface, both can cause hypoxia and death. Additionally, their abundant presence on general body surface of larvae may interfere with locomotion, feeding moulting, etc. Parasite burden increases until ecdysis provides relief	Malachite green, 1.0 ppm	Z
Ciliate infection (Zoot*hamnium* sp., *Epistylis* sp.)			and Formalin combined, 25 ppm	M
			Quinacrine hydrochloride, 0.8 ppm	PL
			Chloramine-T, 0.5 ppm	
			Methylene blue, 8. 0 ppm	
			Saponin, 10 %, 5.0 ppm	
Viruses	Hepatopancreas, anterior midgut	Penaeid baculoviruses infect epithelial cells of the hepatopancreas and, less commonly, anterior midgut, causing high mortality in the postlarval stage		PL
Penaeid baculoviruses (PB MBV, BMN)				

Disease	Affected parts	Symptoms	Treatment	Life stages affected
Infectious hypodermal and haematopoietic necrosis (IHHN)	Hypodermis haematopoietic organs	Prawns dying from acute IHHN show massive destruction of cuticular hypodermis and often of the haematopoietic organs, of glial cells in the nerve cord and of loose connective tissues such as the subcutis and gut serosa. Only prawns within a size range of 0.05–1.0 g have been observed to have these epizootics, resulting in massive mortalities (often 80–90 % within 2 weeks of onset)		PL
Miscellaneous diseases	Appendages	Occur as a result of poor quality of spawner		N
Abnormal nauplii				
Amoebiasis of larvae	subcutis, muscles	Invasion of muscles and subcuticular tissues located in the abdomen. Cephalothorax, antenna and eyestalks, by unclassified amoeba		ZZ
Larval encrustation	Exoskeleton	Brown to black encrusted deposits which contain iron salts affect larval penaeids		ZPL

N nauplius, *Z* zoea, *M* mysis, *PL* postlarva

Recording Physical Parameters

Physical parameters like salinity, temperature and pH are to be monitored twice a day, before filling the tanks.

Population Estimation

Four to eight samples of known volume (250 ml) should be collected at random from different places of the tank, and number of animals present per litre is estimated, from which total population in the tank can be computed by using the formula

$$\text{Total population in the tank} = \text{No. of larvae per litre} \times \text{tank volume.}$$

Table 4 Daily routines of larval section in chronological order

Time	Technical	Non-technical
06.00–07.00	–	Treflan treatment, feeding encapsulated diets, preparing tanks for stocking Nauplii
07.00–08.00	Walk through, population estimation	Recording physical parameters, draining the tanks
08.00–9.00	Microscopic examination, algal and *Artemia* feed calculations	Acclimation and stocking nauplii
09.00–10.00	Feeding *Artemia*	Feeding with algae and *Artemia*, refilling the tanks
10.00–11.00	Estimating the population of postlarval transfers	Harvesting and transferring, postlarvae
11.00–12.00	Data maintenance	Treatment with drugs
12.00–13.00	Data maintenance	Disinfecting tanks, screens, filter bags and washing the floor
13.00–14.00	–	Treflan treatments, feeding with encapsulated diets
14.00–15.00	Walk through	Cleaning
15.00–17.00	Algal and *Artemia* counts and feed calculations. Microscopic observations and data maintenance	Temperature readings, feeding algae and *Artemia*
		Washing screens, filter bags and plastic ware; feeding with encapsulated diets
21.00–22.00	Walk through, *Artemia* counts and feed calculations	Feeding with *Artemia* and encapsulated diet. Treflan treatment

Microscopic Observation

A sample of five to ten larvae from each tank should be observed first under the stereo-zoom dissecting microscope and then under compound microscope. The observation on:

- Swimming activity
- Feeding
- Developmental stages
- Morphological characters
- Symptoms of stress
- Presence of ectoparasite protozoans
- Presence of specific disease like necrosis, larval mycosis, etc. should be recorded on the daily sheet.

The suitable prophylactic (Table 2) or therapeutic treatment (Table 3) should be given to the tank, if necessary, based on the microscopic observations.

Water Exchange

As water exchange plays a vital role in the success of larval rearing operation, it needs intensive care. The water exchange schedule for each state is given in Table 2. A strainer suitable for the larval stage present in the tank is selected and checked thoroughly for any possible damage and leakage. Then it is disinfected in 200 ppm chlorine bath and then thoroughly rinsed in freshwater till chlorine smell is removed. Then it is fixed to the flexible hose of the inner standpipe; it is lowered to the bottom of the tank. Then outer standpipe is lowered to drain water from tank into a harvesting bucket of suitable mesh size to prevent the escape of animals, if any, but the water flows off. While draining is going on, seawater should be splashed on the strainer once in every 10 min to wash the adhering larvae into the tank. The flow rate should be adjusted in such a way that the animals are not forced in to the strainer. When the draining is completed, the exposed walls of the tank should be wiped with clean sponge pad to remove the adhering debris and dirt. Then algae and *Artemia* in calculated quantities should be given before refilling the tanks.

Feeding

The feeding regime for each larval stage is given in Table 2. Generally three types of feeds, viz., algae, *Artemia* and microencapsulated diets, are used in larval rearing. Zoeal stages require only unicellular algae. Mysis and PLs need *Artemia* nauplii along with algae.

Algal Feeding

Algae have to be fed immediately after water exchange and in the evening at 3 P.M. The residual algal cell density is estimated by using haemocytometer. Then the volume of algal water to be added can be computed by using the formula

$$AW = \frac{(DD - RD) \times TV}{AD}$$

where

AW = Algal water to be added to the tank (in litres)

DD = Desired cell density (cells/ml)

RD = Residual cell density (cells/ml)

TV = Larval tank volume (in litres)

AD = Cell density in algal culture tanks (cells/ml)

The algal feeding pump and flexible hoses should be disinfected with 200 ppm chlorine, and then rinsed with freshwater before and after every feeding.

Feeding *Artemia* nauplii

The *Artemia* nauplii requirements should be estimated and informed to the *Artemia* hatching division on the previous day itself. After harvesting, the density of *Artemia* nauplii in the storage tank should be estimated. Counting the number of *Artemia* nauplii present per ml and thereby computing the number per litre can do this. The volume of *Artemia* concentrate to be added to larval tank can be estimated by using the formula

$$AC = \frac{(DD - RD) \times TV}{AD}$$

where

AC = *Artemia* concentrate to be added to the tank (in litres)

DD = Desired density (nauplii/ml)

RD = Residual density (nauplii/ml)

TV = Larval tank volume (in litres)

AD = Density in *Artemia* concentrate (nauplii/ml)

Care should be taken not to allow the entry of cysts or cyst shells along with *Artemia* nauplii. The cysts may bring bacteria in to the larval tank. Freshly hatched *Artemia* nauplii are good for larval feeding.

Feeding of Microencapsulated Diets (Optional)

The microencapsulated diets are given as supplementary feed along with natural diets for faster and healthy growth. Sometimes, they are used as substitutes to natural feeds during the periods of scarcity. Capsule size of 5–30 μm is suitable for zoea, 40–90 μm for mysis and 90–150 μm for postlarvae. Required quantities of the feed (Table 2) is weighed out into a beaker and mixed in water and fed after decanting the debris and scum. This feed is given at 6 A.M., 1 P.M., 5 P.M. and 10 P.M.

Drug Treatments

The drug treatments should be avoided at transition stages. Every new batch of any drug should be tested on larvae by conducting a bioassay for 24 h in the laboratory. Only aquatic grade drugs, which are soluble in water, should be used.

Harvesting and Transferring Postlarvae

The postlarvae are harvested and transferred to postlarval section at PL 3 stage, for further rearing. The tank water is drained to 3 t level, and the postlarvae which are concentrated will be scooped out by using scoop nets and transferred to 50 L plastic bins, which are aerated continuously. This vigorous aeration makes the postlarvae distribute themselves homogenously in the container. Then a few aliquot samples of known volumes (35 ml) are collected and the PLs present in those samples are counted; from this, number of PLs per litre is estimated, and total PLs present in the container can be computed by using the formula

Total PLs in the container
= No. of PLs per litre
× Total volume of the container in litres.

This will be repeated for all the transfers from that tank and added to compute total population harvested. This is recorded on the tank data sheets, and percentage survival from N6 to PL 3 is calculated.

Day-to-day observations and activities should be recorded immediately on the relevant data sheets. This data will help to trace the culture history of different batches of production and to study the economics of the operations.

The following data sheets are used in larval section:

1. Larval rearing daily work sheet
2. Larval rearing tank data sheet

Algal Production

Introduction

Shrimp larvae at their early stages such as zoea and mysis are planktonic filter feeders. At these stages, they should not be fed with an artificial diet: the diet must be planktonic. A typical diet for these stages consists of marine phytoplankton (algae), small unicellular plants 3–30 μm in diameter. In hatchery rearing, a reliable and efficient production of this unicellular algal food species is of paramount importance. Algal cultures must be started several days prior to the zoeal stages so that the mass culture reach maximum cell numbers just before the larvae begin feeding.

Although large numbers of algal species have been used as larval food, only a few have been cultured in quantities and concentration sufficient for large number of larvae. The species cultured for the feeding of shrimp larvae are *Skeletonema costatum*, *Chaetoceros gracilis* and *Isochrysis galbana*. Among these *Chaetoceros* sp. is the suitable feed for *P. monodon* hatchery. *Chaetoceros* sp. is identified by a pair of thin long spines at each end, which fusion with those of neighbouring cells form into chains.

Facilities Required for Algal Growth

Culture room: The culture room should have shelves of different heights to accommodate various sizes of vessels. The room should be isolated from the other hatchery facilities to avoid contamination. Light is generally supplied by banks of white cool fluorescent bulbs (40 W) fitted in chambers, preferably painted white. All light banks should be arranged properly to meet the light intensity requirements of different culture vessels. Light intensity varies from 1,000 to 8,000 lx depending on the volume of culture. Temperature is maintained between 24 and 26 °C by using air conditioners. In addition, the culture room should have a provision for air supply. A simple binocular microscope and a haemocytometer are required for evaluating cell quality and for estimating cell density. A balance for weighing chemicals and a magnetic stirrer for mixing chemicals are required.

Culture vessels: Algal culture vessels should be transparent to allow light and easy to wash. Conical flasks of 50 ml, 100 ml and 2 L capacity and carboys of 20 L capacity are best suited for algal culture. For mass culture, transparent FRP cylinders of 200 L capacity and fibreglass tanks of 1 MT and 2 MT are required.

Table 5 Culture media composition and preparation

S. No.	Nutrient	Primary stock solution (1 L of distilled water)	Working stock solution (1 L of distilled water)	Dosage (1 L of seawater) *F*	*F*/2	*F*/4
1.	Nitrate and phosphate	–	75 g of sodium nitrate and 5 g of sodium phosphate	2 ml	1 ml	0.5 ml
2.	Silicate	–	55 ml of sodium silicate	2 ml	1 ml	0.5 ml
3.	Trace metals	(a) 10 g of copper sulphate	1 ml of primary stock solutions of a, b, c, d, e and 4.56 g of EDTA and 5.15 g of ferric chloride	2 ml	1 ml	0.5 ml
		(b) 22 g of zinc sulphate				
		(c) 10 g of cobalt chloride				
		(d) 180 g of manganese chloride				
		(e) 6 g of sodium molybdate				
4.	Vitamins	(a) 20 g of thiamine hydrochloride	5 ml of primary stock solution of a, b and c	2 ml	1 ml	0.5 ml
		(b) 100 mg of biotin				
		(c) 100 mg cyanocobalamin				

Seawater: Seawater should be good quality to get the best results in algal culture. It should be clean and devoid of pollutants and other live organisms. Pre-filtered seawater may be chlorinated (5–7 ppm) and dechlorinated (with sodium thiosulphate) and filtered through sand and cartridge filters to make it clean and free from all other organisms.

Environmental Conditions Required for Algal Culture

Water temperature	24–26 °C
Salinity	25–29 ppt.
Water filtration	1 μm for outdoor and small volume indoor cultures
	0.22 μm for stock culture (test tubes and 250 ml flasks)
Aeration	Continuous
Light	Artificial fluorescent tubes for indoor
	Sunlight for outdoor 1,000 lx for stock culture 1,000–8,000 lx for Indore culture

Preparation of Nutrient Solutions

Algae need various macro- and micronutrients (Table 5) for their healthy growth. The stock solutions of the nutrients should be prepared in distilled water. Primary stock solutions of trace metals and vitamins can be stocked for 1 month in the refrigerator. Working stock solutions of nitrate, phosphate, silicate, trace metals and vitamins should be prepared every 10 days.

Algal Culture Operations (Plate 12)

Isolation of Pure Algal Strain from Raw Seawater

Twenty litres of seawater collected during high tide should be enriched with nutrients and left under light till algae bloom. The nutrients added to the seawater are suitable for the growth of algal species and favour their dominance in the culture. Hence, subculturing should be repeated till *Chaetoceros* or other required algae dominate in the bloom. The pure strain can be obtained from the bloom by serial dilution. Pure strain should be cultured in 50 ml sterilised flasks with enriched seawater till a density of one million cells per ml is obtained. This is stock or starter for mass culture.

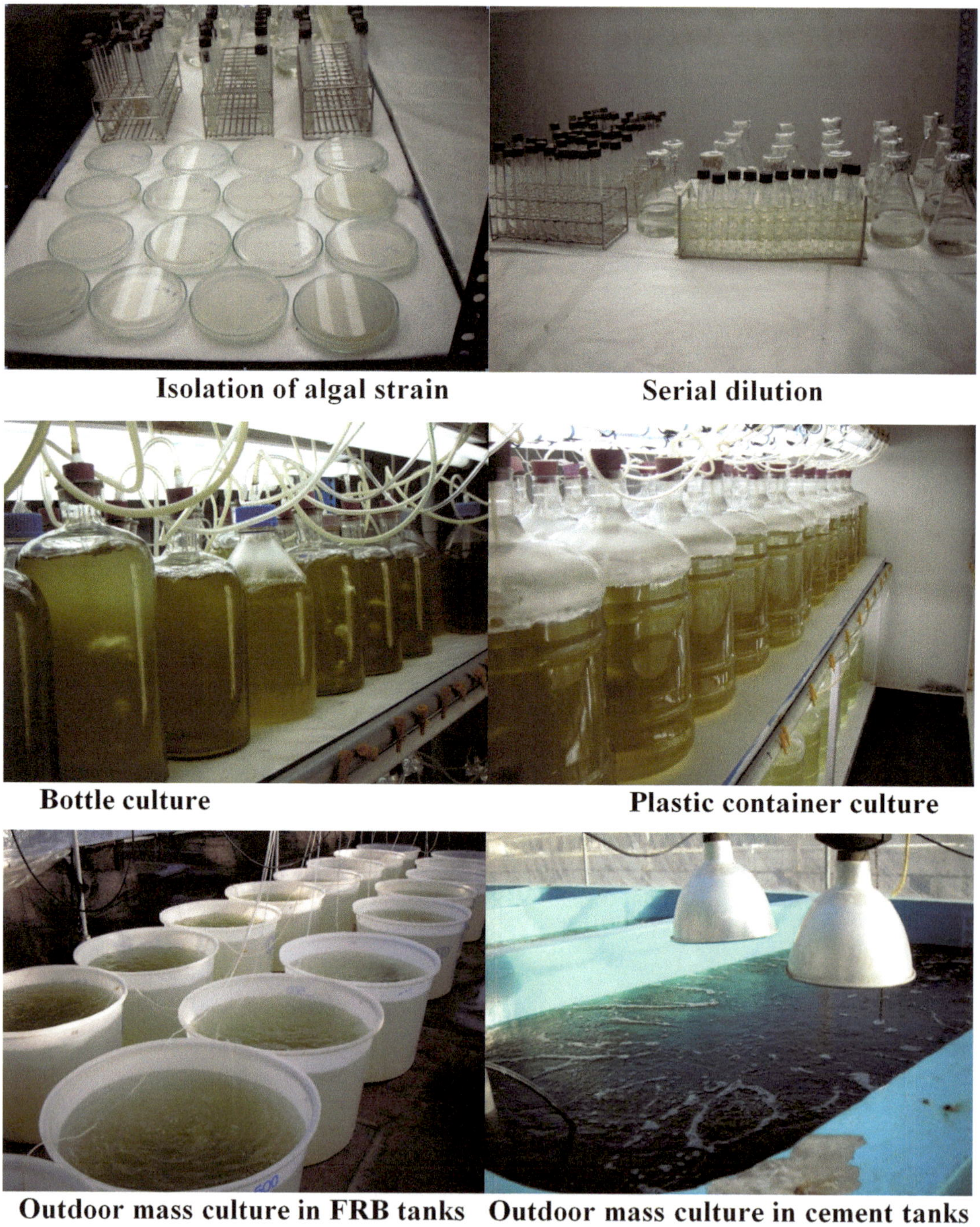

Plate 12 Algal culture

Preservation of Stock Culture

Small flasks (50 ml) or test tubes (20 ml) filled with enriched seawater are inoculated with 0.1 ml of stock culture and incubated in light with a photoperiod of 12 h. In this method, the algae can be stored and maintained for 15 days. Afterwards, the above procedure should be repeated to keep the algae in the active growth

Table 6 Algal culture operations

Culture type	Container	Volume	Medium	Inoculum Source	Inoculum Volume	Incubation period (days)	Comments
	Flask	200 ml	F/2	Master culture or 200 ml flask culture	20 ml	2	Select good culture and transfer to 200 ml or 2 L flask
		2,000 ml	F/2	200 ml flask culture	200 ml	2	Select good culture and transfer to 20 L glass carboy
Indoor							
	Carboy	20 L	f/2	2 L flask culture	2 L	2	Transfer to 200 L FRP cylinder
	FRP tank	200 L	f/4	20 L carboy culture	20 L	5	Transfer to larval rearing tank
	Bucket	10 L	f/4	2 L flask culture	2 L	2	Transfer to 100 L tanks
Outdoor							
	FRP tank	100 L	f/4	20 L culture	20 L	2	Transfer to 1,000 L tanks
	FRF tank	1,000 L	f/4	200 L culture	200 L	2	Transfer to larval rearing tank

phase. The stock culture can be stored in the refrigerator for 1 month.

Flask Culture

The stock culture maintained in 50 ml flasks passes through many progressive culture steps before it reaches mass culture phase. Twenty millilitres of stock culture is inoculated into 250 ml sterile flasks with enriched seawater and incubated in light for 2 days with continuous aeration, to get a density of five million cells per ml. Some of these 250 ml flask will be used for inoculating the small flasks. After 2 days, these 250 ml flasks should be transferred to 2 L flasks with enriched seawater and incubated in light with aeration for 2 days to get five million cells per ml density. Then this is inoculated into 20 L carboys with enriched seawater and incubated in light with aeration for 2 days to get a density of five million cells per ml. Once in 15 days, a new starter culture will be introduced into the system to maintain the vigour of the culture.

Mass Culture

The carboy culture of five million cells per ml density will be transferred into FRP cylinders (200 L) filled with enriched seawater. The cylinders are aerated and incubated in light for 5 days. When it reaches one million cells per ml density, it is transferred to the larval rearing tanks. Where there is no facility for indoor mass culture, indoor culture can be stopped at 2 L flask level. And outdoor mass culture can be started with inoculating 20 L buckets with 2 L flask culture. The buckets should be incubated in sunlight with aeration for 2 days till the density reaches five million cells per ml. Then the culture is transferred to 100 L cans with enriched seawater which are then aerated and incubated in sunlight for 2 days to get a density of two millions cells per ml. Subsequently, the culture is transferred to fibreglass tanks (1 MT) filled with enriched seawater. The tanks should be aerated and incubated in sunlight for 2 days, to get a desirable density of one million cells per ml. Then the culture is transferred to larval rearing tanks (Table 6).

Growth Characteristics in Mass Culture

Growth passes through the following five phases:

1. Lag or starting phase
2. Exponential or growth phase
3. The phase of declining relative growth
4. Stationary phase
5. Death phase

In mass culture, algae will be in the first phase on the first day of inoculation, and for the next 2 days (2nd and 3rd day), it will be in exponential phase. On the fourth day, it reaches the declining phase of relative growth and then stationary and death phase. Prolonged culture results in a decrease in size and nutritive value.

Harvest and Feeding

The algal water from indoor and outdoor mass culture tanks is pumped directly to the larval rearing tanks in desirable volume using a submersible pump. The pump and pipe should be disinfected and washed thoroughly before and after use.

Daily Routine

Preparation of Culture Medium and Culture Vessels

The F/2 medium is prepared by adding 1 ml of the working stock solutions of nitrate, phosphate, silicate, trace metals and vitamins to 1 L of filtered seawater. Thoroughly washed and clean flasks (250 ml; 2 L) should be filled with media (except vitamins) and autoclaved at 120 °C and 15 Psi for 15 min. The flasks should be allowed to cool down to room temperature and vitamins added on the next day just before inoculation. The thoroughly washed and clean carboys and FRP cylinders should be filled with filtered seawater and enriched with nutrients on the next day before inoculation. All culture vessels with media or filtered seawater should be ready by the evening for the inoculation of the next day.

Microscopic Observations

All the transferable cultures are observed under microscope by taking a sample from each container to estimate cell density and cell quality.

Cell Density Estimation

The cells present in 5 squares (4 corner and 1 centre) out of 25 squares of the counting grid of haemocytometer are counted, and the cell density is estimated by using the following formula:

$$\text{Number of cells in 5 squares} \times 5 = \text{Total number of cells in 25 squares}$$

$$\text{Number of cells in 25 squares} \times 10^4 = \text{Total number of cell per 1 ml of culture.}$$

Cell Quality

The cell should be uniform in size and normal in shape and healthy in appearance without vacuoles. Usually culture with dead algae (aggregated cells) and contamination should be discarded. Only the best algae should be transferred.

Inoculation

The best selected algal cultures are inoculated into culture vessels filled with enriched seawater. Ten percent inoculum is required at all culture levels to get desirable density. Inoculation process at the flask level should be carried near spirit lamp to avoid bacterial contamination. 2 L flasks and 20 L carboys can be directly transferred to carboys and fibreglass cylinders, respectively.

Cleaning

Cleaning Glassware, Hoses, etc.

All glasswares used should be treated with mild acid (20–50 % commercial HCl) first, then washed with liquid detergent and finally rinsed thoroughly with freshwater. Air hoses and air stones should be soaked in disinfectant (chlorine 200 ppm) for 1 h. and then removed and cleaned with lab detergent and washed with freshwater.

Cleaning Mass Culture Tanks

Immediately after algal harvest, the mass culture tanks should be thoroughly rinsed with freshwater. Any visible algal residue should be removed by scrubbing the tank walls and cleaning with liquid detergent. Then rinse thoroughly with freshwater and allow to dry for some time.

Cleaning Floors

At the end of every day activity, the entire facility should be flushed with freshwater. The bench tops and shelves should be cleaned with disinfectant and freshwater. The floor should be flushed with chlorine water (200 ppm) once or twice in a week.

Hygiene

It is very important to maintain hygiene in the algal section to avoid any contamination. The section should be highly restricted to the section personnel only. Glass windows can be used for showing the facility to the selected visitors. All personnel should walk through disinfectant baths and dip the hand in chlorine baths while entering the section.

All equipments should be kept as clean as possible and at all time. Movement of equipment like glassware, buckets, etc. should be highly restricted. The workers should be trained to disinfect every equipment after use.

Record Keeping

Microscopic observation, cell counts and other details such as salinity (ppt) and temperature (°C) should be recorded in a daily data sheet. A record of material consumption should also be maintained.

Shutdown Procedure

To reduce or eliminate disease-causing organisms and other contaminants which have become established in the system, a thorough cleaning and drying period is essential to make improvements and modification and to give the personnel a rest period before the next cycle.

The Following Disinfection Should Be Performed

- All equipments (glass ware, fibreglass tanks, buckets, tubs, etc.) should first be scrubbed with a strong chlorine solution (200 ppm).
- After a thorough washing, all small equipments should be placed in a tank filled with chlorine water (200 ppm) and allowed to soak for 5 days. All large tanks should be filled with chlorine water (200 ppm).
- Then they should be washed thoroughly with detergent and freshwater and dried.
- Spray the walls, floors and ditches with chlorine water (200 ppm) and allow to soak for 5 days and then wash with freshwater and dry for 10–14 days.

Production of *Artemia* Nauplii

Introduction

Artemia, a crustacean, found in salt lakes and brine ponds is called 'Brine Shrimp'. It can survive in the salinities of 150–200 ppt. But under these circumstances, it resorts to the encystment of the embryo at gastrula stage. The embryo in the dry cysts can be in dormancy for 8–12 months, and these cysts can be stored in tins and preserved for a long time. When the cysts are hydrated, the embryonic development resumes, and within 24–36 h the nauplii will hatch out from the cysts. The newly hatched nauplii are used as feed for shrimp larvae. *Artemia* have two main advantages: they are live particles and do not pollute the rearing environment and they have a high energy content and well suit the nutritional requirements of larvae.

Artemia hatching tank: *Artemia* hatching tank is an FRP cylindrical-shaped tank with transparent conical bottom and lid. Its capacity ranges from 400 to 500 L (Plate 13). It should be provided with continuous aeration and with two florescent lights at 20 cm height above it. It should be provided with a central standpipe, which is fitted into a drainpipe, which can be regulated with a ball value.

Optimum Conditions for Hatching

Stocking density	Maximum 1 g cysts/L
Water temperature	28–32 °C
Water salinity	25–35 ppt
Water filtration	50 μm minimum
Aeration	Strong to keep cysts in suspension
Light	2,000 lx (to trigger metabolism)

The time necessary for hatching depends on the strains, the batch and the brand of provided cysts. Generally the first nauplii collection can be made about 18–24 h after preparation, the second harvest after about 27–30 h and the third one after about 44–48 h. The hatching time as well as the hatching rate must be checked each time a new batch is supplied.

Decapsulation of Cysts

- Hydrate cysts for 1 h in 10 L of aerated freshwater.

Plate 13 *Artemia* hatching

Artemia hatching section

Hatched *Artemia* nauplii

Meanwhile:

- Dissolve 50 g of soda in 1 L of freshwater, caution with liberation of heat. Place the flask in the refrigerator.
- Dissolve 20 g of sodium thiosulphate in 1 L of freshwater. Collect the eggs on a 100 μm filter.

Prepare the decapsulation solution:

Soda solution:

- Chlorine solution containing 150 g of active chlorine.
- Make up to 6.5 L with freshwater.
- Place the eggs in the decapsulation solution with strong aeration and check the temperature.
- There is a possible need to add ice to avoid temperature increase above 40 °C; stir continuously.

If the reaction is too slow, add a little more chlorine solution.

- Decapsulation is over when the egg suspension becomes orange in colour, after 10–15 min. Observe the eggs under binocular microscope.

- Rinse the eggs with seawater on a 100 μm mesh.
- Place in thiosulphate solution for neutralising; make up to 6.5 L seawater.
- After 10 min, stop aeration so as the eggs to settle. Remove floating material (empty shells, non-decapsulated eggs).
- Rinse the eggs with seawater.
- Place in brine: 1.2 kg of table salt in 3 L of seawater. Slight aeration for 2–3 h.
- Count a few 1 ml samples. Store in brine. Do not expose to direct sunlight:

Storage

Stocking density	400,000–500,000 N/L maximum
Water temperature	25–30 °C
Water salinity	Same as for larval rearing
Water filtration	Same as for larval rearing
Water exchange	None
Aeration	Strong to obtain oxygen saturation
Light	None
Feeding	None
Treatments	Formalin 10 ppm

Harvest of Nauplii

Three harvests can be made from the same batch of cysts placed to hatch. They are based on the fact that nauplii are attracted by light and that eggs, shells and other particles float. Therefore, nauplii swim down to the cone base of the tank where light enters, while eggs and shells without aeration rise to the surface. This partition is about 100 % reached and much care is required for harvesting nauplii without eggs.

Remove the central pipe and airstone. Place the black cover over the tank. No light should pass through. Allow partition for roughly 45 min for 1st and 2nd harvest and 30 min for 3rd harvest. Then, start harvest on 100 μm concentrators through a 207 μm filter. Nauplii will pass through 207 μm, while eggs will not. Draining flow during the collection must be regular to avoid creating water currents in the tank.

When outflowing water becomes clear (as opposed to orange colour), frequently check with a transparent beaker the density of existing nauplii. When the density reduced sharply, stop the harvest. Close the outlet valve. Remove the black cover and set up the central pipe and airstone. Drain the outlet pipe slowly through 207 μm filter into the concentrator. Harvest is over. Incubation of hatching can be continued.

After the 3rd harvest, drain all water from the tank. Wash it (and its equipment) with soap and chlorine. Rinse thoroughly with seawater. Rinse the collected nauplii well with seawater in the concentrator and then transfer with a very small quantity of clean water to the storage tanks. Seawater in this storage tank should be added with 10 ppm of formalin. By sampling with pipettes, nauplii can be counted in this tank.

Daily Working Schedule

07.30 A.M.

- Prepare the 3rd harvest from tanks initiated 2 days before.
- Remove airstone and central pipe cover.
- Fill both *Artemia* storage tanks with 150 L of seawater.
- Add 2 ml of formalin to each tank.

8.00 A.M.

- Start the 3rd harvest in concentrators by using 100 μm through separator with 207 μm.
- Rinse the collected nauplii with seawater.
- Place nauplii in the storage tank for feeding.
- Count by sampling report.

8.30 A.M.

Prepare the 1st harvest from tanks initiated 1 day before.

According to requests, collect necessary quantities of nauplii from the storage tank for larval feeding. Wash all equipments with soap and chlorine (40 ppm) water.

9.30 A.M.

- Start the 1st harvest in concentrators by using 100 μm through separator 207 μm.
- Rinse collected nauplii with seawater.
- Place them in storage tanks for 1st harvest. Add more seawater according to density.
- Count by sampling.

09.30 A.M.

- Prepare the required quantity of cysts for hatching.
- Place in the tanks with freshwater only (about 50–80 L).

10.30 A.M.

- Fill new tanks for hatching with seawater. Check the salinity and adjust to required value.
- Set up airstone.
- According to requests, collect the necessary quantities of nauplii from storage tanks (1st harvest) for the larval feeding.

11.00 A.M.

- Wash the storage tanks with soap and seawater for 3rd harvest.
- If nauplii remain in the storage tank for 1st harvest, fill with seawater to a minimum of 150 L.
- Before leaving, check that all tanks are in good working order: aeration, valves, etc.

01.00 P.M.

- Prepare the 2nd harvest from tanks initiated 1 day before.
- Rinse the storage tank for the second harvest well (used in the morning for 3rd one).
- Fill with 100 L of seawater and set up aeration.
- Add 2 ml of formalin.

02.00 P.M.

- Start the 2nd harvest in concentrators by using 100 μm through filters 207 μm.
- Rinse collected nauplii with seawater; place them in storage tank for 2nd harvest. Add more seawater according to density.
- Count by sampling report.

2.30 P.M.

- According to requests, collect the necessary quantities of nauplii from storage tanks (1st and 2nd harvest) for the larval feeding.

04.00 P.M.

- Wash storage tanks as soon as empty with soap and chlorine water (40 ppm) and also the equipment: airstones, pipes, buckets, concentrators, filters, etc.
- Wash the *Artemia* room and storage room with chlorinated water (40 ppm).

Postlarval Rearing

Introduction

Postlarval rearing technique includes the rearing of 5-day-old postlarvae till 20-day-old postlarvae. The operation in this section warrants meticulous observations and management as the viability of the hatchery depends much on the cost of production of postlarvae.

The larvae at PL5 stage are harvested and stocked in the postlarval section at the rate of 25–50/L and reared till harvest. During the rearing period, the postlarvae are continuously monitored for their health. The various operations during the rearing period include water management, feed management and disease management. The postlarvae are subjected to regular prophylactic treatment to check bacterial, fungal and other parasitic infections. Apart from this, the postlarvae are monitored on daily basis microscopically and infection is treated accordingly. *Artemia* and egg custard are two different types of feeds used. Postlarvae of 20 days old or 13 mm in size are harvested and marketed. The day-to-day performance of various batches is recorded.

Hatchery Facilities

Rearing of postlarvae of shrimp in the hatchery needs the following facilities:

Seawater

The optimum salinity and temperature for the postlarval stages are 31.5 ± 1.5 ppt and 28 °C ± 1, respectively; pH should be 8.2. Water should be free from pollutants like heavy metals, pesticides and turbidity.

Aeration

Since the stocking density is high in tanks, the aeration has to be efficiently maintained to keep the optimum dissolved oxygen levels and to maintain proper circulation of water for facilitating availability of feed to the postlarvae. A standby air blower and a generator should be kept ready.

Postlarval Rearing Tanks

Reinforced concrete tanks of 20 t capacity are recommended for the rearing of various postlarval stages of *P. monodon*. The inner surface of the tanks should be coated with epoxy paints (optional).

Live Feed Culture Facility

Fibreglass tanks of 450 L capacity are used for the culture of *Artemia*. The details of culture technique and feeding rate are given elsewhere.

Laboratory

Laboratory may be equipped with microscope, refractometer, thermometer, weighing balance, water testing kits (chlorine, pH, nutrients, dissolved oxygen), detergents, glassware, chemicals and antibiotics.

Portable Water for Cleaning

Portable water for cleaning and washing the tanks, floors and other installations is essential.

Postlarval Rearing

Preparation of the Tank

First, the postlarval rearing tanks (20 t) should be scrubbed to remove invisible residue (organic matter) by bleaching it (25 ppm of $CaOCl_2$) or by applying detergents and later rinsed with freshwater. Later the tank should be dried for 24 h.

The treated seawater of salinity 31.5 ± 1.5 ppt is pumped into the tanks and aerated. The water should be treated with antibiotics and Treflan 0.05 ppm as prophylaxis (the second treatment of 0.025 ppm Treflan is to be given in the night). Now the tank is ready for stocking.

Stocking of Postlarvae

The postlarvae should be acclimated to the water in the postlarval tank. This is done by floating the plastic bin in the postlarval rearing tank and adding water from the postlarval tank at frequent intervals of 2–5 min, so that, the water characteristics in the plastic bin and the tanks are almost same. Then the postlarvae should be slowly released into the tank.

Rearing Methodology

The postlarval rearing tanks are carefully monitored for water quality and disease prevention. The animals are to be fed at regular intervals with utmost care. The physicochemical parameters like salinity, temperature and tank hygiene are to be checked regularly, and water management is to be done accordingly. The animals have to be examined for their feeding levels, general health and disease. Then prophylactic or therapeutic treatments are to be given accordingly.

Daily Activities

Water Parameters

Salinity, temperature, cloudiness, dirt and left overfeeds, etc. should be recorded.

Population Estimation

Population estimation is an important index for evaluating the status of postlarval health in the tank. Samples of 500 ml are taken and numbers of animals are counted. This should be repeated for five times. Calculate the mean per ml from which total population in the tank can be estimated.

Examination of Animals

Walk Through

The morning walk through should be made on regular basis, as in the larval rearing operation. All the parameters listed below are carried out on individual tank basis.

(a) Activity
(b) Feeding rate by observing the gut or residue
(c) Water colour – clean or turbid (due to high-level organics dissolved in water)
(d) Moulting
(e) Mortalities

Microscopic Observation

Postlarvae should be assessed for their health conditions. Healthy postlarvae appear to be clean, active, with full guts and well-developed tail muscle. Unhealthy postlarvae appear to be less active, with empty or partially filled guts and dirty (contaminated with protozoans, bacteria, fungi and dirt) with tails grainy, skinny and diffused chromatophores.

Aeration

All the tanks should be checked to make sure that the airflow is uniform along the entire length of the tank. Aeration should be gentle.

Data Keeping

Physical parameters, feeding levels, stage of postlarvae, population estimation, problems and remedial action taken on individual tank should be recorded in daily data sheet.

Water Management

Water exchange is an important daily operation for proper maintenance of the stock. Before starting water exchange, make sure the screens to be fixed to the drain pipes are not damaged. Screened tubs are placed in drainage ditch, so that when outside stand pipe is pivoted down towards the floor and water will spill into the tub and then flow out through screened cutouts into drainage ditch. Check contents of screened tubs two or three times during the first 5 min of drainage cycle to see if any postlarvae are escaping through the screen outlet of the tank. Lower the water level in the tank down to two metric tons, and then flush the tank for 15 min and refill. Use 5–10 μm filter bags for filling the tanks.

Siphoning

Siphoning should be done from postlarval stage onwards. First the aeration system is turned off and allowed to stand for 5–10 min so that the suspended matter consisting of *Artemia*, and remains of supplementary feed, settles to the bottom.

Precautions During Water Management

- Spray water along the sides of the walls during lowering of water to wash down the adhering PLs.
- Care should be taken to eliminate most of the waste matter during siphoning.
- Clean all the airlines and sides of the walls with sterilised cloth.
- Turn on the aeration system immediately after siphoning.

Feeding

Feed is one of the important factors determining the growth of larvae.

Artemia

Artemia is an important source of animal protein for shrimp postlarvae. *Artemia* requirement for a day should be estimated and informed to the *Artemia* hatching division the previous day itself. After harvesting, the density of *Artemia* nauplii in the storage tank should be estimated. Counting the number per litre can do this. The volume of *Artemia* concentrated to be number per litre. The volume of *Artemia* concentrate to be added to postlarval tank can be estimated by using the formula as earlier.

Care should be taken not to allow the entry of cysts or cyst shells along with *Artemia* nauplii. Freshly hatched *Artemia* feeding schedule for postlarvae is as follows:

PL 5 to PL 6	–	2 naup/ml × 3 feedings/day
PL 7 to PL 12	–	3 naup/ml × 3 feedings/day
PL 15 to PL 20	–	3 naup/ml × 1 feedings/day

Artemia is fed as soon as the tanks are refilled, at the desired density as in the larval rearing section. Subsequent feeding of *Artemia* is done after estimating the residual level.

Microencapsulated Diet

Microencapsulated feed is used as supplementary diet, mainly as a substitute for live feed.

Preparation of Feed

Known weight of the feed is dissolved in 1 L of water and stirred well and allowed to stand for 5–10 min. Before applying the feed, top layer consisting of broken capsules should be removed. The feeding rate is 0.5 g ton which is given twice in a day.

Egg Custard

Egg custard is another important diet used in the shrimp hatchery. Application of egg custard is done on weight basis. Feeding with egg custard is done generally from PL 8 onwards. The feeding rate ranges from 15 g/feed to 40 g/feed depending on the stage. The scheduled feeding rate is as follows:

PL 8 to PL 10	–	15 g/feeding × 2 times
PL 11 to PL 15	–	25 g/feeding × 5 times
PL 15 to PL 20	–	40 g/feeding × 5 times

However, feeding rate with egg custard is done on demand and care should be taken to siphon out daily the left overfeed.

Preparation of Egg Custard

Ingredients Required

Eggs	56 nos
Cod liver oil	75 ml
Yeast	75 ml
Beef liver	150 g
Polychaete worms	150 g
Vitamin drops	20
Flesh of squid or prawn	200 g
Milk powder	75 g

Procedure

Mix all the ingredients into a custard form by using a mixer. Cook it in a pressure cooker for about an hour. Take the required quantity of the cooked material, and sieve it through screen (500 μm) which gives fine granules. Wash the granules till all the fat content is removed. Squeeze out the water and mix it with water and broadcast it in the PL tank. Remaining part can be stored in a refrigerator for further use.

Drug Treatment

In postlarval tanks, an antibiotic and an antifungal drug are used as prophylaxis on the first day during preparation of the tank and then therapeutic treatments can be administered as and when need arises. The following aquatic antibiotics are usually used for postlarval treatment.

1.	Erythromycin (EM)	2–4 ppm
2.	Prefuran (PREF)	1 ppm
3.	Chloramphenicol (CP)	4 ppm
4.	Neomycin (NM)	5–10 ppm

(continued)

5.	Furazolidone (FZ)	5–5 ppm
6.	Oxytetracycline (Oxytet)	4 ppm

Every new batch of drug should be tested for its effect on postlarvae by conducting a bioassay test before treating the tanks. Usually the drug treatments are administered immediately after water exchange.

A few common diseases of postlarvae and treatment are given below:

Disease	Treatment
1. Bacterial infection like necrosis	Any of the broad spectrum antibiotics listed above
2. Luminescent bacteria	FZ (5 ppm), CP (4–10 ppm), NM (5–10 ppm), Pref (1 ppm)
3. Vibriosis	FZ (5 ppm), CP (4–10 ppm), EM (2–4 ppm), Pref (1 ppm)
4. Fungal diseases	Treflan (0.05 ppm), malachite green (0.0075 ppm), formalin (10 ppm)
5. Ectoparasites like protozoans, *Vorticella* or ciliates	Malachite green and formalin (10 ppm, PL3-8; 25 ppm, PL9-20)

It is advised to keep track of the sensitivity of the antibiotics on the microbial species at frequent intervals by a microbiology test.

Sanitation and Hygiene

- Disinfect and wash all glassware, pitchers, buckets, screened stand pipes, screened tubs, cleaning pads and other equipments used and store in proper location.
- If floors have been flooded as a result of water exchanges, harvesting or other activities, remove standing water by draining.
- Cool and store all hoses.
- Clean and dry all counters and containers in the laboratory.
- Make sure no specimen have been left on the microscope.

Harvesting and Packing of Postlarvae

Harvesting

- Make sure timing is well coordinated with the packing section, i.e. the PL section should be informed at least 24 h. in advance so as to organise manpower and setting of packing area (arrangement of basins as per the quantity of shipment setting of oxygen cylinders, regulators and boxes).
- Once water volume has been reduced to desired level, install harvesting apparatus in drainage ditch in position to catch outflow from outside stand pipe of the culture tank and allow it to fill with water coming out of the tank. Also start a flow of clean-filtered seawater such that a circular flow of water is generated inside the inner harvesting chamber.
- When harvest chamber is filled, remove screened inner stand pipe of tank and allow PLs to flow to harvest chamber.
- Transfer the PLs from the harvesting chamber into 60 L trash can. See that continuous mild aeration is maintained.
- Five 25 ml samples were collected at random from the 60 L trash can and kept under vigorous aeration (mild aeration is maintained once the sampling is finished).
- Population of harvested postlarvae can be estimated as follows:

Total PLs counted from

$$\frac{\text{Samples}}{\text{Total volume of sample (ml)}} \times \text{Total volume of trash can} = \text{Total no. of PLs harvested}$$

Packing

- After arriving at the overall population in the known volume of water in the trash can, PLs per litre can be calculated.
- Distribute the PLs into the tubs containing 8 L of water. Care should be taken that the basins are also provided with gentle aeration.
- The temperature in the tubs is then reduced to 5 °C below the normal water temperature. This is important because lower temperature reduces the normal activity of the PLs.
- Once the temperature is reduced to the desired degree, the PLs along with the water in the tub are transferred into double-layered polythene bags and packed at the ratio of 1:5 (water/air).

- The bag is sealed in a cardboard carton lined inside with thermocol sheet to maintain temperature. Few bags of ice are also placed adjacent to the polythene bag in the box to maintain the low temperature.

Notes
(a) No feed is given in bag during transportation, as this spoils the water quality.
(b) Distribution of PLs per bag is decided on the transport time and size of PLs.

Shrimp Hatchery Diseases

Introduction

Disease has been recognised as one of the several biological factors, which can limit or hinder the development of shrimp aquaculture.

Disease has been defined as a definite morbid process having a characteristic train of syndrome, and it may affect the whole body or any of its parts, and the aetiology and pathology and prognosis may be known or unknown. Disease may be infections or noninfectious.

Bacterial Diseases

(A)	Common name	Luminous bacterial disease
	Causative agent	*Vibrio alginolyticus*, *Vibrio harveyi*
	Stages affected	Eggs, larvae (even juveniles and adults)
	Gross signs	Larvae become weak and opaque
		Infected larvae exhibit a continuous greenish luminescence when observed in total darkness
	Microscope observation	The internal tissues of these larvae are densely packed with highly motile bacteria
	Effects on host	Systemic infection results in mortalities of larvae and postlarvae reaching up to 100 % of affected population
	Preventive methods	Prevent the entry of luminous bacteria into the hatchery system by using ultraviolet irradiated water or by employing a series of filtration equipment (sand filters, filter bags, cartridge filters, 0.5 μ pre-sized microfilter, etc.) and chlorination procedures prior to and during the larvae rearing
		Siphon out the sediments and debris from the tank bottom since these could serve as a base for bacterial growth
		Disinfect infected stock before discarding
	Treatment	Water change must be 80–90 % replacement daily

Antibiotic	Stage	Concentration
Chloramphenicol	Larval/PL	15 ppm
Furazolidone	L/PL	3/10 ppm
Malachite green	L/PL	0.075 ppm
Neomycin	L/PL	8/15 ppm
Formalin	PL	25 ppm
Static bath		50 ppm
Erythromycin	L/PL	2–4 ppm

(B)	Common name	Shell disease, brown/black spot, black rot/erosion, blisters, necrosis of appendages
	Causative agent	Bacteria belonging to *Vibrio*, *Aeromonas* and *Pseudomonas* groups
	Stages affected	Larvae and postlarvae (even juveniles and adults)
	Gross signs	Appearance of brownish to black erosion of the carapace, abdominal segments, rostrum, tail, gills and appendages
	Additional information	Blister containing cyanotic gelatinous fluid may develop on the carapace and abdominal segment
		The blister may extend to the underside of the ventrolateral section of the carapace creating a bulge on the underside
		In larval and postlarval stages, the affected part shows a cigarette, butt-like appearance
	Effects on host	Infection is usually initiated at sites of punctures or injuries made from either telson or rostrum, cracks on the abdominal segment from sudden flexure of the shrimp body or from

(continued)

		other damage caused by cannibalism
		Progressive erosion of these exoskeletal lesions follows upon entry and multiplication of bacterial pathogens. The infection may lead to loss of the affected appendages or musculature. When these occur, normal locomotion or moulting is hampered and may result in mortality
		The affected shrimp becomes susceptible to cannibalism or dies from stress or energy exhaustion
	Preventive methods	Maintain good water quality
		Keep organic load of the water at low levels by removing sediments, especially dead shrimps and moulted exoskeletons, which harbour high numbers of bacteria on the lesions
		Provide adequate diet. Minimise handling and avoid overcrowding
		Avoid injuries to the exoskeleton of the shrimps to prevent the development of primary portals of entry
	Treatment	Malachite green 0.0075 ppm and formalin 25 ppm in static condition for 24 h
(C)	Common name	Filamentous bacterial disease
	Causative agent	*Leucothrix* sp.
	Stages affected	Larvae and postlarvae (even juveniles and adults)
	Gross signs	Presence of fine, colourless, threadlike growth on the body surface and gills as seen under a microscope
	Effects on host	Infected eggs show a thick mat of filaments on the surface, which may interfere with respiration or hatching
		In larvae and postlarvae, filamentous growth on appendages and body surface may interfere with normal locomotory process and with moulting and may entrap other microorganisms (like fungal spores), which may initiate a new infection
	Preventive method	Maintain good water quality with optimum dissolved oxygen levels and low organic matter levels
	Treatment	Prefuran 1 ppm, erythromycin 4–6 ppm and malachite green 0.075 ppm in static treatment for 1 day. Can be administered for another day if situation demands. Potassium permanganate 2.5–5 ppm for 4 h

Fungal Diseases

(A)	Common name	Larval mycosis
	Causative agents	*Lagenidium* sp., *Haliphthoros* sp.
		In *Lagenidium*, 300 spores are developed in a vesicle, which is formed at the end of discharge tubes. These aquatic fungi produce motile zoospores that can easily invade other hosts
	Stages affected	Eggs, larvae and early postlarvae
	Gross signs	Infected eggs, larvae and postlarvae appear whitish, become weak and eventually die
		Signs are readily apparent when the disease is already widespread
	Effect on host	Mortality up to 100 % soon after infection is observed
		The fungal hyphae replace the internal tissues of the shrimp and extend outside the shrimp body to form discharge tubes
		Infected eggs do not hatch and larvae lose equilibrium and exhibit respiratory difficulties
	Preventive	Siphon sediments and dead shrimps
		Reduce stocking density
		Increase water circulation
		Disinfect materials and tanks with 100 ppm detergent for 24 h
		Observe rigid water management and sanitation
		Disinfect eggs with detergent at 20 ppm for 2 h long before hatching. For spawners use Treflan at 5 ppm for 1 h or dip in 500 ppm formalin for 5 min
		In areas where larval mycosis is known to occur, Treflan or trifluralin may be used as prophylactic levels of 0.1 ppm every 2–3 days and malachite green 0.075 ppm

(continued)

	Dispose infected stocks only after disinfection with 100 ppm detergent or 1 ppt chlorination or 500 ppm formalin
	Regular monitoring of the stock species through microscopic examined

Protozoan Diseases

(A)	Common name	Protozoan infestation
	Causative agents	*Vorticella* and *Zoothamnium* sp.
		Vorticella is solitary and has a contractile stalk
	Stages affected	Eggs, larvae, postlarvae (even juveniles and adults)
	Gross signs	Fuzzy mat on shell and gills of heavily infected juveniles and adults
		Reddish to brownish gills
	Effects on host	Microscopically, protozoan may be observed attached to any external part of the shrimp
		The protozoans cause locomotory and respiratory difficulties when present in large numbers on the appendages and gills, respectively. Loss of appetite
	Preventive methods	Maintain good water quality
		Avoid high organic load, heavy siltation, turbidity and low oxygen levels
	Treatment	Protozoan infestation in all stages is observed to be eliminated by formalin 12–25 ppm and malachite green 0.0075 ppm in static condition or formalin bath at 50 ppm for 1 h during postlarvae stage

Nutritional, Toxic and Environmental Diseases

(A)	Common name	Black gill disease
	Causative agents	Chemical contaminants like cadmium, copper, oil, zinc, potassium permanganate, ozone, ammonia and nitrite in rearing water
	Stages affected	Larvae and postlarvae (even juveniles and adults)
	Gross signs	The gills show reddish, brownish to black discolouration and atrophy at the tip of the filaments
		In advanced cases, most of the filaments are affected and the gills become totally black
		Physical deformities. Loss of appetite. Mortalities
	Effects on host	Microscopic observation show that the blackening of the gills may be due to the heavy deposition of black pigment at sites of heavy haemocyte activity (inflammation)
		Extensive accumulation of blood cells in the gill filaments may result in respiratory disturbances
		Secondary infections by bacteria-fungi and protozoans via the dying cells of the gills
	Preventive methods	Avoid overfeeding
		Change water frequently
		Avoid heavy metal discharges nearby factories from getting into rearing facilities
	Treatment	Adequate water exchange
		Methylene blue 8–10 ppm, prefuran 1 ppm and malachite green 0.0075 ppm in static conditions
(B)	Common name	Muscle necrosis
	Causative agents	Stressful environmental conditions like low oxygen levels, temperature or salinity shock, overcrowding and severe gill fouling
	Stages affected	Postlarvae (even juveniles and adults)
	Gross signs	Opaque white areas on the abdomen
		Blackening on edges of the uropod followed by erosion
		Liquid-filled boils at the tip of uropods in advanced stages
		Weakness and eventually died
		The disease causes gradual death of cells of affected parts such as uropods and musculature leading to erosion especially in the tail portion. This condition may serve as portals of entry for a secondary systemic infection by bacteria
	Preventive method	Reduce stocking density
		Give adequate feed
		Adequate water exchange

Hatchery Disinfection and Shutdown Operations

The hatchery should be disinfected and shutdown at least once in 3 months of continuous production in order to ensure non-settlement of microbes in the system and to provide rest for the staff and the machinery. Environmentally unfavourable and lean demand periods have to be advantageously utilised for this purpose.

The following works will be undertaken during shutdown period:

Disinfection of culture tanks and the building: All the culture tanks, walls, roofs and floor of the building and reservoir should be disinfected with 200 ppm chlorine water for 3 days. Then it should be cleaned thoroughly with detergent water and freshwater and then allowed to dry at least for 7 days.

Disinfection of PVC lines: The PVC lines of air, seawater and freshwater should be filled with 1,000 ppm chlorine for 3 days and then rinsed with freshwater thoroughly, and all the PVC lines should be filled with 50 ppm formalin for 3 days. Then they may be rinsed thoroughly with freshwater and dried. Aeration lines should be dried by blowing dry air continuously for 2–3 days.

Maintenance works for building and machinery: All the maintenance works for building, PVC lines and other facilities should be taken up. All the machineries like pumps, generators, blowers, air conditioners, refrigerators and deep freezers should be repaired and serviced.

Procuring the inputs: All the necessary inputs like feeds, chemicals, antibiotics, *Artemia* and packing material should be procured and stored in sufficient quantity. Drying every inch of the hatchery is essential to ensure the better production in the next cycle.

General Maintenance Equipments Genset

- Check the engine oil level every day morning.
- Check the diesel level every day morning and evening.
- Check valves (both inlet and exhaust) once in a week. If required remove, clean and reinstall.
- Clean the oil filters once in a week.
- Clean the outer surface of the genset with oily cloth once in a week to avoid rusting.
- Check the power connection once in a month and rectify the problems when necessary.
- Check the diesel consumption once in a week by measuring amount of diesel utilised per running hour.
- Clean the diesel pump once in a month and reinstall.
- Grease all moving parts like shafts, bearings, wheels, etc. once in a month.
- Do over oiling once in every 6 months.
- Dismantle the generator completely, check for defective parts, replace the defective parts and reassemble once in a year.
- Check output of the generator regularly.

Air Blower

- Check air inlet filters every day in the morning. If necessary remove, clean and reinstall to avoid chocking.
- Check output air pressure using pressure gauge fitted with delivery line. If it is not showing correct reading, identify the problem and rectify it.
- Check power input and motor RPM everyday to avoid the problem of overloading.
- Check the blower for overheating. If felt, stop the blower and start the standby.
- Grease the moving parts once in a month.
- Check the fan periodically to avoid reverse action.
- Do not run any particular blower for more than 8 h continuously.
- Do over oiling once in every 6 months.
- Dismantle the blower completely, check and replace defective parts, and reassemble during shutdown once in a year.

Pumps

- Check inlet and outlet pipes and valves for leakages.
- Check the impeller for chocking when found pump delivering less water.
- Grease the moving parts regularly.
- Check for shaft packing materials every week. If found burnt, dismantle, replace and reassemble it.
- Check electrical connections once a week to avoid short circuits.
- Do not run pump more than 8 h continuously; use standby.
- Change the impeller periodically as per the manufacturers' specification continuously.
- Replace defective parts and do overall servicing once in a year.

Heaters

- Clean the heaters with soap water before and after use.
- Wipe the heaters using clean cloth or tissue paper after washing.
- Check electrical circuits periodically.
- Check the heaters for its waterproof quality.
- Check thermostat regularly.
- Wind the cables around the heater and keep it in a safe place when it is not in use.

Microscope

- Clean the microscope with tissue paper before and after use.
- Check the eyepiece and object piece for their cleanness and correct magnification.
- Cover microscope with a clean cloth when it is not in use.
- Avoid unnecessary and rough handling of the microscope.
- Send the microscope for servicing to manufacturers or to authorised dealers once in a year.

Balances

- Read the instructions given by the manufacturers before using the electronic balance.
- Clean weighing plate before and after use.
- Do not touch the weighing plates when the instrument is under calibration.
- Do not give overloading.
- Check the balance regularly for its sensitivity.
- Cover the balance with clean cloth when it is not in use.
- Send the balance to the manufacturers or to authorised dealers for yearly servicing.

Cooling Equipments (Refrigerator, Deep Freezer, A/C Machine)

- Check the compressor and fans regularly.
- Clean inner sides of the refrigerators and deep freezers once in a week.
- Check the stabiliser for power fluctuation.
- Avoid any part of the equipment making direct contact with seawater.
- Calibrate and check thermostat regularly.
- Call authorised service man when problem occurs.

Underwater Torch

- Clean torch with soap water and freshwater before and after use.
- Keep all threaded parts tightly to avoid water entry.
- Avoid direct contact of inner parts with seawater.
- Remove battery from the torch when it is not in use.
- Clean the inner parts of the torch once in a week.

Seawater Systems

Sand Filter

- Do back washing at least once in a day.
- Avoid forcing of water flow into the filter more than the recommended rate.
- Operate pressure release valve every hour.
- Give acid washing to the filter and filter bed once in a week.
- Remove valves and fittings, give hydrochloric acid washing and reassemble once in a month. Remove total sand, wash and refill and filter once in every 3 months.
- Change the sand once a year.

Cartridge Filters

- Do back washing at least once a day.
- Assure only sand-filtered water is passing through cartridge filter.
- Operate pressure release valve frequently to get relief from excess pressure development.
- Remove cartridges, valves and fitting once in every 10 days and give acid washing.
- Replace cartridges once in a moth or depending upon the condition of the cartridge.

Water Exchange Filters

- Clean the filters with freshwater before use.
- Check the filters for damages before use.
- Clean the filters with 50 ppm chlorine water after use.
- Drain the water from the filter and dry it in a dust-free place after washing.
- Hang the filters on the wall when it was not in use.
- Change the filter mesh cloth once in three cycles or if damaged.

Valves

- Check the valves everyday for leakages.
- Remove all water valves, wash with hydrochloric acid, check for damages and refit once in a month.
- Handle the valves by using both the hands (hold the valve in one hand and operate with another hand).

PVC Pipe Lines

- Check every day all pipelines and fitting for leakages.
- Do not allow water to get stagnated in the pipelines.
- Clean and disinfect all waterlines with 40 ppm chlorine once after each cycle.
- Avoid exposing nets to heat.

Electrical Installations

- Check the connection of all tube lights and fans once in a weak.
- Replace faulty tube lights immediately.
- Check all cables and joints once in a week to avoid the problem of burning due to overload.
- Check all wires every day for damages and leakages.
- Check load factor at main panel every hour.

Civil Structures

Truss and Asbestos Roofing

- Check all joints, bolts and nuts once in a month. Faulty parts can be corrected or replaced.
- Replace asbestos sheets immediately when found broken.
- Repaint all truss pipes once in a year to avoid rusting.

Reservoir and Storage Tank

- Clean reservoirs and storage tanks with soap powder once in a week.

- Disinfect tanks with 50 ppm chlorine once in a month.
- Check tanks frequently for the development of fungus infection.
- Check tanks painting once in every 6 months and if required repaint it.
- Give dry up for 2 days between each cycle for better maintenance.

Drainage System

- Clean all drains and floors with freshwater every day.
- Disinfect with chlorine powder once in 3 days.
- Medicate once in a week to avoid mosquito problem.
- Remove any waste materials and debris from drains to avoid blockages.
- Maintain continuous flow of water in the drain. Do not allow water to stagnate.

Economics of a Shrimp Hatchery

Establishment of shrimp hatchery requires investment which includes both nonrecurring and recurring expenditure. Nonrecurring capital investment in shrimp hatchery includes cost of land, land development, buildings, reservoirs, tanks, seawater intake, aeration system, electrification, machinery, equipments, vehicles, consultancy fee, etc. These components may vary depending upon the capacity of the hatchery. The operational cost of recurring nature for shrimp hatchery includes broodstock, feed, antibiotics and chemicals, fuel and administrative expenses, repairs and maintenance, insurance, etc. The hatchery can be operated almost 10 months in a year giving an allowance of 2 months for shutdown due to repairs, maintenance and unfavourable environmental conditions. The economics worked out for a shrimp hatchery of a production capacity of 50 million per annum is given.

Analysis of the various statements infer that to start a 50 million capacity shrimp hatchery, a sum of Rs. 160.00 lakhs is required as capital cost and Rs. 52.00 lakhs are required as operational cost. A net profit of Rs. 25.00 lakhs is possible at the end of the first year of operation after depreciation, interest, etc.

Capital Costs

S. No.	Particulars	Estimated cost (Rs. in lakhs)
1.	Cost of land	0.75
2.	Land development including site survey	1.50
3.	Buildings	52.95
4.	Tanks	16.00
5.	Seawater supply system	14.85
6.	Freshwater supply system	1.75
7.	Aeration system including supply	5.50
8.	Internal and external electrification	8.00
9.	Machinery and equipment	10.90
10.	Office furniture	0.50
11.	Vehicles	4.00
12.	Expenses before commencement of production	4.74
13.	Contingencies	6.06
14.	Technical consultancy fee	10.00
15.	Interest on term loan	10.70
16.	25 % of the operational cost	12.80
	Total (say Rs. 160 lakhs)	161.00

Operational Cost

S. No.	Particulars	Estimated cost (Rs. in lakhs)
1.	Broodstock	2.00
2.	Feed	12.50
3.	Antibiotics and chemicals	7.50
4.	Fuel and power	7.50

(continued)

S. No.	Particulars	Estimated cost (Rs. in lakhs)
5.	Supplies and consumables	2.00
6.	Salaries and wages	12.00
7.	Vehicles maintenance	0.50
8.	Administrative expenses	3.00
9.	Regular maintenance and repair	3.00
10.	Insurance	1.16
	Total (say Rs. 52 lakhs)	51.16

Financial Analysis of First Year of Operation

1	Total production	50 million PL 20s/annum
2	Sale price	Rs. 250/1,000 PL 20s.
3	Total income	Rs. 125 lakhs
4	Expenditure (operational cost)	Rs. 52.00 lakhs
5	Profit before depreciation and interest	Rs. 52.00 lakhs
6	Depreciation (10 % of the capital cost of Rs. 160 lakhs)	Rs. 16.00 lakhs
7	Profit after depreciation but before interest	Rs. 57.00 lakhs
8	Interest at 20 % on capital cost, i.e. Rs. 160 lakhs	Rs. 32.00 lakhs
9	Net profit after depreciation and interest	Rs. 25.00 lakhs

References

Chakraborty C, Sadhu AK (2001) Biology, hatchery and culture technology of tiger prawn and giant freshwater prawn. Daya Publishing House, New Delhi, pp 1–98

Handbook on shrimp hatchery operation and management by the marine products export development authority, pp 1–78

Operation manual for tiger shrimp hatchery by Sampath Associates, pp 1–110, 1994

Paul Raj S (1999) Shrimp farming techniques, problems and solutions. Palani Paramount Publications, Palani, pp 34–53

Semi-intensive Culture Techniques for Shrimp Farming

B. Gunalan

Introduction

Aquaculture, though considered to have over a 2,500 year history, was mostly practiced as an art. It began to be transformed into a modern science in the second half of the twentieth century. Within a period of 25 years or more, it had begun to impress upon as a major food production sector, having recorded an annual average growth rate of nearly 8 % in the last two decades, as often purported to be the fastest growing primary production sector. Currently, aquaculture accounts for 50 % of the global food fish consumption (De Silva and Davy 2010). The sector has been and continues to be predominant in developing countries, particularly in Asia, which accounts for more than 85 % of the global production. Asian aquaculture by and large is a small-scale farming activity, where most practices are family owned, managed, and operated. The sector has provided direct and indirect livelihood means to millions, a significant proportion of which is rural, and for some Asian nations, it is the main source of foreign exchange earnings. Furthermore, it has contributed to food security and poverty alleviation and therefore is considered to be a successful primary food sector globally (De Silva and Davy 2010). Shrimp industry is a key sector in Indian economy because of its significant contributions to export earnings and gainful employment. But in reality, shrimp exports in the country have stagnated since late 1990s. Problems began a few years earlier with the outbreak of white spot disease (WSS). Later on, the judgment by the high court of India on shrimp aquaculture also had profound impact on its advancement. In response to address the rising concerns about the sustainability of the sector, in the year 2000, MPDEA with the technical assistance of NACA initiated the aforementioned project.

The potential brackishwater area available in the coastal regions of the country for shrimp culture is estimated between 1.2 and 1.4 million ha. Presently an area of about 1,50,000 ha is under culture with an average production of about 1, 20,000 t per year. Cultured shrimps contribute about 50 % of the total shrimp exports. The technology adopted ranges from traditional to improved traditional within the Coastal Regulation Zone (CRZ) and extensive shrimp farming outside the CRZ. About 91 % of the shrimp growers have a holding in between 0 and 2 ha and 6 % between 2 and 5 ha, and the remaining 3 % have an area of 5 ha and above (Nazeer 2005).

There are around 290 hatcheries in the country with an installed seed production capacity of 11 billion. There are about 33 feed mills with an annual installed production capacity of 1,50,000 MT, and the current feed requirement is in the range of 1,35,000 MT. Shrimp farming provides direct employment to about 0.3 million people. Major credit for catapulting shrimp

B. Gunalan (✉)
C.A.S. in Marine Biology, Faculty of Marine Sciences, Annamalai University, Parangipettai 608502, Tamil Nadu, India
e-mail: aquagunal@gmail.com

S. Perumal et al. (eds.), *Advances in Marine and Brackishwater Aquaculture*,
DOI 10.1007/978-81-322-2271-2_13, © Springer India 2015

farming to its present importance goes to the promotional efforts of Marine Products Export Development Authority (MPEDA) and the Ministry of Agriculture (Nazeer 2005).

Tamil Nadu is having about 56,620 ha of brackishwater lands, and the potential for shrimp culture is identified as about 3,684.18 ha. During the year 2004–2005, the state produced 6,674 MT of shrimp through aquaculture, which accounted for just 5 % of the country's cultured shrimp production. The shrimp production for the last 10 years in Tamil Nadu is presented in Table 1.

Table 1 Shrimp production in Tamil Nadu for the last 10 years

Year	Area (ha)	Production (MT)
1995–1996	2,003.89	1,102.00
1996–1997	2,053.00	1,156.31
1997–1998	2,246.33	1,197.00
1998–1999	2,504.00	1,802.00
1999–2000	1,835.00	2,856.00
2000–2001	2,537.00	3,712.00
2001–2002	2,478.00	4,511.00
2002–2003	3,615.00	4,986.00
2003–2004	3,213.00	6,070.00
2004–2005	3,684.18	6,674.00

Shrimp farming technology is gaining considerable importance in India in recent years. Amidst problems such as environmental issues and dreadful shrimp virus-related diseases, the shrimp farming sector is growing by leaps and bounds. Sustainable shrimp farming is the need of the hour in India.

Pond Renovation

Most of the existing traditional shrimp ponds are large (1.5–0.9 ha each), irregular in shape, and relatively not much deep (usually 70–80 cm), often resulting in great variation in water temperature and salinity. These ponds could be easily improved by renovation, making them more regular in shape, uniform in size, and sufficiently deep for the installation of proper inlet and outlet gates to facilitate water exchange through supply and drainage canals. To renovate a pond for semi-intensive culture, farmers are advised to carry out the following procedures.

Renovating the pond (using tractor dozer)

Readjust the size and shape of the pond for better management at an appropriate expense. The pond should be rectangular in shape and cover about 0.5–1 ha in size. Dig up the pond to 150–180 cm in depth to hold more water and to prevent abrupt change of water temperature during the day. The paddle wheels, when installed and operated, will not stir up dregs at the pond bottom; otherwise, they would cause water to become turbid. The pond dike should be made wider and stronger with soil compressing to prevent water leakage and dike rupture when water depth is high up. If there are pieces of tree roots at the pond bottom, take all of them away because they can cause leaks in the pond and also be a sanctuary of shrimp predators. Moreover, rotten roots can easily deteriorate the water quality. After clearing the roots, farmers should smoothen the pond bottom to make a slope toward the outlet gate.

When we modify to a semi-intensive pond, we have to make two separate water gates, one for letting water in and the other for discharging water and harvest. The size of water gates should be of a suitable proportion with the pond size for adequate water exchange and harvest shrimp within a proper time. Its width should be about 1 m for 1 ha pond area. Farmers should install water pumps which are powerful enough to pump water into the pond at any time. There must be reservoir ponds in which water is kept and improved to a required quality before it is pumped into the raising pond. Moreover, the water must be filtered through a sieve or a cloth filter to prevent shrimp predators from entering into the pond which may reduce shrimp production.

Pond Preparation

The preparation of pond before introducing the seed is important to ensure higher production. Eradication of unwanted organisms is carried out by draining the water and drying the pond until the soil cracks in the pond bottom, killing the predator fishes and other competitive organisms in the pond. Drying the pond bottom is the cheapest way of eradication of unwanted species aside from allowing the release of obnoxious gases from pond bottom. Let the pond bottom dry in the sun for a period until the black color and bad smell in soil are gone. This process of drying can help in loosening hard bottom and in mineralization of pond bottom.

When pond bottom is fully dried, scrape 2–4 cm of the top soil from the pond bottom (Soundarapandian and Gunalan 2008). Lime must be broadcast over the pond bottom, which is wet and smelly, to kill disease caused by the accumulation of dead algae. Lime can also help to decompose organic matter and kill predators or other undesirable aquatic organisms living at the pond bottom. After lime application plough the pond bottom horizontally and vertically a

Drying the pond

Scraping the pond bottom (using man power)

Plowing the pond bottom

depth of 30 cm to remove the obnoxious gases, oxygenate the bottom soil, discoloration of the black soil to remove the hydrogen sulfide odor and to increase the fertility, smoothen it and make a slope toward the outlet gate. Repair leaks on the earthen dike and around the water gates. The sieve at the inlet gate must be strongly fitted and without any holes to prevent escape of fry and entry of predators.

Biosecurity Methods

Biosecurity has been defined as "...sets of practices that will reduce the probability of a pathogen introduction and its subsequent spread from one place to another..." (Lotz 1997). The basic elements of a biosecurity program include the physical, chemical, and biological

Lime application in the pond bottom

Bird fencing

methods necessary to protect the pond from the consequences of all diseases that represent a high risk. Effective biosecurity requires attention to a range of factors, some disease specific, some not, ranging from purely technical factors to aspects of management and economics. Various levels and strategies for biosecurity may be employed depending on the pond facility, the diseases of concern, and the level of perceived risk. The appropriate level of biosecurity to be applied will generally be a function of ease of implementation and cost, relative to the impact of the disease on the production operations (Fegan and Clifford 2001).

Biosecurity measures in the shrimp industry can be seen as a two-pronged approach: excluding pathogens and eliminating pathogens when they are present. Lightner (2003) discussed ways of excluding pathogens from stock (i.e., postlarvae and brood stock), especially through the use of quarantine and specific pathogen-free (SPF) certified stocks, and restricting imports of live and frozen shrimp. Excluding vectors and external sources of contamination and preventing internal cross contamination were suggested methods for excluding pathogens from hatcheries and farms.

Stocking pathogen-free postlarvae alone does not guarantee a disease-free culture since the pathogens could still enter the culture environment horizontally and infect the shrimps during the culture. Viral pathogens can still enter the culture environment through the following means, and a better understanding of these can help in the prevention of horizontal transmission.

*By persisting in the soil, *intake water, *aquatic vectors introduced through intake water, by crabs and other animals*, besides the abovementioned carriers, viral particles can also enter the farming system by mechanical carriers like:

- Contaminated land animals and birds
- Contaminated farm inputs – through live feed and semimoist feed
- Contaminated farm implements, nets and vehicles, etc.
- Contaminated personnel

Crabs are one of the carriers of viral pathogens, and providing crab fencing in shrimp farms is considered as one of the important biosecurity requirements. Carriers like crabs could also move from pond to pond over land barriers. To prevent such movements, fencing made of 0.5 m plastic sheet should be put around the culture pond. Live feed and moist feed are more likely to contain pathogens because their ingredients are either in a raw state or subject to insufficient processing.

Birds such as eagle, crow, or water crow pick up the dead and moribund shrimps affected with viral disease from ponds and may drop in unaffected ponds, thereby transmitting the virus mechanically. This could be avoided by using

Water filter system (filter bag)

bird scares and bird fencing over the pond. Similarly land animals like dogs, cats, and cattle can mechanically carry the virus from one pond to another pond. Preventing entry of stray animals and unauthorized personnel into the farming area through fencing is the only way to address this problem. Pond to pond transmission of virus within a farm could easily occur through the use of farm implements and farm workers. Workers move from pond to pond attending to their work. So restriction on the movement of farm workers from pond to pond is necessary.

Bloom Development and Probiotic Application

Fill the pond up to 40–50 cm and then apply tea seedcake at the rate of 150–200 kg/ha to kill common fish or other aquatic organisms which may escape into the pond with entering water. The pond is further enriched with organic fertilizer (dry cow dung, rice bran, and groundnut cake) at a rate of 50–100 kg/ha or inorganic fertilizer (urea, DAP, and super phosphate) at 10–15 kg/ha. The pond is left to provide time for the growth of natural food about 3–4 days with water mixing by paddle wheel aerators. When the water color in the pond turns green or brown, fill more fresh seawater until the desired depth is attained. Once the bloom get stabilized need to apply the probiotic. This will help for better water quality. Now, the pond is ready for shrimp larvae stocking. For semi-intensive shrimp farming, probiotic application should be every 10 days.

Shrimp Seed Stocking

Postlarvae of shrimp should be supplied by a realizable hatchery where chemical or drugs are not used. Moreover, the hatchery must not increase water temperature to higher than 32 °C to stimulate the larvae's growth. The suitable stocking density for semi-intensive pond is >15 pcs/m^2, depending on the condition of the pond. Good quality of shrimp larvae can be noticed as follows:

1. Their rostrum or head is not crinkled.
2. Their uropod or tail is well spread while swimming.
3. By stirring the water slowly with larvae in a small bowl in circular motion, healthy larvae will swim in an opposite direction against the water movement and quickly hold tight to the bowl bottom.
4. The larvae size is particle even.
5. The larvae acclimate themselves quickly to the new surroundings (in different water temperature and degree of salinity).

Prior to purchasing the larvae, farmers should go to the hatchery by themselves. They must emphasize importance to packing and transporting of the larvae. They should count the larvae before packing and recount after arriving at the farm to make sure that the larvae are of a required number. The larvae should be strong enough and acclimatized before stocking which should be done in the early morning so that the water temperature fluctuation is minimal. If the pond salinity differs significantly from the hatchery, acclimatization should be gradual.

Feed Management

During the first 2 days, supplementary or pellet feed is usually not given as the shrimp can eat natural food. However, farmers should also give them some supplementary of pellet feed to ensure a higher survival rate. In semi-intensive system, stocking density is more than the carrying capacity of the pond; hence, the natural food in the pond is not sufficient until harvest. The following is the recommended feeding program of semi-intensive culture at stocking rate >15 pcs/m^2, for guidelines.

Age (day)	ABW (g/pcs)	% Feed	Feed code	% Lift net	Time checking (h)
1	PL 15–.01	2 kg/100,000 pcs/day	001	–	–
2–10	.01–.8	2.4–5.6	001	–	–
11–30	.8–2.5	6.2–17.6	002&003	–	2.5
31–50	2.5–6	18.1–27.6	003&004 s	–	2.5
51–60	6–8	5.5–5.1	004 s	2.4–2.6	2.0
61–70	8–10	5.1–4.5	004	2.6–2.8	2.0
71–80	10–13.5	4.5–4.3	004	2.8–3.1	2.0
81–90	13.5–16	4.3–4.0	004	3.1–3.2	2.0
91–100	16–20	4.0–3.5	005	3.2–3.4	2.0
101–110	20–24	3.5–3.2	005	3.4–3.6	2.0
111–120	24–28	3.2–2.9	005	3.6–3.9	2.0
121–130	28–33	2.9–2.5	005	3.9–4.0	2.0
131–140	33–38	2.5–2.2	005	4.1	2.0

Feeding Program

It is also necessary to use a lift net to find out if the amount of feed is given properly. If the shrimp eat all the feed within a certain time, farmers may give them more feed. But if the feed is not all eaten within the given time, farmers have to reduce the feed to prevent overfeeding.

Leftover feed can cause the pond bottom to decay and the water to become deteriorated easily, causing the shrimp to be weak and stressed. As a result, they will also not feed easily, get sick, and die. Feeding frequency is recommended 4 times/day for the entire culture period. The following are the guidelines for feed checking in lift net.

Round check tray

Boat feeding

Shrimp weight (g)	Feed should be eaten within (h)
Up to 4 g	3.0
5–7	2.5
8–12	2
>12	2

Water Quality Management

The secret of success in semi-intensive shrimp farming mostly depends on the proper management; after stocking, periodical observations on the water quality, soil condition, and growth of the shrimps are to be made to decide on supplementary feeding maintenance and proper salinity.

Parameters	Range
Temperature (°C)	25–33
Salinity (ppt)	10–35
Transparency (cm)	25–60
Dissolved oxygen (ppm)	3–8
pH	7.5–8.5
Total ammonia (ppm)	Less than 1.0
Nitrite (ppm)	Less than 0.25
Hydrogen sulfide (ppm)	Less than 0.25

* *ppm* parts per million, *ppt* parts per thousand, *ppb* parts per billion

Water exchange is essential for the removal of the biological wastes and maintenance of adequate dissolved oxygen level. The purpose of water exchange to maintain water quality is also to stimulate molting of the shrimp, resulting in acceleration of growth and production. Important steps in water quality management are:

Water Exchange

Farmers must have water pumps with a capacity to change as much as 5–10 % of water in the pond every day. During the first 45 days of culture, water exchange is not required. After those periods, check the water quality and other parameters; then if necessary, start water exchange and maintain the good water quality. The oxygen content and water transparency are also factors to help farmers decide to change the pond water. If the oxygen content is lower than 3 ppm in the morning or water transparency is less than 30 cm, more water volume must be required for change than usual. The water exchange will be more efficient if farmers have water storage pond with high-capacity pumps.

Bloom crush

Lab lab (Algal formation)

(A, B, C, and D show the poor water quality management)

Aeration (Paddle Wheels)

Paddle wheels are necessary from the first week itself. For a 1 ha pond with 15/m^2, use six 2-HP paddle wheels (long-arm aerator 10–12 wheels per unit engine). All paddle wheels are usually run 10 h in 40 days. After the 40th day of culture, expect that during feed time, there is a need to run the aerator properly. Otherwise, more possibilities for oxygen depletion in the pond may occur.

Shrimp Health Analysis

Cast net should be used once in 10 days to measure the growth rate of shrimps. *P. monodon* normally grows at the rate of 0.2 g/day after the first 30 days; weekly growth rates range between 1.5 and 2.0 g, depending on the stocking density.

Aerator

Cast net

Harvest

Harvesting should be done during low tide after 5–6 days of changing water to ensure that the majority of shrimps have molted and the shells hardened. A bag net is installed at the outlet gate during harvesting. The shrimp are drained into the bag by releasing the pond water and are collected periodically until the pond water is completely drained out. The remaining shrimps in ponds are collected by hand. The shrimp should be harvested while they are still alive. Harvesting should be made as quickly as possible so that the shrimp are still fresh before they are delivered to cold storage.

Semi-intensive Culture Performance

With the use of semi-intensive culture technique, a higher shrimp production can be obtained. The following are examples of successful farmers in India (Tamil Nadu and Andhra Pradesh) who use the semi-intensive technique with commercial *shrimp* feed (CP feed).

	Ponds		
Particulars	AP	TN	Orissa
Pond area (m^2)	6,000	6,000	6,000
Stoking density/m^2	12	12	12
Initial stock	72,000	72,000	72,000
Average daily growth	0.27	0.28	0.28
Culture period (days)	142	142	143
Total production (kg)	1,874	2,154	2,004
Production (Ton/ha)	3,123	3,590	3,340
Size of harvest (g)	38.46	40.00	40.00
Survival rate (%)	68	75	70
Total feed used (kg)	2,550	2,940	2,725
FCR	1.36	1.36	1.36
Income (Rs)	4,83,492	6,24,660	5,33,064
Total seed cost (Rs)	21,600	21,600	21,600
Total feed cost (Rs)	1,22,400	1,41,120	1,30,800
Other expenses (pond preparation, water culture, probiotics, electrical charges, etc.)	70,000	70,000	70,000
Net profit (Rs)	**2,69,492**	**3,91,940**	**3,10,664**

Conclusion

India has vast potential for coastal aquaculture development. Shrimp farming development holds immense employment potential for the local people. Food safety is one of the major issues in the shrimp produced in India. It is essential that banned antibiotics should not be used in the hatchery and farm. The extension officers are required to educate the farmers on the responsible use of drug in aquaculture and nonuse of banned antibiotics in the culture systems. *Penaeus monodon* is a very suitable species for intensive culture with the availability of pathogen-free seed. The major issues to be considered are biosecurity and maintenance of water quality through constant monitoring. It also requires higher technical knowledge to achieve higher production in sustainable manner. It is hoped that these semi-intensive shrimp culture techniques will guide farmers/entrepreneurs to take up *P. monodon* shrimp farming in a big way and to earn huge profits for themselves and a billion dollar to the nation through exports. This will also generate large-scale employment opportunities for the rural poor and the unemployed youth.

References

De Silva SS, Davy B (Eds) (2010) Success Stories in Asian Aquaculture. Springer, FAO, 2006. The State of world Fisheries and Aquaculture. Food and Agriculture Organization of the United Nations, Rome

Fegan DF, Clifford HC (2001) Health management for viral diseases in shrimp farms. In: Browdy CL, Jory DE (eds) The new wave: proceedings of the special session on sustainable shrimp farming. The World Aquaculture Society, Baton Rouge, pp 168–198

Flegel TW (1997) Major viral disease of the black tiger prawn (*Penaeus monodon*) in Thailand. World J Microbiol Biotechnol 13:433–42

Lightner DV (1996) A handbook of shrimp pathology and diagnostic procedures for diseases of cultured Penaeid shrimp. World Aquacult Society, Baton Rouge, p 304

Lightner DV (2003) Exclusion of specific pathogens for disease prevention in a penaeid shrimp biosecurity program. In: Lee C-S, O'Bryen PJ (eds) Biosecurity in aquaculture production systems: exclusion of pathogens and other undesirables. The World Aquaculture Society, Baton Rouge, pp 1–116

Lotz JM (1997) Disease control and pathogen status in an SPF-based shrimp aquaculture industry, with particular reference to the United States. In: Flegel TW, MacRae IH (eds) Diseases in Asian aquaculture III. Asian fish. Fish Health Section, Asian Fisheries Society, Manila, pp 243–254

Nazeer MA (2005) National consultation on shrimp farming. Aqua International, Chennai, p 11

Ravichandran P, Panigrahi A, Kumaran M (2009) Hand book on seed production and farming of *Litopenaeus vannamei*. CIBA special publication 46, p 43

Soundarapandian P, Gunalan B (2008) Int J Zool Res 4:21–27

Breeding, Larval Rearing and Farming of Mangrove Crab, *Scylla serrata* (Forskal, 1775)

Kumaran Ganesh, Yohannan Chellamma Thampi Sam Raj, Santhanam Perumal, Pandiarajan Srinivasan, and Arulraj Sethuramalingam

Introduction

'Mud crab/green crab/mangrove crab' (*Scylla serrata*) commands sustainable domestic and international market. Owing to this fact, this species is widely exploited from all along the coastal belt of India and also in many parts of the tropical/subtropical regions of the world where its natural distribution is ascertained. As such, geographically, this species is widely distributed in Indo-Asian region and also certain tropical parts of Australian continent (Hongyu et al. 2012).

Scylla serrata

K. Ganesh (✉) • Y.C.T.S. Raj • P. Srinivasan
A. Sethuramalingam
Rajiv Gandhi Centre for Aquaculture, MPEDA, Ministry of Commerce and Industry, 3/197, Poompuhar Rd, Karaimedu, Sattanathapuram, Nagapattinam District, 609109 Sirkali, Tamil Nadu, India
e-mail: ganeshkollam@gmail.com; rgcaho@gmail.com

S. Perumal
Department of Marine Science, Bharathidasan University, Tiruchirappalli 620024, Tamil Nadu, India

Natural habitat of mangrove crab is limited to mangrove areas/mudflats adjoining to brackishwater estuarine aqua zones, bays and saltwater creeks/lagoons. During spawning, it migrates towards sea for ambient water quality conditions which facilitate spawning and hatching. Larval stages are completed in seawater, and the baby crabs migrate to brackishwater areas for further growth.

The male crab normally grows faster and attains bigger size than the female. Mangrove crabs of sizes more than 350 g only command acceptance in conventional live crab market and fetch good value depending on the weight of the individual crabs. Due to high meat content and having a pair of muscle-built claws, it fetches more than Rs. 400/kg (e.g. farm price) in India for an average 500 g crab (BOBP 1992). India exports mangrove crabs in live condition to Singapore, Malaysia, China and Taiwan. Water crabs (molted crab) of sizes varying from 90 to 100 g are having high market potential.

Identifying Features of *Scylla serrata*

1. Polygonal markings found in all legs (including the abdominal flap of female also)
2. Frontal lobe spines high, narrow and bluntly pointed with tendency to concave margins and rounded interspaces

S. Perumal et al. (eds.), *Advances in Marine and Brackishwater Aquaculture*, DOI 10.1007/978-81-322-2271-2_14, © Springer India 2015

3. Wrist of claws with two prominent spines on outer margin, a pair of distinct spines on upper margin of each claw behind the movable finger

Biology of Mangrove Crab (*Scylla* spp.)

Mangrove crabs have flat, broad bodies covered with fan-shaped carapace. One pair of chelipeds, three pairs of walking legs and one pair of swimming legs are the prominent morphological features (Keenan et al. 1998; Sangthong and Jondeung 2006; Jirapunpipat et al. 2008). Walking legs are also used for clasping the female by male and for scrapping the eggs just prior to hatching by female. Under genus Scylla, *Scylla serrata*, *S. tranquebarica*, *S. olivacea* and *S. paramamosain* are the main species (Estampador 1949; Imai et al. 2004). Out of the four species, in India, only *Scylla serrata* and *S. olivacea* are available (Joel and Raj 1983; Macintosh et al. 2002; Anup Mandal et al. 2014).

Life Cycle

Burrows created in mangrove, mudflats and soft bottom shallow intertidal area are the natural habitat of *Scylla* spp. The *Scylla serrata* prefers salinity ranges of 28–35 ppt. They spawn near seawater lagoons, bays and coastal areas. The spawned eggs placed at abdominal flap of female crab would be released with the help of ciliary action of pleopods. Eggs hatch into zoeae and pass through five larval stages, after which they become megalopa. Megalopa molts into a tiny crab, viz., crab instar which is a miniature crab.

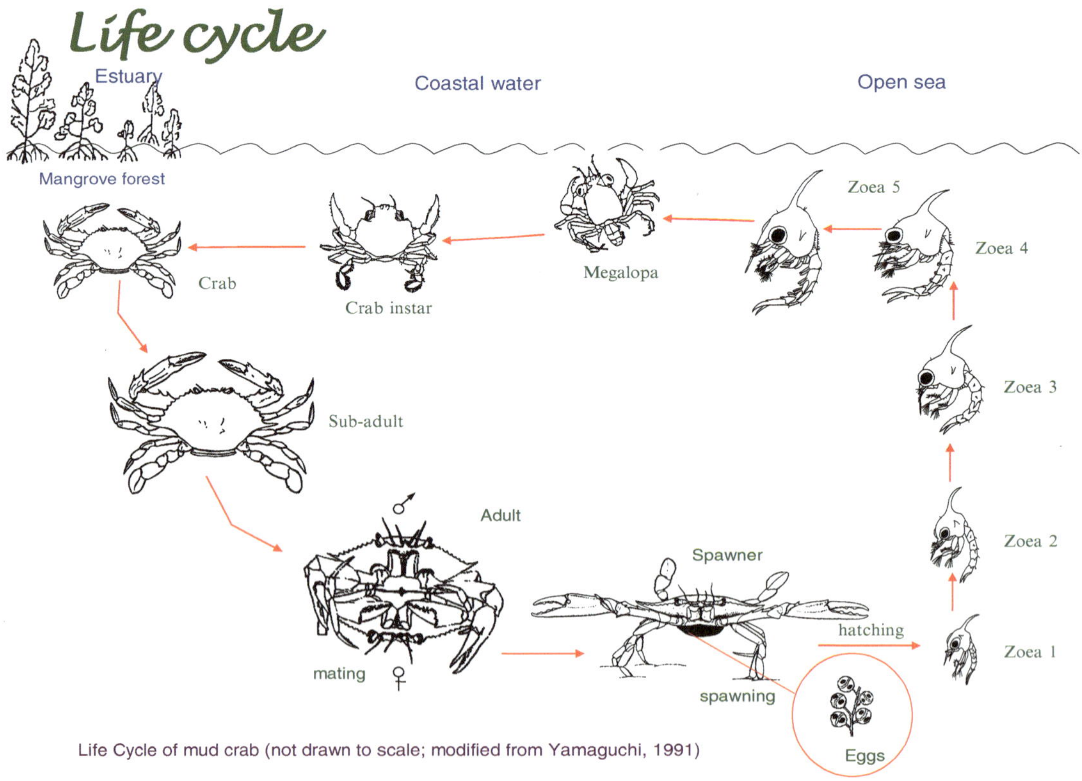

Life cycle of *Scylla serrata*

Hatchery Seed Production of Mangrove Crab

Ambient Water Quality Parameters

The ambient water quality parameters for brood stock maintenance and larval rearing are:

Salinity	28–35 ppt
Temperature	27–32 °C
pH	7.5–8.5
DO	>4 ppm
Ammonia	<0.01 ppm

At the same time, for nursery and grow out, the salinity can be 10–35 ppt. All other parameters should be same as above.

Brood Stock Collection and Maintenance

Male and female crabs can be distinguished based on the structure of the abdominal flap.

For the purpose of hatchery operation, only matured females (crabs with fully matured ovary) or berried crabs are sourced from the wild or from crab rearing ponds. Fertility percentage seems to be higher in brooders collected from the brood stock pond compared to wild collection.

Brood stocks collected from the wild/crab rearing pond are transported to hatchery in wet condition supported with oxygen supply, if necessary. On arrival at hatchery, crabs are subjected to quarantine protocols. During quarantine, crabs are screened for white spot syndrome virus (WSSV), as crab is reported to be a carrier of this virus. Once the brood stocks are qualified in screening, prophylaxis treatment and conditioning are conducted before allowing them to brood in stockholding facilities in the hatchery (FAO 2011).

Spawning and Hatching

A fully matured female weighing more than 500 g can be selected for induced spawning by eyestalk ablation. During spawning, the female crab will attach eggs to the inner portion of the abdominal flap. Hatching of egg takes place within 10–12 days. Routine observation is needed during the embryonic development stages as well as in different larval stages for ascertaining the quality of eggs and larvae.

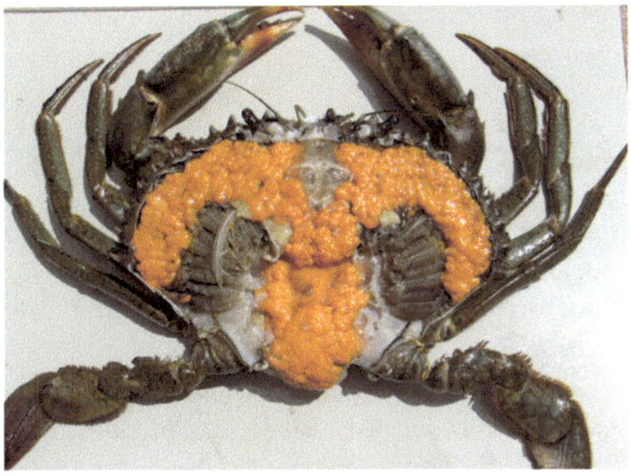

Photo: fully matured female crab

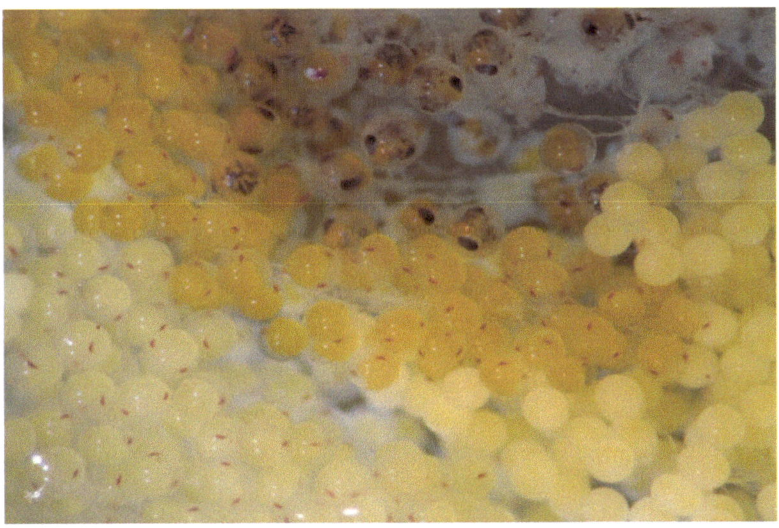

Photo showing different eye stages of developing crab embryo

Larval Rearing

In mangrove crab, there are three larval stages, viz., zoea (1–5), megalopa and crab instar. Generally 27–30 days are required to complete the larval stages so as to get a baby crab (crab instar). Similarly, *Artemia* nauplii and *Artemia* biomass production is also to be maintained as these are the exclusive feed for the later part of the larval stages (Table 1). No artificial or supplementary diets in commercial scale are available at present for mangrove crab larval rearing (Shaji et al 2006).

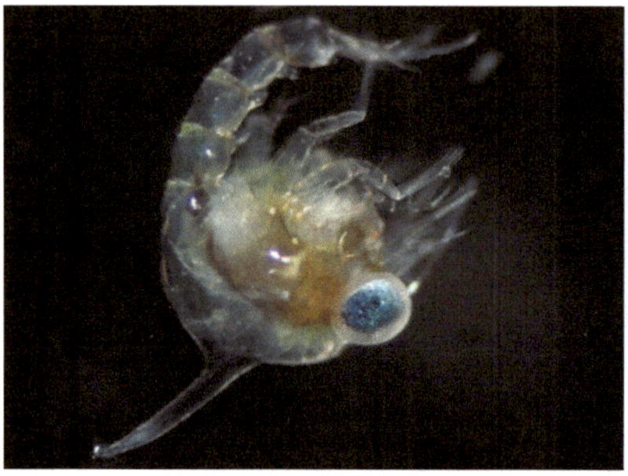

Zoea – 4 stage

Megalopa stage

Success of larval rearing relies on the live feed production and supply in a timely manner as per the demand, based on the feeding efficiency of larvae. This should be a synchronised activity, and therefore, any lag in synchronisation attributes to poor survival. The main live feeds are rotifers and *Artemia* nauplii for the entire larval rearing period. In order to sustain the production of rotifers, initially microalgal species, viz., *Nannochloropsis salina*, *N. oculata*, *Chlorella marina*, etc., are cultured in large scale for feeding the marine rotifer, *Brachionus plicatilis*.

Nursery Rearing

Nursery Rearing of Crab Instars

Nursery phase covers rearing megalopae/crab instars into crablets (in brackishwater ponds/creeks of ambient salinity range of 15–35 ppt.) within a period of 30–40 days, i.e. rearing a megalopae/crab instar of 0.3–0.4 cm carapace width (*CW*) to crablet of 2.5 cm or above *CW*, commonly called as *matchbox size*. It is noticed

Table 1 Food and feeding schedule for mud crab larval rearing

	Natural feed (live feed)			Wet feed
Larval stage	*Nannochloropsis* sp. (feed for rotifer)	*Brachionus* sp. (rotifer)	*Artemia* nauplii	(fish, mussel, killed Artemia biomass)
Zoea 1	50,000 cells/ml	10–20 ind./ml	-	-
Zoea 2	50,000 cells/ml	10–20 ind./ml	-	-
Zoea 3	-	-	0.5–1 ind./ml	-
Zoea 4	-	-	0.5–1 ind./ml	-
Zoea 5	-	-	0.5–1 ind./ml	-
Megalopa	-	-	Adult Artemia (1 ind./ml)	Based on consumption
Crab instar	-	-	-	Based on consumption

that cannibalism is a serious issue in obtaining good survival in indoor nursery facilities. Hence, nursery rearing is recommended only by using nylon/HDPE happas lined with Silpaulin sheet installed in brackishwater ponds or brackishwater creeks. Hideouts are provided to minimise cannibalism during the nursery rearing period. PVC pipes, garden nets, seaweeds (*Gracilaria*), etc. can be used as hideouts in nursery happa. If sufficient hideouts are provided and if grading is practised to maintain even-sized crab instars in happas, a minimum of 60 % survival can be achieved from crab instars to crablets.

HDPE/nylon happas with a total area of 20 m^3 (5 m × 4 m × 1 m) are recommended as ideal size for nursery rearing. Recently, the technology of rearing megalopae stage to crablet in nursery happa is also being practised with a survival of 30–40 %. 25–30 nos/m^2 of crab instars is the ideal stocking density in the nursery happas.

regular intervals at 50–20 % body weight. The average growth rate in terms of carapace width (*CW*) in 30 days of rearing period is 2.5 cm. Sixty-percent survival can be attained in nursery phase. Crablets of >2.5 cm *CW* are suitable for grow-out farming (Quinitio et al. 2009).

Nursery Pond Preparation

Prior to stocking of megalopae, it is necessary for pond conditioning and natural food production. Pond conditioning refers to the following management measures:

1. Draining/flushing of pond bottom.
2. Drying of pond bottom till cracks are developed.
3. Control of pests and competitors by using inorganic/organic eradicators such as ammonium sulphate + CaO combination, Mahua oil cake (MOC)/tea seed cake, etc.

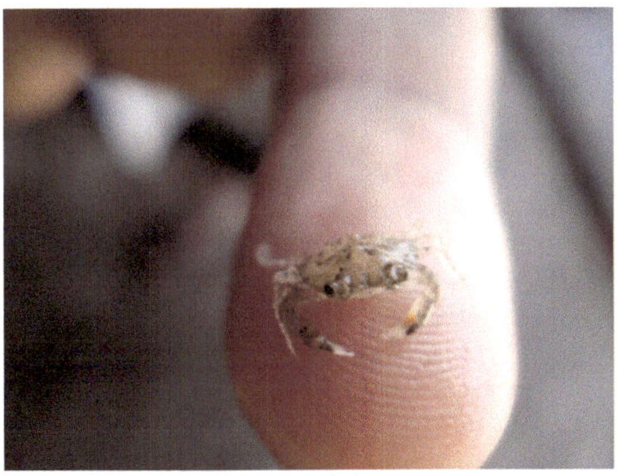

5 mm crab instar

Crablets of match box size

Food and Feeding

Megalopae are fed with either *Artemia* biomass or with chopped fish meat at 50–30 % of biomass/day. Succeeding weeks feeding rate has to be reduced to 20 %. Two times of feeding (40 % morning and 60 % evening) is recommended. The rearing period from megalopa to crab instars is 35–40 days.

The crab instars are fed with low-valued minced fish having good flesh (trash fish) at

4. Induce the natural food production by the application of inorganic/organic fertilisers such as urea/DAP/superphosphate or manure.
5. If properly fertilised, there will be abundant food for megalopa, which will help to control cannibalism and hence improve survival.
6. The salinity range should be 22–32 ppt, and water depth is maintained at 100 cm.
7. Install happas (1.5 mm mesh size having a dimension of 4 m × 5 m = 20 m^2) with support of casuarinas or bamboo poles.

8. Happa nets are installed prior to filling of water in pond.
9. Catwalk of 0.6 m width is constructed for monitoring and feeding of crab instars.

Market Potential for Crab Instars

The conventional crab fattening farmers procure juvenile crabs (100–200 g) or large-sized adult water crabs (350 g and above) for further rearing as per the market requirements. Therefore, no scientific farming techniques are applied in this sort of crab fattening practice. Presently 50–300 g crab (which doesn't have much export demand) is taken to Andhra Pradesh state for further rearing into marketable size. Thousands of acres of farms are operating where crab rearing is exclusively based on wild collected juvenile crabs. Similarly, some entrepreneurs in India started rearing 80–150 g of juvenile mangrove crabs collected from the wild and stocked in individual plastic boxes till molting stage for soft-shell crab export business. But the overdependence on the wild crabs for export, fattening and soft-shell crab production attributes excess pressure on wild population. Hence, farming with hatchery-produced mangrove crab seed reduces the pressure on the natural mangrove crab population, which is presently the only source of mangrove crab for export (Quinitio and Samaj 2007).

Grow-Out Technology

Grow-Out Culture in Earthen Pond

Mangrove crabs are generally tolerant to a wide range of temperature and salinity. They can withstand water temperature from 12 to 35 °C, but their feeding efficiency falls rapidly when the temperature is below 20 °C. They are able to tolerate salinity range of 5–34 ppt. but the optimum salinity range is 15–30 ppt. Pond eradication and water culture are to be done similar to shrimp farm preparation.

Existing or abandoned shrimp culture ponds/ponds in and around mangrove area can be utilised for culturing mangrove crab. However, ponds situated in isolated zones/area are recommended. Net enclosures should be installed along the inner side of the pond dike to prevent the escape of stock.

Nursery reared crablets are used for grow-out culture. Normally 6–7-month period is required to attain the marketable size of 500 g (average). Another way of grow-out culture can be attempted by growing juvenile sizes in small brackishwater earthen ponds for 2 months to attain an average size of 50–75 g. These juveniles can be further grown in grow-out ponds up to marketable size of >500 g within 4–5 months of culture. The second method mentioned facilitates to select even-sized juvenile crabs for grow-out,

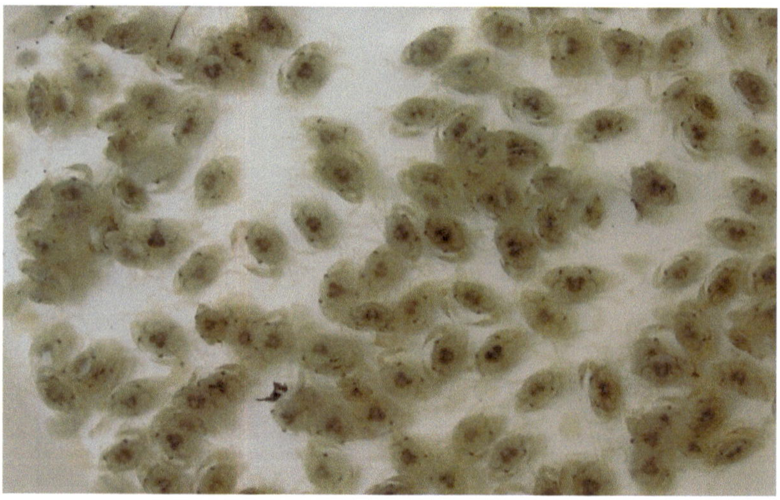

Hatchery-produced crab instars

and as such it would yield successful crop with less differential growth. The size of grow-out pond can vary from 0.25 to 1 ha with proper inlet and outlet for water management. Soil texture with high percentage of silt is not suitable for crab farming. Earthen bund of the pond is to be fenced with HDPE net to avoid escape of crab. Minimum of 1 m water depth is required, and the recommended stocking density is 0.5–0.7 juvenile crab/m^2. Stocking with uniform-sized crablets can yield an average survival of 60 %, whereas stocking with juveniles can provide more than >70 % survival. However, if stocking density is maintained at 0.5 No./m^2, the survival rate can be increased further.

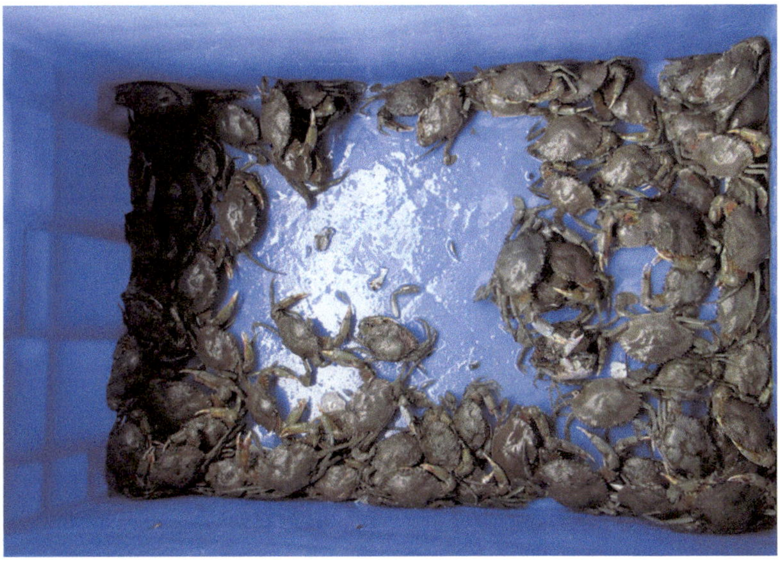

Juvenile crab ready for stocking

In grow-out culture, the cannibalism can be minimised by the use of macrophytes (viz., *Gracilaria* sp.), mangrove twigs and shelters such as sand heaps, old tiles, earthen/cement/PVC pipes, hollow blocks, etc., which will help to attain normal growth and good survival with reduced cannibalism.

Transport, Acclimation and Stocking of Mangrove Crab Juveniles

Mangrove crab juveniles are transported in ventilated bamboo baskets, plastic trays, straw bags and carton boxes. Each container is provided with wet *Gracilaria*/mangrove leaves/some material which can hold water. This would help to minimise fighting among crabs and keep the temperature cool in the container during transport.

Crabs are acclimatised before being released into ponds to prevent thermal and salinity shock that leads to sudden mortality. Avoid stocking in hot sunny time.

Water Management

Daily monitoring of water quality parameters such as temperature, salinity, DO, pH, water colour and transparency is important as a tool for the management of good water conditions in pond. Water exchange can be done according to the need, followed by application of probiotics which will save pumping cost.

Food and Feeding: Grow-Out Culture

Feed comprises 50–60 % of the total cost of production. The use of cost-effective feeds and the right amount will prevent feed wastage and water pollution. General thumb rule for feed calculation based on the carapace width is given below. Also providing feed in several feed trays will help to monitor the crab daily.

Carapace width (cm)	% of feed
<6	10
>6–15	8
>15	6–4

1. Facilitate stocking of different sizes of crab in different pens which may ultimately increase the survival and yield.
2. This type of culture can be practised even in mangrove areas.
3. Easy growth monitoring and harvesting.

Mangrove crab pen culture pond at RGCA farm, Karaikal, India

Feed is given twice, 40 % in the morning and 60 % in the evening. Generally while using trash fish, the FCR varies from 6 to 7 per kg of biomass. Though artificial diets are available for crab farming, its standardisation is under progress.

Pen Culture

Crablets/juveniles can also be reared successfully in pens (e.g. 20 m × 10 m × 1.2 m of HDPE net with mesh size of 10 mm). The stocking density of crablet in a pen can be 0.5–1 No/m^2. The advantages of the pen culture are:

Harvest and Marketing

Selective harvesting can be practised to remove the harvestable-sized, hard crabs continuously during the culture period. This also allows smaller crabs to grow faster, thus helping to shorten the culture period. This type of harvest is done with lift net or scoop net.

Complete harvest can be done by the end of 120–150 days by lift net/scoop net and handpicking after draining the pond.

Conclusion

Presently, mangrove crabs are exploited from the wild and marketed for export and domestic

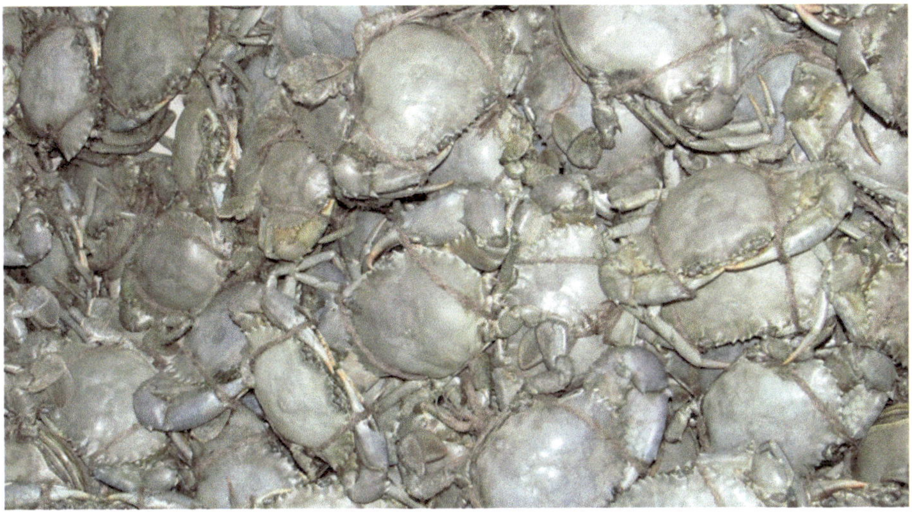

Photo: Hatchery-produced crab instars reared up to 700 g of marketable-sized crabs

consumption. Most of the wild-caught mangrove crabs are exported in live form. The fishers collect mangrove crabs of different sizes starting from juvenile to adult from mangrove mudflats and estuarine areas. Hard-shell crabs of sizes more than 350 g are sold to the exporters directly, and soft-shell crabs of different sizes and juvenile crabs are sold to mangrove crab fattening groups by the fishers. But as the natural stock is depleting due to the overdependence on wild stock, most of the fattening centres are not getting sufficient quantity of crab for their routine operation. As such there is no organised farming prevailing in India to grow mangrove crabs in land-based earthen ponds/pens/cages using hatchery-produced crab seeds. Now MPEDA-RGCA has taken up this challenge to popularise crab farming using hatchery-produced crab larvae, and as a result, a state-of-the-art hatchery facility is built up at Thoduvai, Nagapattinam District, Tamil Nadu, India, for the large-scale production of disease-free crab instars. Meanwhile several trials are progressing at the Aquaculture Demonstration Farm of RGCA at Karaikal for increasing the survival during farm operation while feeding with low-value fish as well as artificial diets.

Photo: The mangrove crab hatchery of RGCA, MPEDA, Thoduvai, Tamil Nadu

Indicative Economics Based on Recurring Expenditure for a Crab Grow-Out Farm

Basic assumptions	
Pond area	0.5 ha
No. of seed stocked (at 0.5 pcs/m^2)	2,500 nos
Period of culture	7 months
Survival (at 50 %)	1,250
ABW at harvest	500 g
Total biomass at harvest	625 kg
Selling price for 500 g crabs	Rs. 400/kg
Feed used (trash fish) at 1:6 FCR	3,750 kg
Recurring expenditure	**(Rupees)**
Pond lease amount for 1 year	20,000
Sluice net and fencing net	10,000
Cost of crablet at Rs. 5 for 2,500 (0.5/m^2)	12,500
Feed -FCR 6:1 at Rs. 15 for trash fish (50 % survival at 500 g)	56,250
Labour salary at 7,000 per month for 8 months – 1 person	56,000
Chemicals, lime/probiotics, etc.	10,000
Power/diesel for pumping and lighting	25,000
Miscellaneous expenses	10,000
Total	**1,99,750**
Revenue	
Sale price of 1,250 nos at 500 g with sale price at 400/kg	**2,50,000**
Profits	
Gross profit per crop	**50,250**

Acknowledgment The authors are thankful to Ms. Leena Nair, IAS, Chairman, MPEDA and President, RGCA, for her wholehearted support for the completion of this work.

References

BOBP (1992) The mud crab: a report of the seminar on the mud crab culture and trade, held at Swat Thani, Thailand, 5–8 November 1991. Angell CA (ed) Bay of Bengal Programme, Madras, India

Estampador EP (1949) Studies on *Scylla* (Crustacea: Portunidae). I. Revision of the genus. Philipp J Sci 78:95–108

FAO (2011) Mud crab aquaculture: a practical manual. Fisheries and aquaculture technical paper no. 567. Shelley C, Lovatelli A (eds). FAO Fisheries and Aquaculture Department, Rome, Italy. Food and Agriculture Organization of the United Nations, Rome, 100 pp

Hongyu M, Chunyan M, Lingbo M (2012) Molecular identification of genus *Scylla* (Decapoda: Portunidae) based on DNA barcoding and polymerase chain reaction. Biochem Syst Ecol 41:41–47

Imai H, Cheng JH, Hamasaki K, Numachi KI (2004) Identification of four mud crab species (genus *Scylla*) using ITS-1 and 16S rDNA markers. Aquat Living Resour 17:31–34

Jirapunpipat K, Aungtonya C, Watanabe S (2008) Morphological study and application of multivariate analysis for the mud crab genus *Scylla* in Klongngao mangrove Ranong province, Thailand. Phuket Mar Biol Cent Res Bull 69:7–24

Joel DR, Raj PJS (1983) Taxonomic remarks on two species of the genus *Scylla* De Haan (Portunidae: Brachyura) from Pulicat Lake. Indian J Fish 30:13–26

Keenan CP, Davie PJF, Mann DL (1998) A revision of the genus *Scylla* de Haan, 1833 (Crustacea: Decapoda: Brachyura: Portunidae). Raffles Bull Zool 46:217–245

Macintosh DJ, Overton JL, Thu HVT (2002) Confirmation of two common mud crab species (genus *Scylla*) in the mangrove ecosystem of the Mekong Delta, Vietnam. Mud crab aquaculture: a practical manual. J Shellfish Res 21:259–265

Mandal A, Varkey M, Sobhanan SP, Mani AK, Gopalakrishnan A, Kumaran G, Sethuramalingam A, Pandiarajan S, Samraj YCT (2014) Molecular markers reveal only two mud crab species of genus *Scylla* (Brachyura: Portunidae) in Indian coastal waters. Biochemical Genetics 52:338–354 (Published online)

Quinitio ET, Samaj YCT (2007) Mud crab culture: challenges and opportunities. In: Vishnu B, Thomas K, Muthuraman AL, Vincent D, Kumar PA (eds) Compendium of the proceedings of Technical Sessions of Aquaculture and Ornamentals at INDAQUA-2007, 11–13 January 2007, Chennai, India, pp 70–80

Quinitio ET, Rodríguez EM, Estepa FD (2009) Nursery and grow-out of mud crab. In: Training handbook on rural aquaculture. SEAFDEC/AQD, Iloilo, pp 87–95

Sangthong P, Jondeung A (2006) Classification of three crabs morphs in genus *Scylla* using morphometric analysis. Kasetsart J (Nat Sci) 40:958–970

Shaji M, Quinitio ET, Samraj T, Kandan S, Ganesh K, Kumar D, Arulraj S, Pandiarajan S, Ismail S, Dhandapani K (2006) Seed production of mud crab *Scylla serrata* at the Rajiv Gandhi Centre for Aquaculture, Tamil Nadu, India. Aquacult Asia 11:17–21

Review of Prospects for Lobster Farming

E.V. Radhakrishnan

Introduction

Capture fisheries have reached a peak in production, hence stock enhancement or aquaculture appears to be the only hope of meeting the ever increasing demand for high-value seafood like lobsters. For aquaculture, the pressing need is to solve the dilemma of promoting expansion while at the same time demanding the development of environmentally sound technologies and farming practices (Mace 1997). Lobsters have excellent market demand and price, and especially live lobsters are the most preferred. Customers in affluent countries are ready to pay more for fresh seafood. Lobsters are exported in different forms: live, frozen tails, whole frozen, whole chilled, whole cooked, frozen and as lobster meat. The live lobster trade increased from 1.3 % during 1993–1994 to 12.7 % during 2003–2004 periods. The foreign exchange revenue from the export of lobsters alone amounts to Rs. 110 crores during 2009. Due to the increasing demand, lobsters of all sizes are caught and marketed, and the resource is under extreme fishing pressure. Though the Government of India has brought a regulation on the size of export of four species of lobsters, undersized lobsters are caught in the absence of any regulation for fishing and are still exported illegally. They fetch very low price and therefore result in a loss of revenue to the fishermen. These lobsters if fattened can fetch a higher price and can be legally exported. Until hatchery technology is commercialised, value addition to the lobsters is possible through short-term fattening. Holding low-value lobsters with a view to value addition and harvesting wild pueruli for commercial grow-out appears to be technically as well as economically feasible (Mohan 2001).

Interest in aquaculture of lobsters has grown worldwide in recent years as the lobster fisheries in many countries are overexploited (Kittaka 1997; Phillips and Evans 1997). In addition, the high value of live lobster (US$ 40) has increased intense interest in lobster farming. Among the spiny lobsters, tropical species have more favourable characteristics and are amenable to farming conditions. Tolerance to high stocking in controlled conditions, communal living without cannibalism, acceptance of pelleted feed and strong market demand are some of the characteristics, which make lobster a widely accepted aquaculture species. However, for a sustainable aquaculture practice, hatchery production of seeds is vital, which is yet to be achieved, though the larval phase of a few species has been completed. The biological and technical feasibility of lobster farming is discussed in this paper.

E.V. Radhakrishnan (✉)
Crustacean Fisheries Division, Central Marine Fisheries Research Institute, Cochin 682018, Kerala, India
e-mail: evrkrishnan@yahoo.com

S. Perumal et al. (eds.), *Advances in Marine and Brackishwater Aquaculture*,
DOI 10.1007/978-81-322-2271-2_15, © Springer India 2015

Growout of Low-Value and Undersized Lobsters for Product Enhancement

Ongrowing wild-caught seed lobsters are widely practised in Vietnam, the Philippines and Indonesia. Farming is typically on a subsistence scale (limited by the availability of seed) although the magnitude of production in Vietnam is collectively very large (1,000 t annually). In Australia and New Zealand, seed lobster cannot be taken for aquaculture except under strict and limited pilot licence conditions. However, some farmers hold legal-sized lobsters for weight gain and/or more favourable (niche) marketing. In India, there being no restriction on fishing, large quantities of juveniles and undersized lobsters are caught and marketed. Though there is a good potential for ongrowing lobsters to a legal size, there is a very little attempt, probably due to non-availability of adequate number of lobsters for stocking. While some entrepreneurs have shown interest, the availability of healthy, quality seed is a major constraint.

Ongrowing involves holding undersized lobsters, which fetch low price or are not accepted legally for export, for a short period until they attain the legal size. These lobsters could be held in tanks, ponds or cages and fed with natural or artificial feed. Growth could be further enhanced remarkably through eyestalk ablation and by proper feed and water management. Since live lobsters fetch high market value, these can be marketed to targeted markets in Southeast Asian countries.

Seed Availability

Among the shallow water species occurring along the Indian coast, *Panulirus homarus homarus*, *P. ornatus*, *P. polyphagus* and *Thenus unimaculatus* are the most promising species. These species can be easily distinguished by the colour and morphological features (Annexure 1). The hatchery production of *T. orientalis* (the species is *T. unimaculatus*) has already been accomplished. The larval culture of the spiny lobster species occurring in India is yet to be successful though some headway has been made. Therefore, farming or growout of lobsters will have to depend upon either the postlarvae (pueruli) or the undersized lobsters caught in artisanal gears. In India, lobsters appear as bycatch in trawls operated in Maharashtra, Gujarat and Tamil Nadu. The quantum of juveniles caught in trawls is low. Artisanal gears such as trammel nets, gill nets and traps are also used for fishing in inshore areas. More than 50 % of the catch in trammel nets are undersized and cannot be legally exported. These are either illegally exported or consumed internally. The secondary holding centres keep the lobsters under highly stressed condition due to paucity of space and seawater facility, and therefore, the lobsters become weak and highly stressed. They may contract disease even if brought and stocked for ongrowing. If proper care is given, these undersized low value lobsters can be used for ongrowing purposes.

Breeding and Hatchery Production

Captive breeding of *P. homarus homarus*, *P. polyphagus*, *P. ornatus*, *P. longipes* and *P. versicolor* has been achieved by different laboratories in India. Adult lobsters maintained in a broodstock holding system mate and breed when optimum environmental conditions and feed are provided. Repetitive breeding is reported in *P. homarus*. Juveniles of *P. homarus homarus* and *P. ornatus* were reared to maturity in captivity and successful breeding achieved. Egg-bearing lobsters procured directly from fishermen are also suitable for egg hatching and rearing. However, berried lobsters procured from secondary holding centres are often infected and not suitable for hatchery operations. Phyllosoma larvae of *P. homarus homarus* have been reared to stage 6 in 60 days. Recently, larvae were reared to stage 8 in 42 days on a mixed diet of *Artemia* and plankton. Similarly, larvae of *P. ornatus*, *P. polyphagus* and *P. versicolor* have also been reared through early stages. Significant advances have been made in Japan, Australia and New Zealand in the culture of phyllosoma larvae of a number of temperate and semi-tropical species of lobsters to settlement. The Japanese were the first to succeed in completing the larval phase of five temperate species and one semi-tropical species. Larvae took 132–319 days to complete the larval phase in different species. The number of instars was 17 for *Jasus* spp., 9 for *Palinurus elephas* and 25 (estimated) for *Panulirus japonicus*. Survival ranged from 0.01 to 10 %. Phyllosoma larvae of *P. cygnus* and *J. verreauxi* were also reared to settlement by the Australians and the New Zealanders. The method developed in northern Japan to culture phyllosoma larvae of five species of cool-temperate spiny lobsters combines the features of upwelling water and cocultured microalgae and the use of mussel gonad as food. The feeding behaviour of phyllosoma larvae shows that they are primarily predators. Recent work has shown that the contamination of culture water by microorganisms such as the fouling protozoans *Vorticella* sp. can greatly reduce phyllosoma survival. Feeding with fish larvae reduced the larval period by half (65 days) in *P. elephas*. The recent success in captive breeding and seed production of *P. ornatus* by Australians has generated high expectations in commercial farming of this promising species. However, commercial feasibility of seed production technology is still doubtful for most of the species because of the prolonged larval phase (>300 days) and poor survival. Until hatchery technology is perfected, lobster farming will have to depend upon naturally available seeds.

Collection and Ongrowing of Pueruli

Spiny lobsters after completing its long larval life metamorphose into postlarva (puerulus), which swims towards the shore and settles in nearshore seagrass/seaweed habitats. They transform into post-puerulus and settle into benthic-dwelling juvenile lobster. In the wild, levels of pueruli settlement vary depending on the species and geographic area, which also depends on the spawning stock. Commercial harvest of pueruli and early juveniles from the wild as suggested by the Australians may be possible where there is a heavy settlement in the inshore habitat. However, the feasibility of such a proposal is to be examined in other areas, as there is apprehension that this will have serious long-term implications on the sustainability of wild resources. Partial harvesting of pueruli from dense settlement areas is suggested as high mortality rates were experienced by juvenile lobsters between settlement and entry into the commercial fishery. This will benefit the wild resource as well as those interested in farming of lobsters. Commercial collection of southern rock lobster pueruli for aquaculture has already commenced in Tasmania. In India, studies conducted in Kovalam near Chennai show that pueruli of three species, *P. homarus*, *P. polyphagus* and *P. ornatus*, settle in rocky areas. However, the intensity of settlement is not very high enough to suggest harvesting for ongrowing. There is no information on the settlement density of pueruli anywhere along the Indian coast.

Indian lobster fishery is poorly managed as this is a low volume resource, and therefore, fishing regulations are not strictly enforced. Gears like trammel nets operated in inshore areas bring in a large quantity of juveniles and subadult lobsters which fetch only low price. The undersized lobsters procured from the fishermen were exported until the Ministry of Commerce and Industry, Government of India, notified the Minimum Legal Size (MLS) limit for export in 2003 (Table 1). Illegal export of lobsters below the MLS is still in vogue. Since there is no regulation for fishing lobsters in any maritime state, fishermen continue to bring the lobsters entangled in the nets and sell it to exporters. Nearly 50 % of the lobsters caught in trammel nets are undersized. In Gujarat and Maharashtra, gill nets operated in the reef areas also bring undersized *P. polyphagus*, which are sold, for a very low price. Along the southwest and east coast of Tamil Nadu, an estimated 25 t of undersized lobsters are landed annually. By catching these undersized lobsters, the total loss of revenue of the fishermen is to the tune of Rs. 1.75 crores. The MLS for *P. homarus* is 200 g and for *P. polyphagus* 300 g, and those below the MLS, which are prohibited from export, can be used for ongrowing to the legal size and fetch a higher price.

Table 1 Minimum legal size for the export of lobsters from India

Species	Live/chilled/frozen (g)	Whole cooked (g)	Tail (g)
Panulirus polyphagus	300	250	90
P. homarus	200	170	50
P. ornatus	500	425	150
Thenus orientalis	150	–	45

Notification No. 16 (RE 2003)/2002–2007 dated 17 July 2003, Ministry of Commerce and Industry, Government of India

Collection, Maintenance and Transportation of Juveniles

Fishermen either keep the lobsters in wet sand until the traders come and pick it or sell it to local merchants. These traders either keep them in holding tanks with recirculation facility (most of them with inadequate water reuse facility) or keep them in cages maintained in rock pools until collected by the exporters. Since only larger lobsters with high price are given more care, the undersized low-price lobsters are kept under high stress in very poor-quality water. Therefore, the lobsters become weak and are easily susceptible to diseases. Such animals are unsuitable for growout purposes. On transportation to long distances, they further become weak and on stocking die gradually. The green colour of the lobster is lost, and their shell becomes reddish in colour, which is indicative of stress. If the undersized lobsters should be useful for growout operations, fishermen and traders should follow certain protocols so that they could supply healthy seed to the entrepreneurs. Juveniles are especially susceptible to stress, and therefore, should be kept in the water soon after they are brought to the shore. It is better to keep them buried in cool moist sand in shade rather than keeping at high densities in poor-quality water. Maximum care should be taken so that their appendages are not lost while removing from the net or they are not injured below the abdomen. During holding in an unhealthy condition, they are infected by deadly pathogens through these injuries, and there are instances of high mortality due to infection by pathogenic bacteria.

Culturable Species

The major species forming fishery in India are *P. homarus homarus*, *P. polyphagus* and *P. ornatus* (Fig. 1). While *P. homarus homarus* is the dominant species along the southwest and southeast coasts, *P. polyphagus* is the major species along the northwest coast. *P. ornatus* form only a small fishery along the southeast coast, though from the culture point of view, this species is the most promising species due to the fast growth rate in captivity and the high demand in the international live lobster market. In India, the focus should be on the most common species, *P. homarus homarus* and *P. polyphagus*. The recent advances in larval culture of *P. ornatus* by the Australians show the high potential of this species as a promising candidate for aquaculture. The culturable characteristics of *P. homarus* are shown in Table 2.

Growout in Land-Based Holding System

Lobsters can be successfully held and grown in land-based holding systems provided suitable environment is provided. Critical environmental parameters include the concentration of dissolved oxygen, ammonia, nitrite and carbon dioxide. Nitrate concentration, pH, salinity and alkalinity levels within the system are also important. Ammonia (unionised) is much more toxic than the ionised ammonium. The concentration of ammonia in culture water is dependent on the pH and temperature of the water. Ammonia should be preferably less than 2 mg/L and nitrite <5 mg/L. Oxygen is generally

Fig. 1 Culturable species of lobsters in India. (**a**) *Panulirus homarus*, (**b**) *Panulirus polyphagus*, (**c**) *Panulirus ornatus* and (**d**) *Thenus unimaculatus*

the major water quality variable limiting the success of live holding systems. The requirement of oxygen will be high during moulting and soon after feeding. Both the processes normally take place during the night, and optimum levels of oxygen shall be maintained during the night to avoid oxygen depletion and mortality. Oxygen-related stress would adversely affect growth significantly. Unlike shrimps, lobsters are oxygen regulators and may not show any sign of low oxygen conditions, as they are capable of adjusting the metabolism even at very low levels of oxygen. Sufficient water needs to be pumped through or the water needs to be aerated to ensure lobsters are supplied with sufficient oxygen. The oxygen saturation should be preferably above 80 % in culture tanks. Sudden salinity fluctuation during rains is a serious situation, and in such circumstances, salinity, alkalinity and pH should be adjusted in a reservoir, before the water is pumped in; though lobsters can tolerate gradual changes in salinity to a limited range, sudden dilution can lead to stress, susceptibility to disease and mortality. However,

Table 2 Culturable characteristics of *P. homarus*

Biological attributes	Advantages	Disadvantages
Morphological features	Hard exoskeleton, robust	Exhibits autotomy
Growth in captivity	Tropical spiny lobster – fast growing	Attains early maturity
Seed production	Hatchery production of few species successful	Long larval phase, 200–400 days; low survival; commercial hatchery yet to be in place; abundance of recruiting seed lobster in grounds along the coast not documented but likely to be comparatively high
Seeds from the wild	Natural seed (juveniles and subadults) available from incidental catch of tangle nets	No adequate information on the level of capture that will not hamper the natural recruitment and fishery
Behavioural attributes	Gregarious behaviour – can be cultured communally	Stocking density should be optimal
Cannibalism	No cannibalism when lobsters are fed adequately	
Tolerance to salinity fluctuation	Can tolerate gradual salinity variation, but not abrupt changes; optimal salinity range, 30–35 ‰	Slow growth in lower and higher salinity regimes
Handling	Can live outside the water for a long duration in certain conditions	To be handled carefully, autotomy – reduction in growth
Food requirements	Feeds on a variety of organisms, mussels, clam, fish; preferred diet, mussels; can be grown on artificial diets	Prefers fresh feeds
Amenability to different culture systems	Amenable to culture conditions both in indoor and sea cage farming systems	Water quality is critical
Shelter for optimal growth	Nocturnal behaviour, intensive feeding activity during night, prefers shelter	Tendency to congregate together
Export potential	Considerable export potential for live lobsters	Rapid price fluctuations, preference for smaller lobsters
Susceptibility to diseases	Relatively free from diseases	Physiological stress and poor holding condition can lead to a disease situation (fungal, bacterial).

Table 3 Various water quality parameters for the culturable species of lobsters

Parameter	Range
Temperature	25–30 °C
DO (% saturation)	Minimum 70 %
	Preferably above 80 %
Salinity (‰)	30–38
Ammonia (mg L^{-1})	<2
Nitrite (mg L^{-1})	<5
Nitrate (mg L^{-1})	100
pH	7.8–8.4
Hardness (ppm)	100–200

juvenile *P. polyphagus* can be acclimated to salinities as low as 17 ‰ and at least as high as 50 ‰ (Kasim 1986).Tolerance limits for various water quality parameters are given in Table 3.

Indoor Tank

Two main systems are currently being used for growout lobsters: flow through and recirculating. In flow through systems, the water that is pumped into a tank is used only once. The water flow is to be decided based on the stocking density and feeding intensity. The incoming water is to be free of sediments and should have water quality parameters required for lobster farming. The water quality should be regularly monitored to avoid wide fluctuations in environmental parameters. In recirculating systems, the majority of the water is reused after each pass through the tanks, first being treated to remove waste products before being returned to the tanks. Even though initial set-up costs may be higher, there is an increasing interest in the use of recirculating systems. In recirculating systems also, the main limiting factor is dissolved oxygen: however, the unionised ammonia concentration becomes increasingly important and is probably the next important limiting factor. Ammonia should be removed from the system at a rate equal to the rate of production to maintain safe concentration. All recirculating systems remove waste solids, oxidise ammonia and nitrite, remove carbon dioxide and aerate the water before returning it to the fattening system. Solid wastes can be removed by mechanical filtration, ammonia and nitrite by biological filtration and carbon dioxide by the provision of an air/water interface. The flow rate calculations must be adjusted according to the species held, the size of the animals, the rate of feeding and the holding temperature. The required estimated flow rate in a tank holding 1000 kg (Crear et al. 2003) of 500 g fed *J. edwardsii* lobsters at 13 °C is 4,500 L h^{-1}. The water management schedule and monitoring frequency of environmental parameters are shown in Table 4.

Tank Design

Raceway, rectangular, square or circular tanks can be used. The most preferred are individual raceway tanks. Circular or square tanks made of brick and cement or concrete are also good. Square tanks will save space when compared to circular and are also less expensive as they could be connected serially (Fig. 2). Raceways are easier to maintain, and with proper slope, the wastes

Table 4 Monitoring schedules for different environmental parameters

Parameters	Schedule	Equipment
Temperature	Daily at 1,000 h and 1,600 h	Thermometer, glass/digital
Salinity	Weekly and before adding new water	Salinity refractometer
pH	Weekly or during water change	pH metre
Oxygen	Weekly (every 2–3 h when new water is added)	Oxygen probe
Ammonia, nitrite, nitrate, alkalinity and hardness	Weekly	Measuring kits

Fig. 2 Indoor lobster culture tanks with self-cleaning facility

can be easily removed through the outlet pipe fixed at the end of the tank. For a complete removal of water, the standpipe can be lifted. Square and circular tanks will have a central drainage system or a self-cleaning two-way waste removal system by which both suspended and settled wastes can be automatically flushed out of the tank. The wastes can be concentrated towards the centre of the tank by creating a vortex by the incoming water. This tank design is used in flow through systems. For recirculating system the wastewater flowing out of the tank is recirculated after removing the waste products and will be a continuous process. Since lobsters grow fast in subdued light, tank covers have to be provided to avoid bright sunlight.

Stocking Density

Undersized lobsters procured from secondary holding centres transported to the growout facility may be kept under quarantine for 48 h to relieve the lobsters from stress. The quarantine facility should be away from grow-out tanks. Healthy lobsters may be stocked at 1.0–1.25 kg/m^2 after segregating into different size groups. The difference in weight between the lower and upper size should not be more than 20 g. Hideouts provided in the tank will help them to congregate around the shelter during day and prevent them from continuously moving in the tank spending lots of energy. Lobsters feed on a variety of natural and artificial feed making them suitable for farming. The natural feed includes mussels, clam, squid, trash fish and smaller crab and shrimp meat. Artificial feed includes shrimp pellets that are suitable for juveniles and subadults. Lobsters feed actively after dusk; feeding during night will reduce feed spoilage and waste.

In Florida, *Panulirus argus* males have been grown from first instarjuveniles to 450 g in 12 months and 1.4 Kg in two years (Lellis and Russell 1990: Lellis 1991). Juveniles of *P. ornatus* stocked at 43/m^2 showed good artificial feed acceptance (>75 %), biomass production (4.7 kg/m^2) and good growth rate (SGR 1.56 %). It is estimated that a weight increase of 1 kg is possible in less than 18 months (Barnard et al. 2011). Juveniles of *P. homarus* in South Africa have attained 60 mm CL in 18 months. In India pueruli of *P. homarus homarus* have been reared to 250 g in 18 months at Kovalam Field Laboratory of CMFRI. *P. homarus* weighing 80 g stocked in a self-cleaning indoor system at Calicut attained 330 g in 12 months with an Food Conversion Ratio (FCR) ranging from 3.5 to 4.0 on feeding with an exclusive diet of green mussel. The minimum exportable weight of 200 g was obtained in 130 days (Radhakrishnan 2004) (Fig. 3). Another study in an indoor grow-out system at Tuticorin showed 172.7 g weight increase in juvenile *P. homarus* in 150 days. In Taiwan, wild-caught animals stocked at 2 g mean weight reach 330 g in 16 months in small 200 m^2 ponds. A growth study on *P. polyphagus* conducted at Northern Territory University showed that lobster juveniles grew substantially to 148.5 g in 9 months.

Growth enhancement of juvenile lobsters by eyestalk ablation has been demonstrated in *P. homarus homarus*, *P. ornatus* and *P. versicolor*. The ablated *P. homarus* gained a weight gain of three to seven times more than the normal (Radhakrishnan and Vijayakumaran 1984). In *P. ornatus*, 100 g lobsters attained 1.5 kg in 8 months. However, food consumption and mortality are higher in ablated lobsters when compared to normal.

Lobster Culture in Intertidal Pits

Fattening of *P. polyphagus* was carried out in intertidal pits along the Bhavnagar coast of Gujarat during the early 1990s (Sarvaiya 1991). The pits are dug in the limestone intertidal areas and are of varying dimensions. No standardised stocking or feeding regimes were followed. 3,000 numbers of lobsters weighing 30–50 g were stocked at a stocking density of 20 lobsters/m^2 in a pit of 21 m × 7 m × 1 m. The pits were covered with nylon nets to prevent

Fig. 3 Harvested *P. homarus* from an indoor farming system

lobsters from escaping during high tide. Lobsters were fed with trash fish, small crabs, marine worms and clams at 10 % body weight and also on compounded feeds. Juvenile lobsters, which were procured at Rs. 20/kg, attained 100–125 g in 90 days, which were sold at Rs. 250/kg in 1991. The harvest of live lobsters packed in bamboo baskets is sold to exporters. The highly profitable practice was discontinued due to intensive poaching and non-availability of enough quantity of seed lobsters.

High-Density Culture of Sand Lobster

Experimental farming of *Thenus orientalis* (*T. unimaculatus*) was carried out at Kovalam Field Centre of CMFRI, Chennai. Juvenile lobsters measuring an average 20 mm carapace length and weighing 5 g collected from the wild were stocked at a rate of 30–35 animals/m^2 in two cement tanks of a 12.5 m^2 floor area and 0.5 m water depth. The environmental parameters in the rearing system were pH 7.8–8.2, salinity 36–38 ‰ and temperature range 27–29 °C. Lobsters attained 140–175 g in 250 days with an overall survival of 90 %. A growth rate of 0.6–0.7 g/day was obtained in these trials (Kizhakudan 2009).

Lobster Culture in Cages

In the Philippines, lobster culture was practised in floating cages made of nylon material. *P. ornatus* weighing 100–300 g stocked at 8/m^2 attained 800 g–1.3 kg in 6–15 months. Survival for larger sizes has been more than 90 %, whereas in smaller juveniles of 30–80 g, survival was less than 50 %. Trash fish is the main feed, which is procured at US$ 0.13–0.22 per kg. The selling price of live lobster was US $ 21–31/kg; hence, a wet Food Conversion Ratio (FCR) of more than 10 can still be profitable.

In Vietnam, mariculture of lobsters in floating, fixed and submerged cages was practised in protected bays or lagoons that have a good tidal flow. *P. ornatus* is the main species, and *P. homarus*, *P. versicolor*, *P. longipes* and *P. stimpsoni* were also cultured. The floating cage is made of nylon net material with a frame and buoy and moored at a depth of 10–20 m. The wooden fixed cage is normally 20–40 m^2 and even large cages of 200–400 m^2 are also used. The off-bottom cage is typically about 0.5 m above the seabed. This kind of cage is suitable for sheltered bays. The submerged cage is made of iron framework with a diameter of 15–16 m. The height is 1.0–1.5 m. The cage has a cover and a feeding pipe and is used for nursery rearing. Seed lobsters of about 25–30 mm TL are stocked (100–200 per cage) and grown to a size of about 50 g (10–12 cm TL). Lobsters are fed exclusively with either a whole finfish or chopped fish or shellfish. Finfish comprised about 70 % of the diet and the preferred fish is lizardfish. Feeding trash fish results in water quality problems. The total nitrogen content in the seawater exceeded the standard level for aquaculture of 0.4 mg/L. The increase in cage farming and using fresh trash fish and shellfish has led to disease problems in some areas. Some common diseases are black gill disease, shell necrosis and red body. Treatment protocols followed are treatment with 100 ppm formalin for 3–5 min. Red tail disease has been reported in spiny lobsters, which is similar to the *gaffkemia* in *Homarus americanus*. The symptoms are lethargy, reddish colour of the underside of the abdomen, spread-eagle posture during late stages, poor food consumption and mortality. The bacteria enter into the body through the injury in the abdomen and multiply in the haemolymph. Isolation of the affected individuals and oral administration of oxytetracycline through feed are protocols for treatment. Antibiotic administration is possible only during the early stages of infection. The disease is highly contagious and spreads through water.

The estimated total annual production of farmed rock lobster was about 1,500 metric tons a year with a farm gate value of US$ 26.75/Kg. The average profit margin was 50 %. Therefore, lobster culture is a profitable industry provided adequate seeds could be collected without disturbing the wild fishery. Suitable artificial diets are to be formulated and feeding regimes established in order to succeed lobster farming in a commercial scale. Disease protection measures should be prioritised. Cage designs should be studied for improved lobster husbandry.

Sea Cage Farming Trials in New Zealand

The potential for using suspended sea cages for the aquaculture of juvenile spiny lobsters was assessed for *Jasus edwardsii* in New Zealand. Lobsters were grown for a year in sea cages at three sites starting from pueruli and regularly fed on opened mussels. Larger juvenile lobsters of two size classes were also held in sea cages with small live mussels for food. Pueruli grew at rates that were close to or greater than those previously recorded from tank experiments. At the most northern site, lobsters grew most quickly to an average of 42.1 mm CL 0.4 s.e. and 36.9 g wet weight 1.0 s.e. in a year. The mortality of lobsters differed with site, but at one site it was lower (14 %) than that was recorded in a previous tank culture experiment (25 %). The differences in growth and mortality among sites appear to be related to differences in ambient water temperatures. Larger juvenile lobsters were found to be unable to feed on the small live mussels. These results indicate that suspended sea cage culture has a considerable potential for the aquaculture of spiny lobster juveniles but will require the careful selection of sites and the development of effective feeding arrangements.

Preliminary experiments in cage culture of lobsters were attempted in India by National Institute of Ocean Technology (NIOT), Chennai. Floating and fixed net cages were positioned in protected areas. Fishermen fattened juvenile lobsters incidentally caught from the wild. Pueruli, post-pueruli, early juveniles and subadults of the spiny lobster, *Panulirus homarus* and juveniles of *P. ornatus* were grown in different dimensions of floating sea cages along the southeast coast of India from May 2003 to May 2007 (Vijayakumaran et al. 2009). The first type of cage had a galvanised iron pipe frame (2.0 m × 2.0 m × 1.2 m) with a steel woven mesh and four inner detachable compartments (0.75 m × 0.75 m × 1.10 m). Fibre-reinforced plastic was used subsequently to fabricate cages (1 m × 1 m × 1 m). Pueruli and post-pueruli of *P. homarus* (1.58 ± 0.62 g *SD*), stocked at 60 individuals/m^2, grew to an average weight of 123.10 ± 26.22 g in 266 days with a survival rate of 70 %. Subadults of *P. homarus* with an average weight (± *SD*) of 123.61 ± 29.26 g reached 341.25 ± 46.22 g in 225 days at a stocking density of 21 individuals/m^2 with a survival of 73 ± 6 %. The post-pueruli grew by 0.46 ± 0.10 g per day with a specific growth rate (SGR) of 1.64, whereas subadults had a growth rate of 0.97 ± 0.20 g per day with an SGR of 0.43. At a higher stocking density of 80 individuals/m^2, juveniles (51.83 ± 10.32 g and 58.20 ± 28.22 g) of *P. homarus* recorded growth rates of 0.86 ± 0.25 (SGR 0.82) and 0.97 ± 0.34 g (SGR 0.96) per day. This study indicates that post-pueruli of *P. homarus* can be grown to over 200 g in 12 months and up to 350 g in 16–17 months in sea cages. Juveniles (average weight 76.35 ± 34.50 g) of *P. ornatus*, reared with *P. homarus* at a stocking density of 80 individuals/m^2, recorded a weight gain of 139 g in 155 days at a rate of 0.89 ± 0.32 g per day with an SGR of 0.67. Marine live clam, *Donax* spp., was the main feed supplemented with the gastropod, *Xancus pyrum*; the green mussel, *Perna viridis*; marine crab (*Charybdis* sp.); squid (*Loligo* sp.); and fish such as clupeids and *Leiognathus* sp. Pueruli and post-pueruli were observed to settle in large numbers (up to 35 individuals/month in one cage) both inside and outside the cages (Vijayakumaran et al. 2009).

Open Sea Cage Farming Trials by CMFRI

The Central Marine Fisheries Research Institute conducted field trials of sea cage farming in open sea at Vizhinjam, Mumbai, Veraval, Chennai and Mandapam. The sea cages with a diameter of 6 m made of 140 mm HDPE with a 1 m high railing at the top and 8 m cylindrical net cage at the bottom attached to a ballast were positioned by attached floats, gabion boxes and shock absorbers close to the coast where the depth is 10 m or more (Fig. 4). At Vizhinjam, juvenile *P. homarus* with an average weight of 114.8 + 25.7 g stocked in the cage grew to 226.0 + 43.0 g in 135 days with an increase in weight of 0.82 g/day and survival of 75 %. Lobsters were fed twice daily with mussel and trash fish. At Veraval and Mumbai, *P. polyphagus* stocked in cages registered a higher growth rate of (1.33g/day) and survival (93%). The suitability of different species to cage farming needs to be studied taking into account the natural habitat of each species. The stocking rates and feeding have to be standardised. For a successful commercial farming, practical formulated diets have to be prepared and its acceptability and conversion rates determined. Studies on these lines are progressing. While there is a potential for high stocking density in floating cages, cannibalism during moulting will have to be minimised by providing adequate shelter.

Results from this study indicate that open sea cage farming is a cost-effective and biologically feasible way of ongrowing juvenile lobsters to market size, in comparison to indoor tanks even with a high biomass of animals. Factors such as

Fig. 4 Harvested *P. homarus* from a sea cage at Vizhinjam

supplemental feed from biofouling, reduced stress, natural light levels and photoperiod are likely to have contributed to the good performance of the sea cages. Nevertheless, this study also shows that the aquaculture potential of a sea cage is highly dependent upon the sea cage structural design.

Economic Feasibility

(a) *Land-based farming systems*
The major inputs for a land-based farming system are the capital investment (land, building, tanks, seawater intake pumps/filters, office, water quality testing equipment, freezers for wet feed storage) and the operating costs (cost of seed lobsters, feed, labour, electricity, wages and maintenance).

The profit will greatly depend upon the seed cost, feed cost and the market price of farmed lobsters. The high demand and attractive price for small lobsters in the export market are likely to inflate the purchase cost of seed lobsters. Feed costs do have the potential to be reduced substantially by using foods cheaper than mussels and clams. Using mussel and clams in lobster culture also generates increased labour costs related to regularly removing the unfed feed and cleaning tanks to avoid bacterial contamination caused by feed decomposition. Feeding with mussels has shown that fresh mussels consistently produce the best growth rates in cultured spiny lobsters. Feeding of lobsters with trash fish in indoor farming is not recommended as this will require thorough cleaning of the tanks and complete replacement of culture water, thereby increasing the cost. Ultimately, the development of a low-cost artificial feed for spiny lobsters has the potential to greatly increase growth rates and reduce labour input and maintenance costs. Recently, practical diets for the ongrowing lobsters have been developed in India, Australia and New Zealand. Labour costs also have the potential to be reduced especially through improving the design of culture tanks so that feeding and cleaning become much more efficient.

(b) *Sea cage farming system*
A preliminary assessment of the feasibility of sea cage farming of *P. homarus* along the southwest coast of India suggests that there is a good potential for this activity to be commercially viable. The major capital inputs are the cost of cage and mooring system and lease amount for the space. The operational cost includes cost of seed lobsters, feed, wages for feeding, cleaning and security, boat charges to visit the cages and harvesting.

The question is whether lobster culture is feasible in the absence of a hatchery technology. In Tamil Nadu, an average 25 t of undersized lobsters are landed at the landing centres. In Maharashtra, an average 180 t of lobsters are landed by artisanal gears and 60–70 % is undersized lobsters. If 50 % of this seed source could be used for farming, this would enable an aquaculture industry of 280 t of cultured lobsters per annum. However, before contemplating aquaculture of this scale, research is necessary to better determine the stock structure of spiny lobster. Such research would enable responsible fishery management policies to be put in place to ensure sustainability of wild fishing stock.

Appropriate controls to regulate the number of cages and culture sites may be necessary to minimise environmental impacts to the adjacent areas. Low-intensity aquaculture of spiny lobsters may be a very profitable export industry in India.

Future Prospects

All marine lobsters are highly considered as fine table food and thus are in high demand. Most wild fisheries are overexploited, with many stocks having already collapsed or catch rates closely regulated to sustain the wild fishery. For these reasons, lobsters fetch high prices. The highest price is paid for a live product; chilled or frozen products bring much lower prices. Aquaculture offers the only prospects by which lobster supplies can realistically and sustainably be increased. Considerable export potential exists for live products in Southeast Asian markets. Packing and transport conditions for the live shipment of lobsters are well developed and may not be a problem. The development of a successful export market would require both continuity of supply and a reasonable volume of production. However, the greatest concern is whether aquaculture production would be sustainable if recruiting seed was taken in large quantities for aquaculture. Attempts in the 1970s and in the late 1990s to establish a large-scale intensive aquaculture of spiny lobsters in the Philippines collapsed within a few years of establishment when seed supplies became insufficient to support the venture. In India the resource is limited to certain pockets along the coast, and lobster landing is drastically declining in all the centres due to indiscriminate fishing. Therefore, before contemplating aquaculture, research is warranted to better determine the stock structures of spiny lobsters and to estimate recruitment patterns and survival rates. Such research would enable responsible fishery management policies to be put in place to ensure sustainability of the wild fishery stocks. Low-intensity aquaculture of spiny lobsters for value enhancement and export strictly adhering to the legal procedures could be a profitable industry in some specific locations along the coast.

References

Barnard RM, Johnston MD, Phillips BF, Ritar AJ (2011) Hatchery production of spiny lobsters: meeting growing demand for premium product. Global Aquaculture Advocate, Sept/Oct, pp 92–95

Crear B, Cobcroft J, Battaglene S (2003) Recirculating systems for holding rock lobsters. Technical report series, (Tasmanian Aquaculture and Fisheries Institute) No. 15

Kasim HM (1986) Effect of salinity, temperature and oxygen partial pressure on the respiratory metabolism of *Panulirus polyphagus* (Herbst). Indian J Fish 33 (1):66–75

Kittaka J (1997) Culture of larval spiny lobsters: a review of work done in northern Japan. Mar Freshw Res 48:923–930

Kizhakudan JK (2009) Larval and juvenile rearing of the sand lobster *Thenus orientalis* Lund, 1793. In: Winter School course manual on recent advances in breeding and larviculture of marine finfish and shellfish. CMFRI, Cochin, pp 145–150

Lellis W (1991) Spiny lobster: a mariculture candidate for the Caribbean? World Aquac 22:60–63

Lellis WA, Russell JA (1990) Effect of temperature on survival, growth and feed intake of postlarval lobsters, *Panulirus argus*. Aquaculture 90:1–9

Mace PM (1997) Developing and sustaining world fisheries resources: the state of the science and management. In: Hancock DA, Smith DC, Grant A, Beumer JP (eds) Developing and sustaining world fisheries resources: the state of the science and management. Second world fisheries congress, Brisbane 1996. CSIRO Publishing, Melbourne, pp 1–20

Mohan R (2001) Tropical spiny lobster - a new mariculture species for the Sultanate of Oman and the Arabian Gulf States. In: Goddard S, Al- Oufi H, McIlwain J, Claereboudt M (eds) Proceedings of the 1st international conference on fisheries, Aquaculture and Environment in the NW Indian Ocean, Sultan Qaboos University, Muscat, Sultanate of Oman, pp 158–164

Phillips BF, Evans LH (1997) Aquaculture and stock enhancement of lobsters: report from a workshop. Mar Freshw Res 48:899–902

Radhakrishnan EV (2004) Prospects for grow out of the spiny lobster, *Panulirus homarus* in indoor farming system. In: Program and abstracts, 7th international conference and workshop on Lobster Biology and Management, 8–13 February, Hobart, Tasmania

Radhakrishnan EV, Vijayakumaran M (1984) Effect of eyestalk ablation in spiny lobster *Panulirus homarus* (Linnaeus). 1. On moulting and growth. Indian Fish 31:130–147

Sarvaiya RT (1991) Successful spiny lobster culture in Gujarat. Fish Chimes 10:30–31

Vijayakumaran M, Venkatesan R, Murugan TS, Kumar TS, Jha DK, Remany MC, Thilakam JML, Jahan SS, Dharani G, Kathiroli S, Selvan K (2009) Farming of spiny lobsters in sea cages in India. N Z J Mar Freshw Res 43:623–634

Mussel Farming and Its Potential in India

K. Sunil Mohamed

Introduction

Marine mussels are bivalve molluscs belonging to the family "Mytilidae" and form one of the most dominant cultivable species all over the world (estimated culture production about 1.5 million tonnes in 2003). Total global trade of mussel is involving 300,000 t, worth about US $ 400 million. They give the highest conversion of primary producers (phytoplankton) to human food, and culture of mussels in the column waters can increase the seafood production several folds. In India, two species of marine mussels (green mussel, *Perna viridis,* and the brown mussel, *P. indica*) are distributed in the rocky coastal areas where they support a traditional sustenance fishery, but scope for increasing natural production from the existing beds is rather limited (Appukuttan et al. 2001).

During post-monsoon period, there is a heavy settlement of mussel spat along the west and east coasts in the intertidal and subtidal rocky areas. During this season, many millions of mussel seeds are attached to the hard substratum, but only a small percentage of these grow to become adults, mainly due to sand deposition and receding tides. The mussel culture technology developed by CMFRI during the early 1970s envisages using these perishing spat for seeding ropes.

K.S. Mohamed (✉)
Molluscan Fisheries Division, Central Marine Fisheries Research Institute [CMFRI], PO Box 1603, Cochin 682018, Kerala, India
e-mail: ksmohamed@vsnl.com

Technology Development

Giving priority to the development of a technique for farming mussels, the CMFRI started a project at Vizhinjam in Kerala in 1971. In 1976, at Calicut, attempts were made to develop open sea farming of the green mussel. The programme was extended to the east coast at Madras during the same year. All these efforts led to the development of technology for mussel culture in the open sea and in protected bays. The mussel seeds collected from the natural beds are wrapped around a thick nylon rope by using a biodegradable cloth, which degenerates within a fortnight. The seeds get attached to the rope by this time and continue to grow there, utilising the natural food. These "mussel ropes" are suspended from grow-out structures like raft or long line deployed at a productive and unpolluted site (Box 1). Harvestable sizes are usually attained within 4–6 months depending on the area of grow-out.

Maritime states along the west coast of India have extensive estuaries, which open to the Arabian Sea. These estuaries are subjected to wide variations in hydrographic condition due to the southwest monsoon during June to September and a less intense northeast monsoon from October to November. The mean of annual rainy days in Kerala has been estimated as 130 days, and of this, 66 % is during June to September, 19 % during October to December and 15 % during January to March. Based on the

S. Perumal et al. (eds.), *Advances in Marine and Brackishwater Aquaculture*,
DOI 10.1007/978-81-322-2271-2_16, © Springer India 2015

Box 1: Methods of Farming

1. *Rack (Trestle) Method*

This method is especially suited for estuaries and shallow seas. Bamboo or *Casuarina* poles are driven into the bottom spaced 1–2 m apart. These stakes are connected horizontally with poles forming a trestle. The horizontal poles should be above the level of water at high tide. Seeded rope can be suspended into the water for farming from these poles. Three seeded ropes can be suspended from 1 m^2 area of the rack.

2. *Long-Line Method*

This method is considered ideal for unprotected open sea conditions. Synthetic rope of 16–20 mm diameter is used for the long line (main line). The main line is supported with 220 l barrels tied to it, spaced at 5 m. The long line and barrels are anchored in position at either ends using concrete blocks and nylon ropes. Seeded ropes are suspended from the long line.

3. *Raft Culture*

Ideal for open sea conditions which are not rough. Square or rectangular rafts are made with sturdy bamboo or *Casuarina* poles. Buoyancy for the raft is given by tying 5 barrels of 200 l capacity (metal oil barrel painted with anticorrosive paint or synthetic barrel). Ideal size of the raft is 5 × 5 m. The rafts are to be positioned at suitable site in the sea using anchors (grapnel, granite, concrete). Three seeded ropes can be suspended from 1 m^2 area of the raft.

hydrographic condition, in most estuaries, two phases, viz., a marine phase during December to May and a brackishwater phase during June to November, have been observed. It is during the marine phase that the ecosystem becomes conducive for mussel culture. The offshore of this coast is subjected to strong currents and upwelling phenomenon during the monsoon season. However, in the fair season, these areas can be used for mussel farming. Long-line units and rafts were utilised for mussel farming in the deeper offshore sites. Both these methods proved to be highly successful.

Transfer of Technology

The constraints which were foremost in stalling the development of mussel culture industry in the 1970s and 1980s were lack of awareness, social inhibitions and finance. Perceiving the drawbacks in extension programmes, in the beginning of the 1990s, an action programme was initiated for location testing as well as for disseminating farming technology in estuaries, backwaters and coastal seas. It was decided to set up demonstration units at several sites along the coastline with the direct involvement of fishermen. This led to the growth and development of mussel farming as a rural development programme especially in the southwestern parts of India, in the states Kerala and Karnataka, especially in the former (Mohamed et al. 1998; Sasikumar et al. 2000, 2006).

Compared to the open sea, the estuarine ecosystems are less turbulent and shallow (<4 m) and, therefore, less prone to risks like loss due to heavy weather and poaching. The demonstrations with fisher participation in north Kerala convinced them about the feasibility of mussel culture, and the local fishers of this area set up their own mussel farms with technical guidance from CMFRI and financial assistance from local governing bodies. From the ensuing season onwards, different fisher groups took up mussel farming as a seasonal avocation, and farmed mussel production in the region was trebled (Appukuttan et al. 2000).

The scientists of the CMFRI in consultation with the district administration created a master plan to transfer the technology to potential women beneficiaries. The DWCRA (Development of Women and Children in Rural Areas) was identified as the most suitable scheme intended for groups of women beneficiaries (Box 2) below the poverty line. The local governing bodies identified the beneficiaries with the help of village extension officers and district administration. The selection criteria took into consideration

(1) primary school as the minimum education level, (2) age of the beneficiaries between 28 and 62 years and (3) fisheries/agriculture as the major occupation. After the selection of beneficiaries, a series of awareness camps on mussel farming were conducted by the institute in each panchayat (village). Beneficiaries were given training in their own farms from seeding to harvesting. One-day workshops were organised in different villages involving bank officials, officers of the district administration and village extension workers. Besides, the National Bank for Agriculture and Rural Development (NABARD) approved mussel farming as a bankable scheme having IRR (internal rate of return) above 50 % and benefit cost ratio of 1:1.34.

Box 2: Technology Adoption by Women

In 1996, groups of women from the North Kerala (Kasaragod district) started their own mussel farms with the financial support extended by the Development of Women and Children in Rural Areas (DWCRA) and Training of Rural Youth in Self-Employment (TRYSEM). The entire farming operation, viz., starting from seed collection to marketing, was done by the women themselves. They were able to pay back the loan within the stipulated period. In succeeding years the farming activities were intensified by the involvement of more groups. Now, mussel farming is a part-time vocation of the coastal women of North Kerala. The local banks and district administration have taken a lead in providing financial assistance to these fishers. Mussel farms are usually set up by November–December, and the crop is harvested before June (to avoid large-scale destruction due to monsoon). Though it is only seasonal, women have recognised that it is something which they can do with minimum effort and financial commitment.

Loans from the government developmental agencies like DWCRA (Development of Women and Children in Rural Areas), IRDP (Integrated Rural Development Programme), TRYSEM (Training of Rural Youth in Self-Employment), BFFDA (Brackishwater Fish Farmers Development Agency) and Farmers Cooperative Banks (such as Malabar Gramin Bank) to newly formed village mussel farming groups (average 13 members in each group) resulted in the start of several mussel farms in this region. All technical help to the farmers was provided by the CMFRI.

These developments paved the way for a consistent increase in farmed mussel production in the state of Kerala. The horizontal spread of the technology in Kerala is shown in the map below (Fig. 1). The major farming area is Kasaragod district, but recently farming has spread in Ernakulam and Kollam districts. The farmers themselves have innovated in the technology used, and many farmers in Kannur, Kozhikode, Malappuram and Kollam districts practise on-bottom farming where seed mussels are spread on the bottom and allowed to grow. In this method the input costs are minimal. Besides, some farmers in Kadalundi and Korapuzha estuaries where the water depth is shallow (about 1 m) practise horizontal rope culture method. In this method unit area productivity is low, but growth is comparable to that in the traditional rack (trestle) system (Velayudhan et al., 2000).

During 2009, the estimated farmed mussel production is more than 18,000 t. About 75 % of this production is from Kasaragod district, and other major districts contributing to the production are Malappuram, Kozhikode and Kannur. Ernakulam, Thrissur and Kollam districts contribute only a small percentage. The growth in production has been particularly steep after 2004 (Fig. 2).

Technological Innovations

Refinements in the technology have been made to reduce capital costs (mainly on nylon ropes) by using alternate core materials (flexible

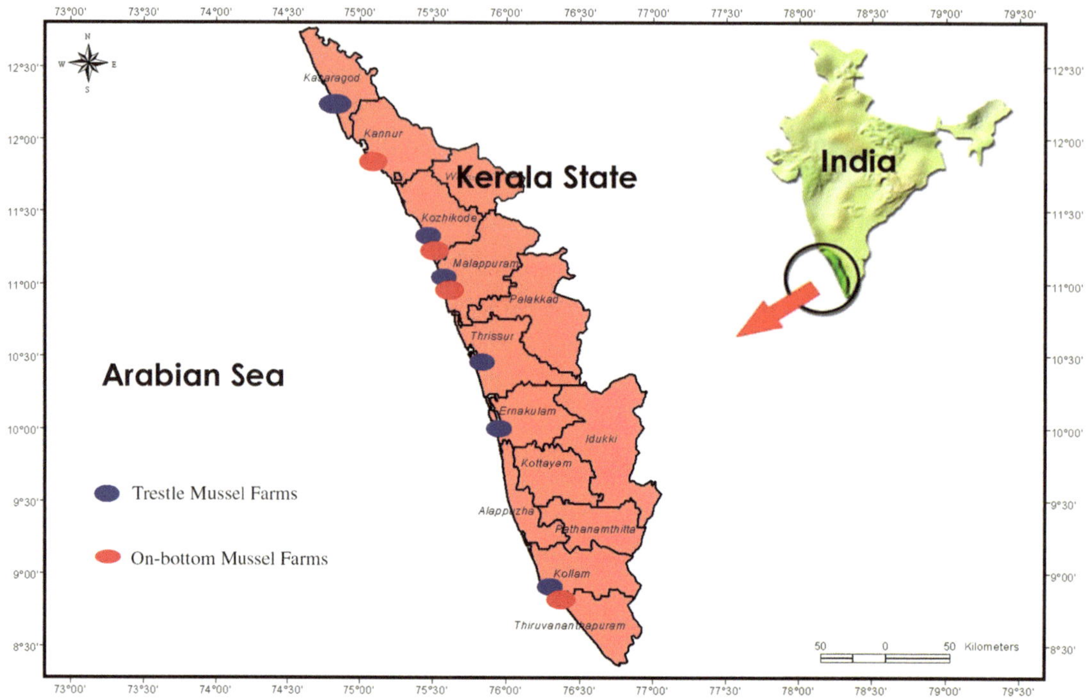

Fig. 1 Map showing the mussel farming technology implemented in Kerala

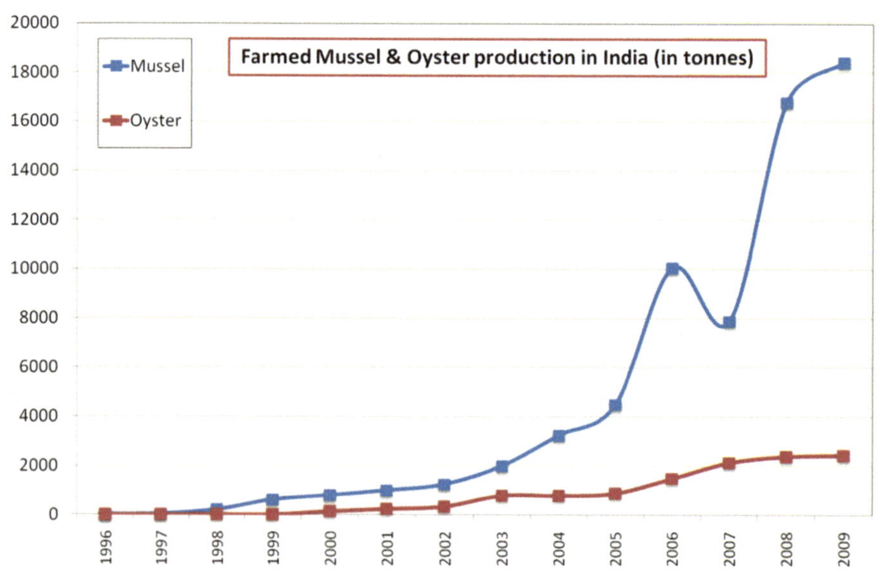

Fig. 2 Farmed mussel and oyster production in India (in tonnes)

plastic strips – FPS) and pre-stitched cotton net tubes. Seeding is one of the most critical activities in mussel farming. The process which is physically demanding (as farmers have to kneel and bend down to do it) is crucial to the success of farming as the uniform attachment of mussel seed around the rope is dependent on how well it is done.

Fig. 3 Details and parts of the mussel seeder (*left*) (*a*) top stand, (*b*) FPS, (*c*) seed holder, (*d*) vertical support, (*e*) PVC pipe, (*f*) notch for PVC pipe and (*g*) base plate. *Right panel* shows the parts of the mussel declumper and its working

Now, to reduce the physical strain and to increase efficiency during this process, a *semiautomated mussel seeder* has been designed (Fig. 3), developed and field tested. The seeder, which has an estimated cost of Rs. 2,500, was successfully field tested and demonstrated to mussel farmers in Kerala. The chief advantages of the seeder are reduction in time taken for seeding resulting in increased efficiency and lower labour costs and reduction in physical strain during the process (Mohamed et al. 2003). The time taken for manual stitching of 1 m rope by the conventional method is 8 min, whereas in the seeder the same can be accomplished in 2 min (Kripa et al. 2001).

Harvesting and declumping (separating mussels from the rope) farmed mussels is done by lifting the mussel ropes and by plucking the mussels from the rope or by stamping if the byssal attachment is very strong. To easily separate the mussels from the rope, the concept of a *semiautomated mussel declumping machine* was developed. The machine had two separate units, a metal drum and a metallic circular, fixed shield with a central opening and with a diameter of 10 mm fixed on a stand and a ramp for placing the harvested rope. One metre mussel rope could be declumped in 2 min. The chief advantages were that physical exertion during harvesting could be avoided and that it was more hygienic and efficient (Mohamed et al. 2005).

Impact of Mussel Farming

The experiences of CMFRI, in the transfer of technology programmes, have clearly shown that fishermen can accept a scientifically proven technology only if they actually observe the benefits from it. Participation by the villagers was found to make the demonstration and adoption processes a sure success. Though the culture technique is the same, to give wider publicity, different small mussel farms were set up in the estuarine systems. The overall social impact was the emergence of mussel farming as a group/community activity with the recognition of scope for income improvement and women empowerment (Kripa and Mohamed 2008). A number of ancillary industries (rope making, seed collection, seeding and marketing) have also opened up. The sector has also become organised in the northern districts with the formation of the Green Mussel Farmers Society (GMFS) at Kasaragod.

Ecological Impacts

Although widely perceived as an environmentally safe technology, mussel farming has also got ecological impacts to the farming system which needs to be seriously addressed. A major impact is the accumulation of sediment (detritus) in the farming areas due to faeces production of mussels resulting in lowering of depths especially in estuaries where flushing and tidal flow is minimal. As a consequence, the benthic ecology can undergo detrimental changes. Based on scientific studies, the CMFRI has recommended that continuous farming for more than 2 years at one site is not advisable and it is necessary to shift the farm to an adjacent site every 2 years.

Prospects for Future Development

It is quite clear from the fast pace of its development in the state of Kerala that mussel farming can develop as a new sunrise mariculture industry in India. Unlike other aquaculture industries, it is not capital intensive and offers great scope for improving the incomes of the rural fishers as an alternate livelihood. But primarily, what has spurred its growth in Kerala is the considerable demand for the produce among the populace. Other mussel-consuming states like Karnataka, Goa and Maharashtra can also be targeted in the next phase of development. Policymakers and planners need to address the following for sustained development of this spanking industry.

1. Promote mussel farming in other maritime states using Kerala as a developmental model.
2. Since farming depends on seed availability from natural sources, development of methods to collect mussel seeds from the wild is necessary.
3. Encourage hygiene (depuration) and value-added products (VAP) for mussels to increase marketing possibilities and to make the farming practice more remunerative.
4. Determine carrying capacity of backwaters/estuaries for mussel farming and restrict farming accordingly.
5. Demark areas for mariculture and create mariculture parks with adequate legal protection and articulate open-access waterbody leasing policies.
6. Make a prospective (5 years) plan to improve hygiene in farming areas using EC guidelines as a criterion.
7. The current international prices for mussels (Fig. 4) are not remunerative enough for India to encourage exports to European markets. Besides, the European markets demand meeting very strict environmental standards which we will be able to fulfil only in the next 5 years. The domestic market retail price for mussels are in the range of Rs. 60–100. The Arabian Gulf markets for expatriate Indians can be explored.

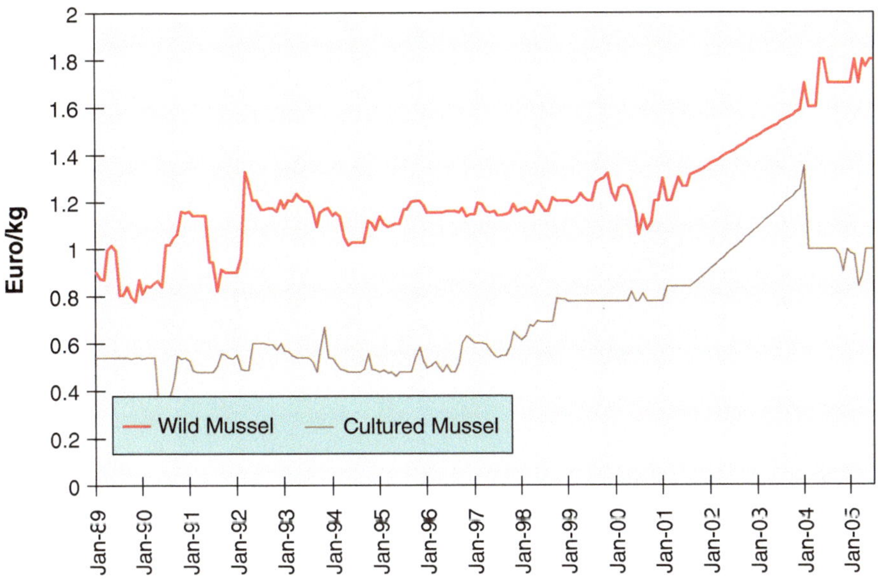

Fig. 4 Price movement in wild and cultured mussels in Barcelona (Spain) markets (Source: Globefish)

References

Appukuttan KK, Kripa V, Velayudhan TS, Mohamed KS, Victor ACC, Kuriakose PS, Laxmilatha P, Muthiah P (2000) In: Pillai VN (ed) Bivalve mariculture in India – a success story in coastal ecosystem development. Published by Asia Pacific Association of Agricultural Research Institutions, FAO, Bangkok, APAARI Publication: 2000/1, 56 p

Appukuttan KK, Mohamed KS, Kripa V, Asokan PK, Anil MK, Sasikumar G, Velayudhan TS, Laxmilatha P, Koya KPS, Radhakrishnan P, Joseph M, Alloycious PS, Surendranathan VG, Sivadasan MP, Nagaraja D, Sharma J, Naik MS (2001) Survey of green mussel seed resources of Kerala and Karnataka. Mar Fish Inform Serv T&E Ser 168:12–19

Kripa V, Mohamed KS (2008) Green mussel, *Perna viridis*, farming in Kerala, India – technology diffusion process and socioeconomic impacts. J World Aquacult Soc 39:612–624

Kripa V, Mohamed KS, Velayudhan TS, Laxmilatha P, Radhakrishnan P, Joseph M, Alloycious PS, Sharma J, Appukuttan KK (2001) Recent developments in edible bivalve farming in India. Aquacult Asia 6:40–43

Mohamed KS, Muthiah C, Nagaraja D, Sampathkumar G (1998) Initiation of marine mussel culture activities in Dakshina Kannada District, Karnataka. Mar Fish Inform Serv T&E Ser 155:10–15

Mohamed KS, Kripa V, Velayudhan TS, Appukuttan KK (2003) Enhancing profitability through improved seeding techniques in the green mussel (*Perna viridis*) farming. J Mar Biol Assoc India 45:214–223

Mohamed KS, Kripa V, Radhakrishnan P, Joseph M, Alloycious PS, Velayudhan TS, Appukuttan KK (2005) Design and development of a semi-automatic seeder for seeding mussels. J Mar Biol Assoc India 47:207–209

Sasikumar G, Muthiah C, Nagaraja D, Sridhara B, Bhat GS (2000) Mussel culture in Mulky estuary, Dakshina Kannada District, Karnataka during 1997–99. Mar Fish Inform Serv T&E Ser 164:14–18

Sasikumar G, Krishnakumar PK, Bhat GS (2006) Monitoring trace metal contaminants in green mussel *Perna viridis* from coastal waters of Karnataka, southwest coast of India. Arch Environ Contam Toxicol 51:206–214 (online)

Velayudhan TS, Kripa V, Appukuttan KK, Mohamed KS, Laxmilatha P (2000) Feasibility of integrated bivalve farming in an estuarine system of Central Kerala, India. In: Tiensongrasmee B, et al. (eds) Mollusk Research in Asia, Asian Fisheries Society, Manila, Philippines, pp 209–215

Present Status and Future Prospects of Sea Cucumber Breeding and Culture in India

P.S. Asha

Introduction

The holothurians or sea cucumbers are commercial echinoderms, which are exclusively marine in their habitat. Holothurians inhabit different depths of the ocean as well as different habitats such as rocky shores, sandy beaches, muddy flats, coral reefs and mangrove swamps. Out of 154 genera and 1,150 species recorded in the world, 62 genera and 160 species are available in India; among them, 88 are recorded in Andaman and Nicobar Islands, 39 in the Gulf of Mannar and 32 in Lakshadweep.

Economic Value

Holothurians are exploited commercially, for the raw body wall or viscera, but mostly for the processed dry product called 'beche-de-mer'. The Chinese are the traditional consumers, and Japanese, Koreans, Melanesians, Micronesians, Polynesians and African people also consume 'beche-de-mer' in significant ways and quantities. In traditional Chinese medicines, the sea cucumbers are used for treating weakness, impotency, debility of the aged, constipation due to intestinal dryness and frequent urination. Recent findings indicated their potential role in the biomedical research, as a rich source of polysaccharides like chondroitin sulphate and glucosamine and other bioactive substances with anti-inflammatory, antitumour and fungicidal activity. Apart from these nutritional and therapeutic values, the sea cucumbers are important components of the food chain, because they play an important role as deposit and suspension feeders. They are often called the earthworms of the sea, because they are responsible for extensive shifting and mixing of substrate and recycling of detrital matter.

Fishery and Trade Status

Sea cucumbers are fished all over the world, particularly in tropical regions. The total global catch of sea cucumbers is in the order of 100,000 t of live animals annually. The leading exporters were Indonesia, the Philippines, Fiji Islands, Japan, Madagascar, Papua New Guinea, Solomon Islands, Thailand and the USA. Due to high demand, ease of harvesting and low cost production technique, the sea cucumber has developed rapidly (Conand 2004).

Indian Beche-de-mer Industry

In India, the 'beche-de-mer' industry was introduced by the Chinese more than 1,000 years ago. It is essentially a cottage industry based in rural areas needing little investment.

P.S. Asha (✉)
Tuticorin Research Centre of CMFRI, South Beach Road, Tuticorin 628001, Tamil Nadu, India
e-mail: ashasanil@gmail.com

S. Perumal et al. (eds.), *Advances in Marine and Brackishwater Aquaculture*,
DOI 10.1007/978-81-322-2271-2_17, © Springer India 2015

The industry consists of the fishermen, who are divers, the processors who act as middlemen and the exporters. Though the sea cucumbers are distributed in Lakshadweep, Andaman and Nicobar Islands and the Gulf of Kutch, the processing of sea cucumbers was going on only in the Gulf of Mannar (GOM), and it played as an important source of income for about 50,000 poor fisherfolk along this area. The cost of the sea cucumber varies with the species and among the seven commercial species; *H. scabra* was the highly priced species followed by *H. spinifera*. The processed 'beche-de-mer' was exported mainly to Singapore from India, because of the lack of domestic 'beche-de-mer' markets. During 1996–1997, India exported 70 metric tonnes of 'beche-de-mer', which has decreased to 3.81 metric tonnes during 2001 (Hong Kong SAR import statistics).

Statutory Regulations

Over-exploitation of sea cucumber resources has caused local extinction of breeding population of some species and the collapse of other stocks in some countries. Owing to the dramatic declining of catch and size of the specimens fished, the Ministry of Environment and Forests, Government of India, imposed a ban on sea cucumbers in 1982, and accordingly fishing of sea cucumber of less than 8 cm was prohibited. The Ministry imposed another ban in June 2001 on sea cucumber fishery and listed the holothurians as protected animals along with 50 other marine species under the Indian Wildlife Protection Act, 1972, which has caused severe impact on the livelihood of several thousands of fishermen populations along the Gulf of Mannar and Palk Bay, who subsist on the fishery of this species, and also reduced the foreign exchange too.

The releasing of hatchery-produced juveniles of commercial sea cucumber species to their natural habitat, a process called restoration, restocking or reseeding, is gaining momentum worldwide, as the only way to replenish the depleted stock of holothurians. The Central Marine Fisheries Research Institute has given subtle importance on the conservation aspects of this fragile resources by implementing projects on the development of seed production techniques of commercial sea cucumbers and succeeded in developing seed production techniques of commercial holothurians like *Holothuria scabra* (Fig. 1a) in 1988 and *H. spinifera* (Fig. 1b) in 2001 for the first time in the world. *Holothuria scabra* commonly called sandfish is one of the most commercially valuable tropical species of sea cucumber with a wide distribution throughout Indo-Pacific areas. The A-grade beche-de-mer processed from sandfish commands some of the highest price on the international market. Since the mass production of juvenile *H. scabra* through hatchery system has been proved, it is being considered as an ideal

Fig. 1 Brood stock of holothurians: (**a**) *Holothuria scabra* and (**b**) *H. spinifera*

candidate for stock enhancement programme in many countries. *H. spinifera*, the second major contributor to the beche-de-mer industry along the south-east coast of India, is already declared as an endangered species in China, and its availability is not yet confirmed in many places of its earlier distribution; hence, sea ranching of its hatchery-produced seed to the natural habitat will have greater importance. Through continuous research effort in the breeding and hatchery aspects of these two species, several improvements have been made in the holothurian hatchery system (Asha 2005; James et al. 1999).

Hatchery Operations

1. *Brood stock collections and management*

The brood stock of the sea cucumbers has been collected from the wild during the breeding season. The quality and liveliness of the specimens are best in those collected by skin divers; they never developed any disease and also often yielded effective spawning. Hence, it can be recommended that the skin diving specimens are best for hatchery operations.

The breeding attempt were carried out during two spawning peaks, i.e. during the major breeding peaks (March–May and October–December), in the case of *H. scabra* and from November to March in the case of *H. spinifera*; hence, the brooders have to be collected prior to the spawning peaks. Care should be taken for the storage of good-quality filtered sea water with adequate salinity. The brood stock collected by skin divers can be acclimatized in the hatchery for 2 weeks before subjecting them for spawning induction by maintaining in a 1 t tank with 6 in. thickness of fine beach sand in a continuous system with a daily exchange of rearing medium, and the brooders can be fed with *Sargassum* spp. powder at 0.5 g/500 l^{-1} (James et al. 1988, 1994).

Spawning Induction

The brood stock was induced to spawn by several techniques. Addition of feed made of rice bran, *Sargassum* spp. and soya bean powder (2:1:0.5) at 50–100 g^{-1} 500 l^{-1} is found to be very effective in the case of *H. scabra* and *H. spinifera* (Asha and Muthiah 2002). On one occasion, the sudden changes in salinity caused effective spawning among the brooders. The egg suspension from an eviscerated female also induced spawning successfully on one occasion. Instead of going for thermal shock, a combination of cold shock followed by hot shock (± 5 °C) is also found to be more effective in *H. scabra*. Since it is difficult to determine the sex of sea cucumbers externally, it is advisable to collect 30–40 specimens having a length and live weight ranging from 20–30 cm and 300–500 g. It is advisable to maintain 15 brooders 500^{-1}.

Larval Rearing

High sperm density will affect the fertilization; hence, care should be taken to filter out the fertilized eggs. Experiment proved that stocking densities of 0.5 ml^{-1} for the eggs and 1 ml^{-1} for the larvae are the optimum hatching and larval rearing conditions (Asha and Diwakar 2013). The fertilized eggs (Fig. 2a) and larvae can be stocked in sterilized sea water filtered through 1 μm with mild aeration to get the effective hatching and development (Asha and Muthiah 2002). The chronological developmental stages with the details of various larval stages up to pentactula with the time of occurrence and mean size are given in Table 1.

1. *Early development*

After fertilization, the first polar body appears within 20–30 min. The first cleavage takes place after 15 min. In 3 h, blastula is fully formed. After 24 h, the gastrula (Fig. 2b) is fully developed, which is oval in shape and motile. After 48 h, early auricularia appears.

2. *Auricularia larva*

Early auricularia (Fig. 2c) is developed within 48 h, which passed through the mid- and later stages within 10 days. It is slipper shaped, transparent and pelagic in habit. It has a preoral loop anteriorly and anal loop posteriorly, which helps in locomotion. The digestive tract consists of mouth, pharynx and sacciform stomach.

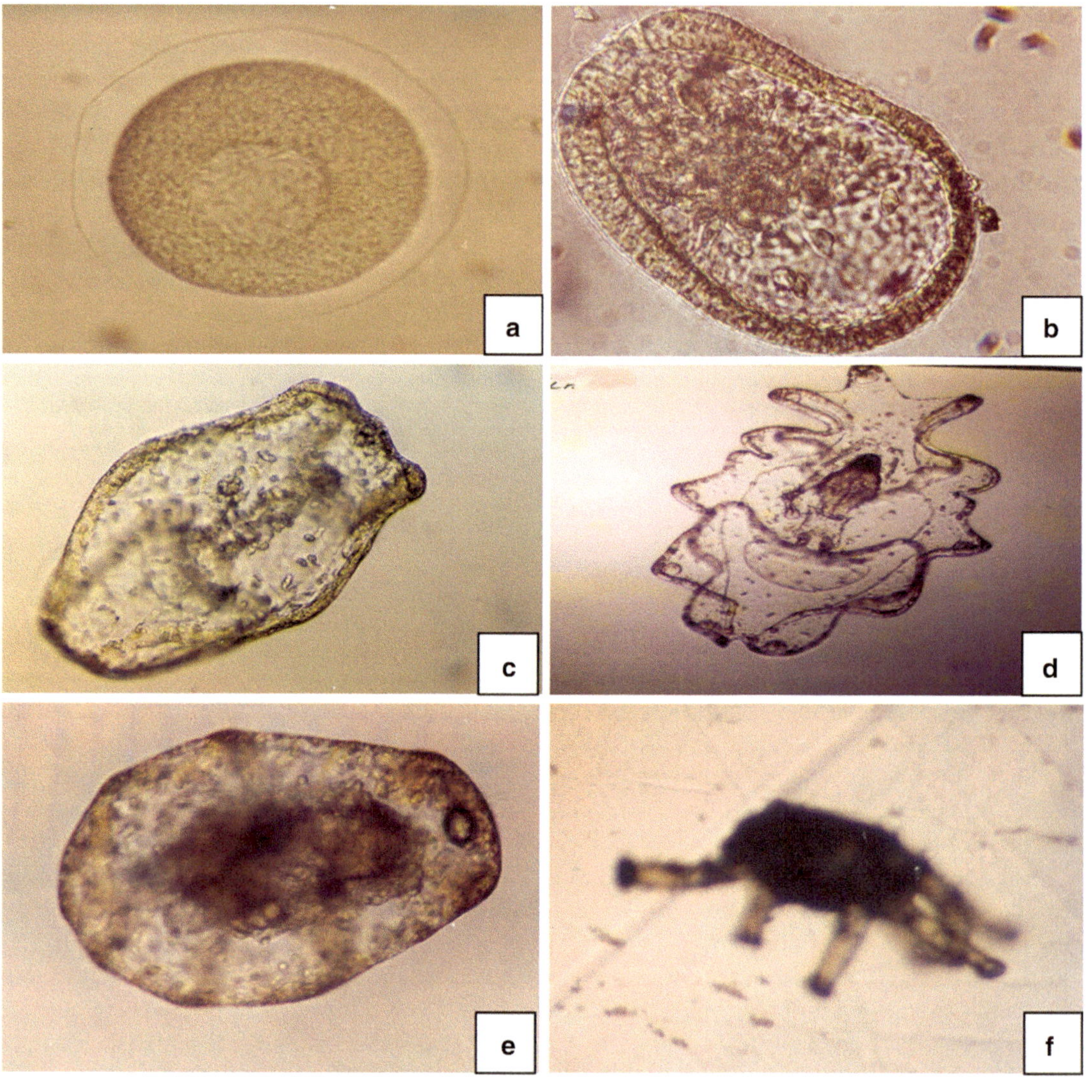

Fig. 2 Developmental stages of holothurians: (**a**) Fertilized egg, (**b**) gastrula, (**c**) early auricularia, (**d**) late auricularia, (**e**) doliolaria and (**f**) pentactula

Table 1 Time after fertilization for the different development stages of holothurians

Development stag	Time after fertilization	Mean size
Blastula	3 h	–
Gastrula	24 h	265.4 ± 6.06 μm
Auricularia (early)	2 days	489 ±14.1 μm
Auricularia (late)	10 days	1.1 mm
Doliolaria	10–12 days	468 ± 23.3 μm
Pentactula	13–15 days	330 ±16.7 μm
Early juveniles	20 days	1 mm
Juveniles	90 days	40 mm

On the tenth day, the late auricularia larva (Fig. 2d) metamorphoses to doliolaria larva.

3. *Doliolaria larva*

The doliolaria is barrel shaped, free floating and in a nonfeeding stage with five bands around the body (Fig. 2e). This stage will last for 2–3 days, and subsequently they metamorphosed to the creeping stage known as pentactula.

4. *Pentactula larva*

The pentactula is tubular with five tentacles at the anterior end and with a single tube foot at the posterior end (Fig. 2f) which helps in the locomotion of the larva. The pentactulae creep over the sides and bottom of the tank.

Larval Feeding

The appropriate concentration of feed is important for the successful larval rearing. The feeding schedule has to be determined by the larval healthiness. The early auricularia for the first 3 days can be fed at 2×10^4 which can be slowly raised to 3×10^4 in the mid-stage to 4×10^4 in the later stage (Asha 2004). The larvae are found to survive well when fed by *Isochrysis galbana* but grow well when fed by *Chaetoceros calcitrans*; hence, a mixture of these two (1:1) is found to be an ideal feed for the auricularia larvae for 10 days (Asha and Muthiah 2006). By periodic assessment of larval growth rate, the feeding regime can be adjusted so that the auricularia larval survival rate can be improved even up to 80–90 %. No feed has to be given to the doliolaria larvae. The doliolaria larvae have to maintain at 2 ml^{-1} in a flow-through system in which equal quantity of water is allowed to let in and let out. The water temperature 28–32 °C, pH 7.8 and salinity 35 ppt are the suitable environmental condition for the metamorphoses and settlement of the larvae (Asha and Muthiah 2005; Asha et al. 2011).

Settlement

The doliolaria can be induced to settle by the daily addition of powdered algae Algamac at a concentration of 0.5 g 500 L^{-1}, which will act as an inducer for the doliolaria to settle and also will serve as a food source for the newly settled pentactulae. The newly settled pentactulae can continue to feed Algamac for 1 month with slowly raising the concentration from 0.5 g to 1 g 500^{-1} (Asha and Muthiah 2007). The periodic thinning out of the pentactulae to reduce the stocking density is found to improve the growth rate (Sui et al. 1986).

Nursery Rearing

One-month-old juveniles of sea cucumbers of both H. *scabra* and H. *spinifera* (Figs. 3a and 4a) were given *Sargassum* spp. extract (<40 μm) for 1 month in a bare tank. When the

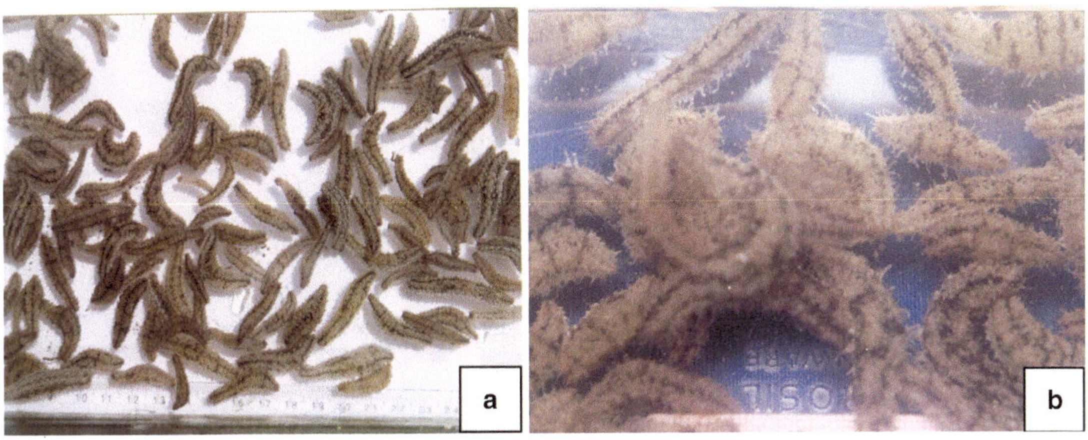

Fig. 3 One-month-old juveniles of sea cucumbers: (**a**) *H. scabra* and (**b**) *H. spinifera*

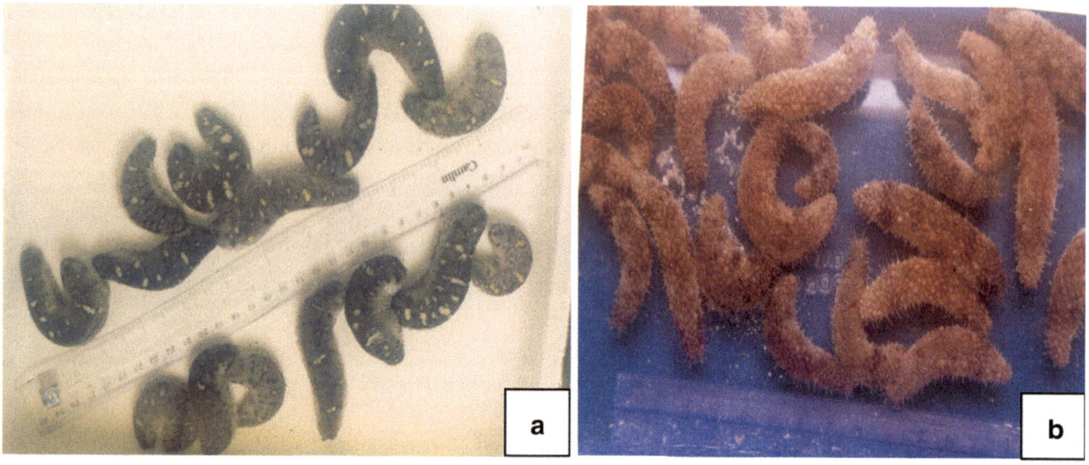

Fig. 4 One-hundred-and-twenty-day-old juveniles of sea cucumbers: (**a**) *H. scabra* and (**b**) *H. spinifera*

juveniles attained an average length of 20 mm, a mixture of *Sargassum* spp. powder and fine sand in a proportion of 1:2 was given at 1 % of the body weight of the juveniles (initially <80; <200 μm as the days progressed) (Asha and Muthiah 2007). Addition of Algamac at 2 % level was provided along with the above feed. Fifty percent water exchange was given daily and the juveniles were taken out using a brush, and the length and number of live juveniles were noted for assessing the growth and survival rate. By maintaining appropriate stocking densities, periodic transferring to new tanks along with size-wise segregation and proper feeding, the growth rate of the juveniles can be improved.

Sea Ranching and Culture

The average size recommended for the release of sea cucumber juveniles to a suitable habitat to replenish the wild population is 20–30 mm. The juveniles of *H. scabra* and *H. spinifera* (Figs. 3b and 4b) produced on several occasions in the hatchery were sea ranched around the Gulf of Mannar area. The culture experiments of juvenile *H. scabra* attempted in a prawn farm using concrete rings yielded encouraging results that indicated their efficiency in cleaning the pond bottom by consuming the feed waste (James et al. 1999). It is advised to stock them at a rate of 30,000 per hectare.

Future Prospects

The continued research effort in the hatchery operation has refined the existing hatchery technology for the mass production of the juvenile *H. scabra* and *H. spinifera*. Future research needs to be focused on the study of the present stock structure of commercial sea cucumbers along the Gulf of Mannar, which is essential to assess the effect of sea ranching in the long run. By upgrading the existing larval and juvenile rearing techniques, cost-effective mass production of holothurian juveniles can be carried out in an effective way which will in turn be helpful in evolving strategies for conservation and sustainable fishery and export of sea cucumbers from India and ultimately will improve the foreign exchange and the economic status of poor fishermen communities along the Gulf of Mannar and Palk Bay area of the southeast coast of India.

References

Asha PS (2004) Effect of feed concentration on larval growth, survival and development of Holothuria

(Theelothuria) spinifera Theel. J Mar Biol Assoc India 46:80–86

Asha PS (2005) Reproductive aspects, larval and juvenile rearing of Holothuria (Theelothuria) spinifera Theel. PhD thesis Manonmanium Sundarnar University, Tamil Nadu, India, 12pp

Asha PS, Diwakar K (2013) Effect of stocking density on hatching rate, larval and early juvenile rearing of edible sea cucumber Holothuria scabra Jaeger. Indian J Mar Sci 42(2):191–195

Asha PS, Muthiah P (2002) Spawning and larval rearing of sea cucumber: *Holothurio* (Theelothurio) *spinifera* Theel. SPC Beche de mer Inf Bull 16:11–15

Asha PS, Muthiah P (2005) Effect of temperature, salinity and pH on the larval growth, survival and development of the commercial sea cucumber Holothuria spinifera Theel. Aquaculture 250:823–829

Asha PS, Muthiah P (2006) Effect of single and combined microalgae on the larval growth, survival and development of the commercial sea cucumber Holothuria spinifera Theel. Aquacult Res 37:113–118

Asha PS, Muthiah P (2007) Growth of the hatchery produced juveniles of the sea cucumber Holothuria spinifera Theel. Aquacult Res 38:1082–1088

Asha PS, Rajagopal M, Diwakar K (2011) Influence of salinity on hatching rate, larval and early juvenile rearing of commercial sea cucumber Holothuria scabra Jaeger. J Mar Biol Assoc India 53(2):218–224

Conand C (2004) Present status of world sea cucumber resources and utilization: an international overview. In: Lovatelli A, Conand C, Purcell S, Uthicke S, Hamel JF, Mercier A (eds) Advances in sea cucumber aquaculture and management, Fisheries technical paper no. 463. FAO, Rome, pp 13–24

James PSBR, James DB (1993) Ecology, breeding, seed production and prospects for farming of sea cucumbers from the seas around India. Fish Chimes 13:24–34

James DB, Rajapandian ME, Baskar BK, Gopinathan CP (1988) Successful induced spawning and rearing of the Holothurian *Holothuria* (matriatyla) *scabra* Jaeger at Tuticorin. Mar Fish Inf Serv Tech Ext Ser 87:30–33

James DB, Gandhi AD, Palaniswamy N, Rodrigo JX (1994) Hatchery techniques and culture of the sea-cucumber *Holothuria scabra*, CMFRI special publication 57. Central Marine Fisheries Research Institute, Indian Council of Agricultural Research, Kochi, pp 1–40

James DB, Asha PS, Ram Mohan MK, Jaiganesh P (1999) Culture of sea cucumbers in shrimp farms – a take off in technology. In: National seminar on development and transfer of fisheries technology. Organized by the Fisheries College and Research Institute, Tuticorin, pp 5–7

Sui X, Hu Q, Chen Y (1986) A study on technology for rearing of post larvae and juvenile of sea cucumber *Apostichopus japonicus* in high density tanks. Oceanol Limnol Sin 17:513–520

Culture and Values of Sea Urchin

K. Maheshkumar

Introduction

In the developed countries, there is a general consumer shift from meat towards seafood. Seafood is an excellent source of protein, fatty acids, low cholesterols and vitamins and minerals. It is also reported that the consumption of seafood especially fish and fish oil produces brainy child and prevents humans from coronary heart diseases. Now more than one billion people worldwide rely on fish as an important source of animal protein. It contributes 180 cal per capita per day (FAO 2002). The world population has been increasing quickly than the total fish food supply from production; this decreases in global per capita fish supply from 14.6 kg in 1987 to 13.7 kg in 2000 (FAO 2002). This has been evidenced in India also, i.e. the annual per capita was very low at 8 kg against the world average of 13.7 kg (Sugunan and Sinha 2001). To overcome the situation we should develop suitable mariculture technology for fin fish and other cultivable organisms in open seas to supplement capture fishery production.

Sea urchins are members of a large group of marine invertebrates in the phylum Echinodermata (spiny skinned animals) that also include starfish, sea cucumbers, sea lilies and brittle stars. All sea urchins have a hard calcareous shell called a test, which is covered with a thin epithelium and is usually armed with spines. The spines are used for protection and for trapping drifting algae for food. Between the spines, they have tube feet that are used in food capture, in locomotion and for holding on to the substrate. The sea urchins are found all along our coastline usually on shallow rocky bottom, although some species live in deep water or in sandy or silty substratum. They are herbivores grazing on attached marine plant and drifted algal fragments whose primary food is kelp and may limit algal distribution.

Why Sea Urchin?

Sea urchins were well known to the ancient Greeks and Romans and have been frequently mentioned in their writing as food, together with oysters, snails and other seafood. Even before Aristotle, the echini were well recognized as a food. In recent years, countries like Japan; California, USA; and Spain, Ireland, France and Greece in Europe are having the regular and regulatory fishery for the sea urchins. Among these Japan, Europe and France, are the world's largest consumers and marketers. Sea urchins are harvested for their internal roe. The gonads of both sexes are equally valuable and are referred to as roe or "uni" in Japanese. Uni is known as a delicacy in Japan and sushi bars worldwide. Sea urchin uni is primarily sold fresh rather than

K. Maheshkumar Ph.D. (✉)
Department of Marine Studies and Coastal Resource Management, Madras Christian College, Tambaram, Chennai 600059, Tamil Nadu, India
e-mail: mh.kumar1976@gmail.com

S. Perumal et al. (eds.), *Advances in Marine and Brackishwater Aquaculture*,
DOI 10.1007/978-81-322-2271-2_18, © Springer India 2015

frozen. Uni has a very sweet flavour as it melts in your mouth.

The riped ova of edible species of sea urchins are prized food items in Japan, approximately ¥14,000/kg, making it one of the most valuable sea foods in the world. Frozen roe pieces are available either in bulk bags or plastic trays. A first-quality A-grade roe costs $15/80 g. In the Tokyo market fresh roe of Chinese, North Korean and US origin cost ¥3,600–3,300, ¥1,700–1,300 and ¥3,600–1,000/tray, respectively, for 280–350 g tray.

Sea urchins that are eaten are distributed among a number of orders of regular echinoids as follows (adapted from Hagen 1996).

Class: Echinoidea
- **Subclass:** Perischoechinoidea
- **Subclass:** Cidaroida
- **Subclass:** Euechinoidea
  - **Infraclass:** Echinoturioidea
    - **Order:** Diadematoida
  - **Infraclass:** Acroechinoidea
    - **Cohort:** Diadematacea
      - **Order:** Diadematidae
        - **Family:** Diadematidae
          - Genus: *Centrostephanus*
          - Genus: *Diadema*
    - **Cohort:** Echinacea
      - **Superorder:** Stirodonta
        - **Order:** Phymosomatoida
          - **Family**: Arabaciidae
            - **Genus:** *Diadema*
      - **Superorder**: Camarodonta
        - **Order:** Echinoida
          - **Family:** Echinidae
            - **Genus:** *Echinus*
            - **Genus:** *Loxechinus*
            - **Genus:** *Paracentrotus*
            - **Genus:** *Psammechinus*
          - **Family:** Echinometridae
            - **Genus:** *Anthocidaris*
            - **Genus:** *Colobocentrotus*
            - **Genus:** *Echinometra*
            - **Genus:** *Evechinus*
            - **Genus:** *Heliocidaris*
          - **Family:** Strongylocentrotus
            - **Genus:** *Hemicentrotus*
            - **Genus:** *Strongylocentrotus*
          - **Family:** Toxopneustidae
            - **Genus:** *Lytechinus*
            - **Genus:** *Pseudoboletia*
            - **Genus:** *Pseudocentrotus*
            - **Genus:** *Toxopneustes*
            - **Genus:** *Tripneustes*

The world sea urchin and other echinoderm production were 97,213 t in 2008, whereas the production in Asia was only 25,782 t. The average worldwide landings are still stable but are obviously not sustainable in the near future, since the commercial harvest of sea urchin population for their gonads is steadily increasing worldwide that leads to a decline in their population. Further, the heavy demand for sea urchin gonad has created opportunities for sea urchin culture.

Echinoculture

The term "echinoculture" refers to the cultivation of echinoderms, i.e. to both sea urchin (Echinoidea) and to a lesser extent sea cucumber (Holothuroidea). Nevertheless, sea urchins are more valuable than sea cucumber and their cultivation is more advanced (Hagen 1996). There are two methods of sea urchin culture: the first involves spawning adult brood stock and rearing resultant larvae/juveniles to marketable size and the second involves enhancing the gonads (i.e. increasing yield and/or quality) of wild-caught adults held in captivity by feeding them with natural or prepared diets (Pearse et al. 2004).

Japan was the first country to address the issue of overexploitation and initiated stock enhancement progress very early (Saito 1992; Hagen 1996). These techniques include habitat enhancement (artificial reef), artificial feeding, translocation and building of hatcheries that produce several million of seeds a year that are transplanted to the field. Hatcheries may be a solution to ensure recruitment where harvesting eliminates adults before they spawn, but good natural habitats are required, like large tide pools, to give enough protection to juveniles released in the field (Grosjean 2001).

The ultimate step in the aquaculture production of sea urchin is independence from natural resources, that is, to control the whole life cycle in culture, from spawning to gonad enhancement (Le Gall 1990; Hagen 1996). Somatic growth of juveniles until they reach marketable size is a process that requires major improvements in current technology and is key to the successful development of closed-cycle echinoculture (Grosjean 2001).

Hatchery Technology

Reduced natural recruitment in many sea urchin fishery countries had led to increased interest in hatchery systems that could provide stock for replenishing natural population and outplanting of juvenile to aquaculture lease sites for sea ranching of sea urchins in Japan (Harris et al. 2003). The Japanese have developed an effective and well-documented hatchery system to produce small animals (sea urchins) for outplanting, and the depletion of wild population was being compensated through the large-scale seed stock release programme to get a sustainable yield of sea urchins from the wild.

Larviculture

Larviculture in Japanese hatcheries commences with the mixing of gametes from several animals. Excess perm is rinsed off, and the fertilized eggs hatch after approximately 20 h (Fig. 1). Three to four days later, they have pluteus stage which requires planktonic microalgae as food. Diatom *Chaetoceros gracilis* is commonly used in commercial hatcheries as food for larvae at the rate of

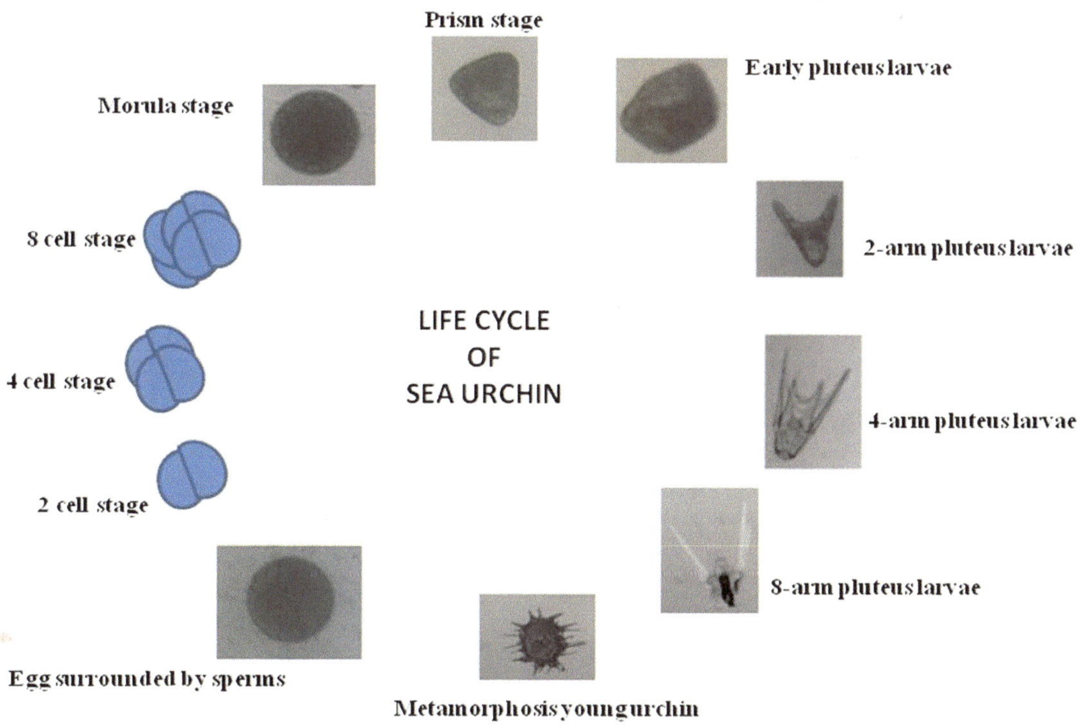

Fig. 1 Life cycle of sea urchin

5,000 cells/ml in which the amount gradually increased to 10 l/tank/day in the final stages of cultivation. The larvae are cultivated in 1,000 l tanks with continuous flow of 1 um filtered seawater. Circulation in the larvae tanks is provided by two large airstones with a gentle flow, one on the bottom and another near the surface. The density is initially 1.5 larvae/ml but is decreased to 0.8 larvae/ml at the time of settlement. The metamorphosed juveniles are approximately 0.3 mm in size.

Early Juvenile Rearing

Larval settlement in sea urchin is induced by introducing placing small rocks taken from the low intertidal area where the parents were collected or transparent polycarbonate plates can be used should be inoculated with the desired algae. Feeding with the soft seaweeds commences when the juveniles reaches 3–4 mm size.

Nursery Culture

The settled juveniles were transferred to nursery tanks with open mesh cages or hanging cages which suspended 1–2 m below the surface. The juveniles were fed with preferred seaweeds. Juveniles can be released to the environment 6 months after fertilization when they are 7–10 mm. The large juveniles are ready to be recaptured 2 years after release when they reach a diameter of more than 40 mm.

Grow-Out System

Closed-Cycle Cultivation

Closed-cycle cultivation requires grow-out facilities. These can be constructed by expanding existing nursery techniques, by adapting technology developed for the intensive cultivation of abalone. The French adopted the last alternative and developed a prototype of multilayered grow-out tanks which consist of four stacked, sloping shelves. Water is pumped up to the top shelf from a reservoir tank under the shelves and then runs down through the stack of shelves in a zigzag pattern. The accumulation of sea urchin faeces in the reservoir tank is siphoned off at regular intervals. The recirculated water is gradually replenished by marine groundwater. The commercial scale grow-out facility would require a stable food supply. Closed-cycle cultivation is capital intensive and has high operational costs but requires only a modest investment in R&D. Full-scale hatchery and nursery technology is well established in Japan (Hagen 1996).

Values of Sea Urchins

Ornamental

Sea urchins were harvested mainly for their gonads and their tests for ornaments all over the world. They are used as decorations for bordering mirrors instead of molluskan shells and used as ash trays and lamp shades, and the spines are used as writing tool for small kids in slates (Fig. 2).

Fig. 2 Ornaments made from sea urchin shell

Nutritional and Food Value

Sea urchin roe is rich in nutrients such as protein, carbohydrate, saturated and polyunsaturated fatty acids and vitamins (Table 1). The nutrient content of sea urchin roe is equivalent to a tuna fish and five eggs.

Uni is known as a delicacy in Japan and sushi bars worldwide. Sea urchin uni is primarily sold fresh rather than frozen. Uni has a very sweet flavour as it melts in your mouth. Whole sea urchins and sea urchin roe are available in the market. Individual roe pieces are available as packed on wooden or plastic trays (Fig. 3).

Table 1 Nutritional composition of sea urchin gonad

Protein	36.14 %
Carbohydrate	4.86 %
Lipid	26.35 %
Saturated FA	46.72 %
MUFA	13.85 %
PUFA	39.41 %
Sodium	35.00 mg/100 g
Zinc	1.042 g/100 g
Vit. B1	24.86 mg/100 g
Vit. B12	6.24 mg/100 g
Vit B6	5.96 mg/100 g
Folic acid	96.9 mg/100 g

Fig. 3 Live sea urchin and packed uni in plastic tray

Ecological Value

Sea urchin plays a major role in the seaweed ecosystem and coral reef ecosystem. It checks the over growth of seaweeds in both the ecosystems. Especially in coral reef ecosystem, it controls the growth of seaweed which competes for the sunlight. Also, it creates new spaces for the coral recruitment. Sea urchin serves as one of the major food item for the lobster, which is key stone species in the seaweed ecosystem. It checks the sea urchin population overgrowth in the ecosystem in order to maintain a balance between seaweed growth and its grazer. So sea urchin is one of the major components in determining the health of the ecosystem in which it lives.

References

FAO (2002) The state of world fisheries and aquaculture. Food and Agriculture Organization, Rome, pp 30–33

Grosjean P (2001) Growth model of the reared sea urchin *Paracentrotus lividus* (Lamarck, 1816). Ph.D. thesis, Universite Libre de Bruxelles, 271 pp

Hagen NT (1996) Echinoculture: form fishery enhancement to closed cycle cultivation. World Aquacult 27:7–19

Harris LG, Madigan P, Waters K (2003) A hatchery system for green sea urchin aquaculture in the Gulf of Maine. World Aquac 34:32–71

Le Gall P (1990) Culture of echinoderms. In: Barnabe G (ed) Aquaculture, vol 1. Ellis Horwood, New York, pp 311–324

Pearse CM, Daggett TL, Robinson SMC (2004) Effect of urchin size and diet on gonad yield and quality in the green sea urchin (*Strongylocentrotus droebachiensis*). Aquaculture 233:337–367

Saito K (1992) Japan's sea urchin enhancement experience. In: Dewees CM (ed) The management and enhancement of sea urchins and other kelp bed resources: a Pacific rim perspective, Publication no. T-CSGCP-028. California Sea Grant College, University of California, La Jolla, pp 1–38

Sugunan VV, Sinha M (2001) Sustainable capture fisheries-based fisheries in freshwater. In: Pandian TJ (ed) Proceedings: seminar on "sustainable fisheries for nutritional security". National Academy of Agricultural Sciences, New Delhi, pp 43–70

Methods of Cultivation and Economic Values of Seaweeds

N. Kaliaperumal

Introduction

Seaweeds or marine algae occur in marine, estuarine and backwater habitats. They are valuable renewable marine living resource. They belong to the primitive group of nonflowering plants and are mostly distributed in the intertidal, shallow and deep waters of the sea up to 150 m depth. They grow on rocks, dead coral stones, pebbles, solid substrates and other plants as epiphytes. They are unique with no distinguishable root, stem and leaves. Based on the type of pigments and morphological and anatomical characters, seaweeds are broadly classified into green, brown, red and blue-green algae. Seaweeds are one of the commercially important marine living renewable resources. They are used as human food, animal feed and fertiliser for several land crops and as medicine. The distribution and resources of seaweeds in India, candidate species and suitable methods of culture are presented. The various economic values of seaweeds are also given in this chapter.

Seaweed Distribution and Resources

About 6,000 species of red seaweeds (Rhodophyceae), 2,000 species of brown seaweeds (Phaeophyceae) and 1,200 species of green seaweeds (Chlorophyceae) occur globally of which approximately 220 species are economically important as food and phycocolloids. In Indian waters, 250 genera and 896 species of marine algae (51 genera and 228 species of Chlorophyceae, 47 genera and 210 species of Phaeophyceae, 151 genera and 455 species of Rhodophyceae and 1 genus and 3 species of Xanthophyceae) have been reported (Umamaheswara Rao 2011). Several species of green, brown, red and blue-green algae with luxuriant growth occur along the south coast of Tamil Nadu from Rameswaram to Kanyakumari covering 21 islands in the Gulf of Mannar. In Gujarat coast, seaweeds occur abundantly at Okha, Dwarka, Porbandar, Veraval, Diu and Gopnath. Rich seaweed beds are present in Lakshadweep, in Andaman and Nicobar Islands and also in and around Mumbai, Ratnagiri, Goa, Karwar, Varkala, Vizhinjam, Visakhapatnam and coastal lakes such as Pulicat and Chilka (Chennubhotla et al. 1987a, 1991).

The world production of seaweeds is estimated more than 15.8 million tonnes, and India is contributing only 3,01,646 t (Chennubhotla et al. 2013a). It has been estimated from the seaweed resource surveys conducted by the Central Marine Fisheries Research Institute, Central Salt & Marine Chemicals Research Institute, National Institute of Oceanography and other government organisations either jointly or separately in the east and west coast, Lakshadweep and Andaman

N. Kaliaperumal (✉)
Regional Centre of Central Marine Fisheries Research Institute, Ramanathapuram District, 623520 Mandapam Camp, Tamil Nadu, India

S. Perumal et al. (eds.), *Advances in Marine and Brackishwater Aquaculture*,
DOI 10.1007/978-81-322-2271-2_19, © Springer India 2015

and Nicobar Islands that the standing crop of seaweeds in the intertidal and shallow waters is 4.33 lakh tons (wet wt.). Of these, the standing crop of Tamil Nadu is 1,01,381 t wet (Subbaramaiah 2010). The Central Marine Fisheries Research Institute and Central Salt & Marine Chemicals Research Institute jointly carried out detailed seaweed resource survey in deep waters of the Tamil Nadu coast from Dhanushkodi to Kanyakumari at the depths ranging from 5 to 22 m during 1986–1991. The total standing crop from 1,863 km^2 sampled area was 75,373 t (wet wt.) consisting of 2,750 t of *Sargassum* spp., 962 t of *Gracilaria* spp., 5,262 t of *Hypnea* spp. and 66,399 t of other seaweeds (Kaliaperumal et al. 1987, 1998; Kaliaperumal 2002).

Now in India, seaweeds collected from wild are used as raw materials for the production of agar, alginate and liquid seaweed fertiliser. The commercially cultivated red seaweed *Kappaphycus alvarezii* is used for manufacture of carrageenan. More than 25 seaweed industries are functioning at different places in the maritime states of Tamil Nadu, Kerala and Karnataka. The red algae *Gelidiella acerosa*, *Gracilaria edulis*, *G. verrucosa* and other species of *Gracilaria* are used for agar production and brown algae *Sargassum* spp., *Turbinaria* spp. and *Cystoseira trinodis* for alginates. All these seaweeds are harvested since 1966 only from the natural seaweed beds occurring in the south Tamil Nadu coast from Rameswaram to Kanyakumari. Data collected by the CMFRI on seaweed landing of Tamil Nadu for 28 years from 1978 to 2005 show that the quantity of agarophytes landed ranged from 248 to 1,518 t (dry wt.) and alginophytes from 529 to 5,537 t (dry wt.) depending on the availability of seaweeds in the natural beds and raw material requirements from the seaweed industries. As the demand for seaweeds, particularly the agarophytes *Gelidiella acerosa* and *Gracilaria edulis*, is more and the natural resources are less, seaweeds are overexploited causing depletion in the biomass of seaweeds in the natural seaweed beds (Kaliaperumal and Kalimuthu 1993; Kaliaperumal 2006a).

Seaweed Cultivation

Seaweeds are cultivated globally for supply of raw material to the seaweed industries and for their use as human food. In India, seaweed cultivation is necessary to meet the required quantities of seaweeds for the production of phytochemicals (Krishnamurthy 1967). The Central Marine Fisheries Research Institute, Cochin, and Central Salt & Marine Chemicals Research Institute, Bhavnagar, have developed viable technologies for commercial cultivation of different agars, carrageenans and algins yielding seaweeds with a view to augment continuous supply of raw materials to the Indian seaweed industries and to conserve the natural seaweed vegetation for sustainable exploitation (Kaliaperumal 2006b). The coral stone method for *Gelidiella acerosa* (Subba Rao and Ganesan 2006; Subbaramaiah 2008), coir rope net and longline methods for *Gracilaria edulis* (Umamaheswara Rao 1974; Chennubhotla and Kaliaperumal 1983; Kaliaperumal et al. 1994, 2003; Kaliaperumal 2006c), *Hypnea musciformis* (Rama Rao and Subbaramaiah 1986; Rama Rao and Ganesan 2006) and *Acanthophora spicifera* (Kaliaperumal et al. 1986) and bamboo raft and nylon net bag methods for *Kappaphycus alvarezii* (Eswaran et al. 2002, 2006; Kaliaperumal et al. 2008) are found suitable techniques for successful large-scale cultivation by vegetative propagation method in the near-shore areas of the sea. The culture technologies attempted, places of cultivation and yields obtained for the above candidate seaweed species are briefly given below.

Gelidiella acerosa Cultivation

The cultivation of *Gelidiella acerosa* was started first in 1971 by CSMCRI in the lagoon of Kurusadai Island by longline rope method. Later, it was expanded as a sponsored project of the agar industry Cellulose Products of India Ltd., Madurai, for several years from 1973. The location was shifted to the open shore at

Ervadi, and the bottom culture technique on coral stone substratum was attempted. During this period, the cultivation technique on a pilot-farm scale was established successfully. Subsequently, appropriate seed material and also its pretreatment with a suitable growth regulator were used. Bottom culture during 1980–1982 over an area of 0.25 ha gave a higher crop yield. Harvest and planting were scheduled at 6 monthly intervals during January and June. A crop production of 0.65 kg (dry)/m^2/year was achieved, and based on these findings, the CSMCRI brought out a farm technology package with an annual harvest yield of 4.5 kg (dry)/ha. During 2000–2001 when this method was evaluated on a 0.5 ha area, lesser crop yield of 4 t (dry)/ha in two harvests was attained (Subba Rao et al. 2004).

During 2004–2006, the CSMCRI has made further improvement on the cultivation of *G. acerosa* in the open shore at Ervadi. Four methods involving different substrates were evaluated. They were floating raft, bottom net, cement block and hollow-cylinder-cement blocks. The floating raft method was found superior based on 1.31 % daily growth rate (DGR) and up to 1,288 g (wet)/m^2 biomass yield. The yield was optimum with the use of a seed material from Sethu Karai, planting density of 500 g (wet)/m^2 and 180-day growth period. The floating raft method being used for *Kappaphycus alvarezii* cultivation in India is found best for *G. acerosa* cultivation (Ganesan et al. 2009). Subsequently, an improved technique known as suspended stone (SS) method was attempted for *G. acerosa* propagation over a 50 m^2 area. In this method, small calcareous stones (15–70 cm^2 surface area and 100–200 g weight) were used as the substrate, and the seed material (2 g wet/plant) was fastened to the stones and hung below the ropes (3 mm dia) on the bamboo frames fixed to the seabed. It differed from the floating raft method as the plants lie 5 cm below the rope. The biomass and growth rate obtained in the SS method were higher with biomass ranging from 528 to 3,645 g (wet)/m^2 and the DGR 2.62 %. Based on these results, the method was recommended for large-scale *G. acerosa* cultivation (Ganesan et al. 2011).

Gelidiella acerosa was cultured by the CMFRI by using small fragments of the plants along with their substratum (coral piece) to coir ropes interwoven on G. I. pipe frames. The frames were tied in submerged condition to poles fixed in the inshore waters. The plants reached the harvestable size after 75 days, and the yield was increased onefold over the seed material used. The fragments of the seaweed were also tied at the mesh intersections of HDPE rope nets and introduced at 4 m depth water in floating condition using plastic buoys and anchors. Twofold increases in yield were obtained after 60-day growth. The fragments were also tied to nylon twines at regular intervals. The seeded twines were wound round on the nails fixed to coral stones, kept in iron cages and cultured at 2 m depth of the sea. The plants reached the harvestable size after 5 months with more than a threefold increase in yield (Chennubhotla et al. 1977).

Gracilaria edulis Cultivation

Gracilaria edulis culture was initiated by the CMFRI using the net method in an open shore environment in the Gulf of Mannar (Umamaheswara Rao 1974). The seed material (4 cm apical fragments) was inserted into the twists of coir rope, and the nets (4 m× 2 m size) were kept at a subtidal level. Harvest after 80 days yielded 4.4 kg (wet)/m^2. Later, it was extended to different localities in the inshore waters of the Gulf of Mannar, Palk Bay and Minicoy (Lakshadweep Islands), and pilot scale cultivation was established at many locations. With a view to find out the feasibility of Lakshadweep lagoons for the cultivation of *G. edulis*, experimental culture was carried out at four sites in the Minicoy lagoon (Kaliaperumal et al. 1992; Chennubhotla et al. 1992a, b), and encouraging results were obtained with about 15-fold increase in yield. Kaladharan et al. (1996) conducted experiments in the Minicoy lagoon during 1990–1992 on the dry matter accumulation, effect of planting density, depth of water above culture nets, periodic cleaning for the removal of epiphytes, grazing, yield and regeneration of *G. edulis* after successive harvests to understand the optimum physical parameters for the successful cultivation of *G. edulis*.

Field cultivation of *Gracilaria edulis* was carried out during 1986–1989 in nearshore areas of the Gulf of Mannar and Palk Bay to determine the various environmental factors which affect the growth of the seaweed (Kaliaperumal et al. 1993). The epiphytes, epifauna, low light density and sedimentation caused by turbulence of water and grazing by fishes were found to be the primary factors hampering the growth of cultured *G. edulis*. The suitable period for field cultivation of this species in the Gulf of Mannar is from December to March. While cultivating *G. edulis* in the nearshore sea at Vadakadu (Rameswaram), the fisherfolk participated were imparted training on the culture technique (Kaliaperumal et al. 2003). The net culture method of cultivation technology developed by the CMFRI was transferred to the fisherfolk under the Lab-to-Land programme of the institute during 1978–1981 (Chennubhotla and Kaliaperumal 1983) and under the Department of Biotechnology sponsored project during 2000–2002 (Kaliaperumal et al. 2003). The CMFRI has also conducted yearly a short-term training course on seaweed culture, processing and utilisation to the fishermen, to the state and central government officials and to private entrepreneurs (Kaliaperumal 2006c).

Gracilaria edulis was cultivated by CSMCRI during 1967 in a sandy lagoon at Krusadai Island near Mandapam. The method consisted of vegetative propagation of fragments by the longline rope method. Fragments of 2.5–3.0 cm were inserted in the twists of ropes. The seeded ropes were tied to bamboo poles planted to the sea bottom and adjusted at a level of 1 ft above the bottom. Three harvests were made at the end of 5, 8 and 10 months after planting, and the total harvest in 1 year was about 3.5 kg from 1 m length of rope (Raju and Thomas 1971). The economic feasibility of its large-scale cultivation was formulated, and the technology was popularised (Krishnamurthy et al. 1975). *G. edulis* was successfully cultivated in the open shore environment at Ervadi by the single rope floating raft (SRFT) technique with an annual yield of 4 kg (wet)/m (Subbaramaiah and Thomas 1990). In the later (1997) cultivation, by using a selected strain as seed material, a high crop yield of 5 kg/m was obtained in four harvests. The cultured plants were 30 cm in length with an average weight of 300 g (wet) giving 30 % agar yield and gel strength of 200–300 g/cm^2 on a laboratory scale (Subba Rao et al. 2004).

Hypnea musciformis Cultivation

Species of *Hypnea* constitute a potential source of carrageenan the world over. *H. musciformis* was cultivated by the CSMCRI in the lagoon of Krusadai Island, the Gulf of Mannar and Tamil Nadu using longline ropes. A yield of 6.34 kg fresh weight from 150 m length of rope in 25 days was obtained (Rama Rao and Subbaramaiah 1986). During June 2001 to May 2002, the average biomass per hectare per month was 1.40 t fresh, and during June 2002 to February 2003, the average biomass per hectare per month was 2.08 t fresh (Rama Rao and Ganesan 2006).

Acanthophora spicifera Cultivation

This edible and agaroid-yielding red alga was cultivated by the CMFRI in the nearshore area of Hare Island near Mandapam at 1 m depth following the vegetative propagation method. Vegetative fragments tied with polypropylene straw were fastened to nylon fishing lines. The plants reached the harvestable size in 25 days, and the yield was increased 2.6-fold over the quantity of seed material introduced (Kaliaperumal et al. 1986). This seaweed was also cultured successfully in the CMFRI fish farm ponds at Mandapam. The pond was connected to the sea through a feeder canal, and hence there was regular inflow and outflow of sea water depending on the high tide and low tide, respectively. The fragments of the plants were tied at the mesh intersections of HDPE rope nets with nylon twines, and the seeded nets were tied in submerged condition to the palmyra poles erected in the pond. The seedlings grew to harvestable size after 45 days yielding a 3.6-fold increase over the quantity of seed material introduced. The remnants of the plants were allowed to grow for another 1 month, and the second harvest yielded more than twofold crop (Chennubhotla et al. 1987c; Kaliaperumal 1993).

Kappaphycus alvarezii Cultivation

In India, initial attempts were made successfully to cultivate the kappa-carrageenan-yielding red alga *Kappaphycus alvarezii* in the Saurashtra region (Mairh et al. 1995). Acclimatisation and commercial cultivation of this alga were achieved during 1995–1997 at the Mandapam coast, southeast India (Eswaran et al. 2002, 2006). Plants were cultivated by vegetative propagation of fragments in perforated polythene bags, net bags, rafts and open culture by the monoline rope method. Raft and net bag culture methods were proved safe and suitable for large-scale cultivation, as they prevented loss by grazing and drifting, produced plants from epiphytes and allowed easy harvests. It showed a daily growth rate between 0.4 and 11.4 % with a mean of 4.0 %, carrageenan yield between 32 and 54.7 % and gel strength of 210–795 g/cm^2. Pilot scale cultivation of this alga was carried out by the CMFRI during June 2004–February 2006 in the nearshore area of the Gulf of Mannar and Palk Bay using bamboo rafts, nylon net cages, perforated polythene bags, HDP longline ropes and pens (Kaliaperumal et al. 2008). The growth and production varied in these methods, and the raft culture method was found more suitable than all other methods, as the plants grown on rafts were robust and thicker with several branches. All the physicochemical factors remained favourable throughout the year for the good growth of this seaweed. The current status of *K. alvarezii* farming and the constraints are given by Johnson and Gopakumar (2011). The empowerment of coastal communities in the cultivation and processing of *Kappaphycus alvarezii* at Vizhinjam village, Kerala, was given by Bindu (2011).

Economic Values of Seaweeds

Seaweeds contain different vitamins, minerals, trace elements, proteins, iodines and bioactive substances. They are the only source for the production of phytochemicals such as agar, carrageenan and alginate. Agar is extracted from red algae like *Gelidiella*, *Gracilaria*, *Gelidium* and *Pterocladia*. Some other red algae, viz. *Kappaphycus*, *Eucheuma*, *Chondrus*, *Hypnea* and *Gigartina*, are used for the production of carrageenan. Algin is manufactured from brown algae like *Sargassum*, *Turbinaria*, *Laminaria*, *Undaria*, *Macrocystis* and *Ascophyllum*. These phycocolloids are used as gelling, stabilising and thickening agents in food, pharmaceutical, confectionary, dairy, textile, paper, paint, varnish industries, etc. Other chemical products such as mannitol, iodine, laminarin and fucoidin are also obtained from marine algae. Many protein-rich seaweeds are used as human food in Southeast Asian countries in the form of soup, salad, curry, etc. Jelly, jam, chocolate, pickle and wafer can also be prepared from seaweeds. Meal can be prepared from seaweeds and used as food for poultry, cattle and other farm animals. Seaweeds are also used as fertiliser for various land crops and as medicine (Chennubhotla et al. 1981, 1987b, 2013b; Silas et al. 1986; Kaliaperumal et al. 1995; Kaladharan et al. 1998; Kaliaperumal 2003).

Conclusion

Now, commercial-scale cultivation of carrageenan-yielding red seaweed *Kappaphycus alvarezii* using the bamboo raft method is going in various places in the coastal districts of Ramanathapuram, Tuticorin, Pudukkottai and

Thanjavur in Tamil Nadu State and in the coastal areas of different districts in Gujarat State. In addition to this species, large-scale cultivation of other economically important seaweeds has to be undertaken by the fisherfolk, seaweed utilizers and private entrepreneurs in the bays and creeks present in the open shore along the east and west coast, lagoons or coral reefs in the southeast coast of Tamil Nadu, Andaman and Nicobar Islands and atolls of Lakshadweep, which are suitable localities for seaweed cultivation, following the technologies already developed and by availing financial assistance from banks and other funding agencies connected with rural development programmes. Seaweed cultivation on large scale will not only augment uninterrupted and continuous supply of raw materials to the Indian seaweed industries, but it will also provide employment to the people living in the coastal areas of mainland, Lakshadweep and Andaman and Nicobar Islands. This will help in improving their economic status and rural development (Periasamy et al. 2013).

Acknowledgement The author is grateful to Dr. P. Santhanam, Assistant Professor, Department of Marine Science, Bharathidasan University, Tiruchirappalli, Tamil Nadu, for his keen interest and encouragement in the preparation of the this paper.

References

Bindu MS (2011) Empowerment of coastal communities in cultivation and processing of *Kappaphycus alvarezii* – a case study at Vizhinjam Village, Kerala, India. J Appl Phycol 23:157–163

Chennubhotla VSK, Kaliaperumal N (1983) Proven Technology. 7. Technology of cultured seaweed production. Mar Fish Inform Serv Tech Ext Ser 54:19–20

Chennubhotla VSK, Kalimuthu S, Najmuddin M, Selvaraj M (1977) Field culture of *Gelidiella acerosa* in the inshore waters of Gulf of Mannar. Suppl J Phycol 13:454

Chennubhotla VSK, Kaliaperumal N, Kalimuthu S (1981) Seaweed recipes and other practical uses of seaweeds. Seaf Export J 13:9–16

Chennubhotla VSK, Kaliaperumal N, Kalimuthu S (1987a) Economically important seaweeds. CMFRI Bull 41:3–19

Chennubhotla VSK, Kaliaperumal N, Kalimuthu S (1987b) Common seaweed products. CMFRI Bull 41:26–30

Chennubhotla VSK, Kaliaperumal N, Kalimuthu S, Ramalingam JR, Selvaraj M, Najmuddin M (1987c) Seaweed culture. CMFRI Bull 41:60–74

Chennubhotla VSK, Kaliaperumal N, Kalimuthu S, Ramalingam JR (1991) Commercially important seaweeds of India, their occurrence, chemical products and uses. Mar Fish Inform Serv Tech Ext Ser 107:11–16

Chennubhotla VSK, Kunhikoya KK, Anusukoya N (1992a) Record production of *Gracilaria edulis* in culture system in Minicoy Lagoon. Mar Fish Inform Serv 118:12–13

Chennubhotla VSK, Kaladharan P, Kaliaperumal N, Rajagopalan MS (1992b) Seasonal variation in production of cultured seaweed *Gracilaria edulis* (Gmelin) Silva in Minicoy lagoon (Lakshadweep). Seaweed Res Util 14(2):109–113

Chennubhotla VSK, Umamaheswara Rao M, Rao KS (2013a) Exploitation of marine algae in Indo-Pacific region. Seaweed Res Util 35(1 & 2):1–7

Chennubhotla VSK, Umamaheswara Rao M, Rao KS (2013b) Commercial importance of marine macro algae. Seaweed Res Util 35(1 & 2):118–128

Eswaran K, Ghosh PK, Mairh OP (2002) Experimental field cultivation of *Kappaphycus alvarezii* (Doty) Doty ex. P. Silva at Mandapam region. Seaweed Res Util 24(1):67–72

Eswaran K, Ghosh PK, Subba Rao PV, Mairh OP (2006) Field cultivation of *Kappaphycus alvarezii* in India. In: Tewari A (ed) Recent advances on applied aspects of Indian marine algae with reference to global scenario, vol 1. CSMCRI, Bhavnagar, pp 250–261

Ganesan M, Thiruppathi S, Eswaranand K, Jha B (2009) Cultivation of *Gelidiella acerosa* in open sea on the south eastern coast of India. Mar Ecol Prog Ser 382:49–57

Ganesan M, Thiruppathi S, Eswaran K, Reddy CRK, Jha B (2011) Development of an improved method of cultivation to obtain high biomass of the red alga *Gelidiella acerosa* (Gelidiales, Rhodophyta) in the open sea. Biomass Bioenerg 35:2729–2736

Johnson B, Gopakumar G (2011) Farming of the seaweed *Kappaphycus alvarezii* in Tamil Nadu coast – status and constraints. Mar Fish Inform Serv Tech Ext Ser 208:1–5

Kaladharan P, Vijayakumaran K, Chennubhotala VSK (1996) Optimization of certain physical parameters for the mariculture of *Gracilaria edulis* (Gmelin) Silva in Minicoy lagoon (Laccadive Archipelago). Aquaculture 139:265–270

Kaladharan P, Kaliaperumal N, Ramalingam JR (1998) Seaweeds-products, processing and utilisation. Mar Fish Inform Serv Tech Ext Ser 157:1–9

Kaliaperumal N (1993) Seaweed culture. In: Handbook on aquafarming seaweed, sea urchin and sea cucumber. MPEDA, Cochin, pp 9–22

Kaliaperumal N (2002) Seaweed distribution and resources in India. In: Anand N (ed) Algological research in India. Bishen Singh Mahendra Paul Singh, Dehradun, pp 177–190

Kaliaperumal N (2003) Products from seaweeds proceedings of the national seminar on marine biodiversity as a source of food and medicine. SDMRI Res Publ 3:33–42

Kaliaperumal N (2006a) Seaweed landings in India. In: Tewari A (ed) Recent advances on applied aspects of Indian marine algae with reference to global scenario, vol 1. CSMCRI, Bhavnagar, pp 174–185

Kaliaperumal N (2006b) Seaweed cultivation. In: Handbook of fisheries and aquaculture. Directorate of Information and Publications of Agriculture. Indian Council of Agricultural Research, New Delhi, pp 430–437

Kaliaperumal N (2006c) Culture of *Gracilaria edulis* and other seaweeds in India. In: Tewari A (ed) Recent advances on applied aspects of Indian marine algae with reference to global scenario, vol 1. CSMCRI, Bhavnagar, pp 94–108

Kaliaperumal N, Kalimuthu S (1993) Need for conservation of economically important seaweeds of Tamil Nadu and time-table for their commercial exploitation. Mar Fish Inform Serv Tech Ext Ser 119:5–12

Kaliaperumal N, Kalimuthu S, Ramalingam JR, Selvaraj M (1986) Experimental field cultivation of *Acanthophora spicifera* (Vahl) Boergesen in the near shore area of Gulf of Mannar. Indian J Fish 33(4):476–478

Kaliaperumal N, Chennubhotla VSK, Kalimuthu S (1987) Seaweed resources of India. CMFRI Bull 41:51–54

Kaliaperumal N, Rajagopalan MS, Chennubhotla VSK (1992) Field cultivation of *Gracilaria edulis* (Gmelin) Silva in Minicoy lagoon (Lakshadweep). Seaweed Res Utli 14(2):103–107

Kaliaperumal N, Chennubhotla VSK, Kalimuthu S, Ramalingam JR, Muniyandi K (1993) Growth of *Gracilaria edulis* in relation to environmental factors in field cultivation. Seaweed Res Util 16(1 & 2):167–176

Kaliaperumal N, Kalimuthu S, Muniyandi K (1994) Experimental cultivation of *Gracilaria edulis* at Valinokkam Bay. In: Paulraj S (ed) Proceedings of the national symposium on aquaculture for 2000 AD. Madurai Kamaraj University, Madurai, pp 221–226

Kaliaperumal N, Kalimuthu S, Ramalingam JR (1995) Economically important seaweeds. CMFRI Spec Publ 62:1–36

Kaliaperumal N, Chennubhotla VSK, Kalimuthu S, Ramalingam JR, Krishna Pillai S, Muniyandi K, Rama Rao K, Thomas PC, Zaidi SH, Subbaramaiah K (1998) Seaweed resources and distribution in deep waters from Dhanushkodi to Kanyakumari, Tamil Nadu. Seaweed Res Util 20(1&2):141–151

Kaliaperumal N, Kalimuthu S, Ramalingam JR (2003) Pilot scale field cultivation of the agarophyte *Gracilaria edulis* (Gmelin) Silva at Vadakadu (Rameswaram). Seaweed Res Util 25(1&2):213–219

Kaliaperumal N, Ramalingam JR, Sulochana B (2008) Growth and production of carrageenophyte *Kappaphycus alvarezii* in pilot scale cultivation in Gulf of Mannar and Palk Bay. Seaweed Res Util 30(1&2):57–66

Krishnamurthy V (1967) Marine algal cultivation-necessity, principles and problems. In: Krishnamurthy V (ed) Proceedings of the seminar on sea, salt and plants. CSMCRI, Bhavnagar, pp 327–333

Krishnamurthy V, Raju PV, Thomas PC (1975) On augmenting seaweed resources of India. J Mar Biol Assoc India 17(2):181–185

Mairh OP, Zodape ST, Tewari A, Rajyaguru MR (1995) Culture of marine red alga *Kappaphycus striatum* (Schmitz) Doty in the Saurashtra region, west coast of India. Indian J Mar Sci 24:24–31

Periasamy C, Anantharaman P, Balasubramanian T (2013) Social upliftment of coastal fisher women through seaweed (*Kappaphycus alvarezii* (Doty) Doty) farming in Tamil Nadu, India. J Appl Phycol 26:775–781

Raju PV, Thoma PC (1971) Experimental field cultivation of *Gracilaria edulis* (Gmel.) Silva. Bot Mar 14(2):71–75

Rama Rao K, Ganesan M (2006) Prospects of mariculture of *Hypnea* species in India. In: Tewari A (ed) Recent advances on applied aspects of Indian marine algae with reference to global scenario, vol 1. CSMCRI, Bhavnagar, pp 288–295

Rama Rao K, Subbaramaiah K (1986) A technique for the field cultivation of *Hypnea musciformis* (Wulf.) Lam. In: Proceedings symposium on coastal aquaculture. MBAI, Cochin, pp 1190–1192

Silas EG, Chennubhotla VSK, Kaliaperumal N (1986) Seaweed resources, products and utilization. Seaweed Res Util 9(1&2):11–24

Subba Rao PV, Ganesan M (2006) Field cultivation of *Gelidiella acerosa* in India. In: Tewari A (ed) Recent advances on applied aspects of Indian marine algae with reference to global scenario, vol 1. CSMCRI, Bhavnagar, pp 296–303

Subba Rao PV, Eswaranand K, Ganesan M (2004) Cultivation of agarophytes in India: retrospect, prospects and future strategies. Seaweed Res Utli 26(Special Issue):181–184

Subbaramaiah K (2008) Whither *Gelidiella acerosa* production – by cultivation and by natural propagation. Seaweed Res Util 30(Special Issue):1–11

Subbaramaiah K (2010) Marine algal distribution and resources in India. In: Souvenir of national symposium on marine plants. Annamalai University, Parangipettai, pp 15–18

Subbaramaiah K, Thomas PC (1990) Raft cultivation of *Gracilaria edulis* (Gmel.) Silva. Proc Indian Acad Sci (Plant Sci) 100:123–127

Umamaheswara Rao M (1974) On the cultivation of *Gracilaria edulis* in the near shore areas around Mandapam. Curr Sci 43(20):660–664

Umamaheswara Rao M (2011) Diversity and commercial feasibility of marine macroalgae of India. Seaweed Res Util 33(1&2):1–13

Water Quality Management in Fish Hatchery and Grow-Out Systems

R. Saraswathy, M. Muralidhar, J.K. Sundaray, N. Lalitha, and P. Kumararaja

Introduction

Aquaculture in India with its immense potential in marine, coastal and brackishwater resources is expected to contribute significantly in protein-rich fish food, rural employment and utilisation of water resources and wastelands. Brackishwater aquaculture, which made quantum jump in the 1980s, had to face a setback in the second half of 1990 due to the uncontrolled disease outbreak. Diversification of species has become unavoidable to sustain the aquaculture. The logic for introducing different alternative species in aquaculture system is that the intensity of the pathogens dependent on a particular host will be reduced and the consequent problems will get reduced. In this context, for India, many finfish like Asian sea bass, grouper, milkfish, cobia, pearlspot, pompano, etc., are considered as suitable alternative fish species farming in all the culture ecosystems. To have adequate amount of quality seed, there is a need to set up fish seed production hatchery with suitable technology. As it is evident the good water quality will ensure the quality seed production, we need to have good water management in fish seed production cycle and in fish grow-out system (Bisson et al. 1992; Qin et al. 1995). This chapter will bring out the best way to manage water quality in a fish hatchery and grow-out system.

R. Saraswathy (✉) • M. Muralidhar • J.K. Sundaray
N. Lalitha • P. Kumararaja
Central Institute of Brackishwater Aquaculture, No. 75, Santhome High Road, R.A. Puram, Chennai 600028, Tamil Nadu, India
e-mail: saraswati@ciba.res.in

Hatchery

A fish hatchery is a place for artificial breeding, hatching and rearing through the early life stages of animals. It consists of different units like egg collection, incubation, larval rearing for culture and nursery rearing facilities. The supporting facilities for live feed algae, rotifer and brine shrimp, nauplii production and enrichment facilities are to be in built into the system.

The efficient operation of a fish hatchery depends on a number of factors such as suitable site selection, soil characteristics, water quality, adequate facility design, water supply structures and water sources. Among these, water quality determines to a great extent the success or failure of a fish culture operation. Physical and chemical characteristics such as suspended solids, temperature, dissolved gases, pH, mineral content and the potential danger of toxic metals must be considered in the selection of a suitable water source.

S. Perumal et al. (eds.), *Advances in Marine and Brackishwater Aquaculture*,
DOI 10.1007/978-81-322-2271-2_20, © Springer India 2015

Parameters Influencing Hatchery Operation

Temperature

No other single factor affects the development and growth of fish as much as water temperature. Metabolic rates of fish increase rapidly as temperatures go up. Many biological processes such as spawning and egg hatching are geared to annual temperature changes in the natural environment. Each species has a temperature range that it can tolerate, and within that range it has optimal temperatures for growth and reproduction. These optimal temperatures may change as a fish grows. Cobia requires 23–27 °C during spawning. Most of the fish require 26–28 °C during larval rearing to promote fast growth of larvae. The growth of cobia larvae is usually retarded at low temperature, and sometimes high mortality also occurs when temperature decreases below 16 °C. Hence, more consideration should be given to select a water supply with optimal temperatures for the species to be reared or, conversely, to select a species of fish that thrives in the water temperatures naturally available to the hatchery. If unsuitable temperatures occur, water has to be cooled or heated based on the requirement.

The major temperature differences between hatchery water and the streams into which the fish ultimately may be stocked can greatly lower the success of any stocking programme to which hatchery operations may be directed. Within a hatchery, temperatures that become too high or low for fish impart stresses that can dramatically affect production and render fish more susceptible to disease. Most chemical substances dissolve more readily as temperature increases; in contrast and of considerable importance to hatchery operations, gases such as oxygen and carbon dioxide become less soluble as temperatures rise.

Photoperiod

One of the factors considered of great importance to the inducement of sexual maturation and spawning is photoperiod. Photoperiod manipulation is now being employed to alter the normal production of a cultured fish species, for example, 16 h light and 8 h darkness is recommended for sea bass when temperature remains below 21 °C, and above this temperature 20 h light and 4 h darkness is recommended. The greatest advantage of altering the spawning time of the cultured species is the availability of fry throughout the year.

Salinity

Some fish spp. migrate from marine to freshwater environment in order to spawn, while others migrate from freshwater to marine environment to complete their reproductive cycle. Hence, salinity may influence gametogenesis but probably does not function as a synchroniser for the timing of maturation.

Dissolved Gases

Nitrogen and oxygen are the two most abundant gases dissolved in water. Although the atmosphere contains almost four times more nitrogen than oxygen in volume, oxygen has twice the solubility of nitrogen in water. Carbon dioxide also is present in water, but it normally occurs at much lower concentrations than either nitrogen or oxygen because of its low concentration in the atmosphere. In general, oxygen concentrations should be near 100 % saturation in the incoming water supply to a hatchery. A continual concentration of 80 % or more of saturation provides a desirable oxygen supply.

Generally, waters supporting good fish populations have less than 5.0 ppm carbon dioxide. Carbon dioxide in excess of 20 ppm may be harmful to fish. If the dissolved oxygen (DO) content drops to 3–5 ppm, lower carbon dioxide concentrations may be detrimental. Both conditions easily can be corrected with efficient aerating devices.

Toxic gases like hydrogen sulphide and hydrogen cyanide in very low concentrations can kill fish. Hydrogen sulphide derives mainly from the anaerobic decomposition of sulphur compounds in sediments; a few parts per billion are lethal.

Alkalinity and Hardness

Alkalinity and hardness imply similar things about water quality, but they represent different types of measurements. Alkalinity refers to the ability to accept hydrogen ions (or to neutralise acid) and is a direct counterpart of acidity. The anion (negatively charged) bases involved mainly are carbonate (CO_3) and bicarbonate (HCO_3) ions; hardness represents the concentration of calcium and magnesium cations, also expressed as the $CaCO_3$ equivalent concentration.

Fish grow well over a wide range of alkalinities and hardness, but values of 120–400 ppm are optimum. At very low alkalinities, water loses its ability to buffer against changes in acidity, and pH may fluctuate quickly and widely to the detriment of fish. Fish also are more sensitive to some toxic pollutants at low alkalinity.

Probiotics

Application of probiotics in the water of tanks has an effect on fish health by improving several qualities of water, since they modify the bacterial composition of water or sediments. Micro-algae (*Tetraselmis suecica*) can inhibit pathogenic bacteria of fish such as *Vibrio anguillarum*, *V. salmonicida* and *Yersinia ruckeri*. It might be due to the presence of bioactive compounds in the algal cells. In hatchery, photosynthetic bacteria (*Rhodomonas* sp.) have been used as water cleaner and auxiliary food. Their results showed that the water quality of the pond treated with the bacteria remarkably improved, the fouling on the shell of the larvae was reduced, the metamorphosis time of the larvae was 1 day or even earlier and the production of post-larvae was more than that of the control.

Recirculation System

Recirculating aquaculture system (RAS) represents a new and unique way to farm fish. This system rears fish at high densities in indoor tanks with a controlled environment. Recirculation systems filter and clean the water for recycling back to the fish culture tanks. RAS has a lot of benefits; it includes a method to maximise the production on a limited supply of water and land, complete environmental control to maximise fish growth year round, complete and convenient harvesting and quick and effective disease control. Most water quality problems experienced in the recirculation system were associated with low DO, high fish waste metabolites, total ammonia nitrogen (TAN), un-ionised ammonia and nitrite–N, CO_2 and total suspended solids (TSS). To overcome this problem, proper water quality is maintained by pumping water through special filtration and aeration equipment. To provide suitable environment, recirculation system must maintain uniform flow rate (water/oxygen), fixed water levels and uninterrupted operation.

Suggested Water Quality Parameters for Water Sources (Concentration in ppm Except pH)

Parameter	Values
pH	6.5–9
Dissolved oxygen	5– saturation
Carbon dioxide	0–10
Total alkalinity as $CaCO_3$	50–400
Un-ionised ammonia	0–0.05
Nitrate	0–3.0
Nitrite	0–0.5
Phosphate	0.01–3.0
Manganese	0–0.01
Iron	0–0.5
Zinc	0–0.05
Lead	0.00
Hydrogen sulphide	0.00

Grow-Out Systems

In fish pond culture system, water quality is affected by chemical, biological and physical factors, which ultimately influence the aquatic environment and productivity. Aquaculture animals adopt themselves to these natural fluctuations to a certain level and fail to survive thereafter due to stress (Culver and Geddes 1993). The most important principle regarding water quality and soil management is that a pond has a finite capacity to assimilate nutrients and organic matter (Hopkins et al. 1994). When capacity exceeds, water and soil quality deteriorate. The maintenance of good water quality is essential for survival, growth and production. There are numerous water quality variables in pond fish culture. Fortunately, only a few of them normally play an important role (Boyd and Zimmermann 2009).

Water quality variables such as salinity and temperature play a decisive role when assessing the suitability of a site for a culture of particular species. Other properties such as alkalinity, turbidity and compounds of phosphorus and nitrogen are also important because they affect plant productivity, which, in turn, may influence aquaculture production. DO, CO_2, ammonia and other factors come into play during the grow-out period, because they are potential stressors for the animal in culture (Colt 2006).

Physical Characteristics

Water Temperature

Temperature of water is obviously very important. All metabolic and physiological activities and life processes such as feeding, reproduction, movement and distribution of aquatic organisms are greatly influenced by water temperature. Temperature also affects the speed of chemical changes in soil and water and the contents and pressure of dissolved gases. The requirement for dissolved oxygen is higher in warm water than cool water. Warm water fish grow best at temperatures between 25 and 32 °C.

The littoral zone of aquaculture pond has high temperature, whereas the benthic zone of aquatic pond has lower temperature, and this unequal temperature distribution in the pond results in thermal stratification in deeper ponds. This degrades the water quality by accumulation of methane, hydrogen sulphide and ammonia. Fish have poor tolerance to sudden changes in temperature. It can tolerate gradual changes in temperature, for example, from 25 to 32 °C, over several hours, but a sudden change in temperature of as little as 5 °C will stress or even kill the fish.

Salinity

Salinity refers to the total concentration of all ions in water (calcium, magnesium, sodium, potassium, bicarbonate, chloride and sulphate). Each species has an optimal salinity range. This optimum range of salinity allows the aquatic animals to efficiently regulate their internal body fluid composition of ions and water by the process of osmoregulation. Therefore, salinity plays an important role in the growth, reproduction and migratory behaviour of the fish as well as its general metabolism through osmoregulations of body minerals from that of the surrounding water. Normal level of salinity is around 10–32 ppt. *Chanos chanos* requires 18–30 ppt and *Mugil cephalus* requires 8–30 ppt. Water exchange with high or low salinity will provide the required salinity. The stress response associated with the sudden decrease in salinity was much reduced when the calcium concentration of the low salinity was increased from 84 to 150 ppm.

pH

The initial pH of pond waters (before biological activity adds to or removes CO_2 from water) is a function of the total alkalinity of the water. During culture, the pH of water is strongly influenced by both photosynthesis and respiration. As a result of respiration, carbon dioxide

(CO_2) is released into the water. Carbon dioxide decreases the pH of water as it is acidic. The rate of CO_2 production and CO_2 consumption depends on the density of animals and phytoplankton density respectively. Diurnal fluctuations of pH occur, depending on the number of aquatic life within a pond. With higher algae concentrations, more CO_2 is removed from the system and hence pH levels will rise. The reverse will occur at night when more CO_2 is produced, therefore leading to a drop in pH levels. pH is also changed by organic acids (produced from protein, carbohydrates and fat from feed wastes by anaerobic bacteria), mineral acid – sulfuric acid (washed down from dikes during rains) – and lime application.

The proportion of total ammonia existing in the toxic, un-ionised form (NH_3) increases as the pH increases. High pH increases algal bloom formation and reduces the swimming performance of fish due to ammonia accumulation. Low pH increases nitrite toxicity and also the fraction of H_2S (toxic form). Chlorine and metals such as copper, cadmium, zinc and aluminium are affected by pH. At higher temperatures, fish are more sensitive to pH changes.

Waters with pH values of about 6.5–9 are considered best for fish production. Daily fluctuation of pH should be within a range of 0.4 differences. Vigorous fluctuation of pH causes stress to culture organisms. However, the pH of brackishwater is usually not a direct threat to the health of the aquatic animal, since it is well buffered against pH changes. Calcium is a particularly important modulator of pH toxicity because calcium affects the permeability and stability of biological membranes. Filter alum and lime may be added to decrease and increase the pH of water respectively.

Turbidity

Turbidity refers to an optical property of water that causes light to be scattered or absorbed rather than transmitted through the water in a straight line. Water turbidity results due to the presence of suspended material, planktonic organisms or from suspended clay particles, and this reduces the light penetration, thereby limiting photosynthesis in the bottom layer. High turbidity can cause temperature and DO stratification in fish ponds. Turbidity caused by plankton is desirable, whereas turbidity resulting from suspended particles of clay is undesirable in aquaculture ponds. It will restrict light penetration, adversely affecting plant growth and destroying benthic organisms. In case of very high turbidity, fish die due to gill clogging. High value of transparency (>60 cm) is indicative of poor plankton density, and the water should be fertilised with the right kind of fertilisers. Low value (<20 cm) indicates high density of plankton and hence fertilisation rate and frequency should be reduced. Turbidity can be measured in terms of transparency using Secchi disc. The optimum range of transparency is 25–50 cm.

Clay particles responsible for turbidity repel each other due to negative charges, and these can be neutralised by electrolytes, resulting in coagulation. Alum (aluminium sulphate) and ferric sulphate might be more effective than hydrated lime and gypsum in removing clay turbidity. Both alum and gypsum can depress pH and total alkalinity. Hence, a simultaneous application of lime is recommended to maintain the suitable range of pH.

Chemical Characteristics

Dissolved Oxygen

Dissolved oxygen (DO) is the most critical water quality variable in fish culture. Dissolved oxygen in water is utilised by an aquatic organism to hold metabolism and is excreted as carbon dioxide (CO_2). Oxygen is regenerated within the pond from the liberated CO_2, which is used by photosynthetic plant forms to restore oxygen within the pond. Much of this oxygen is consumed by the aquatic organism and some is returned to the environment. Changes in the oxidation state of substances from the oxidised to the reduced form can be caused by low levels of dissolved oxygen in the pond environment. The concentration of toxic substances such as un-ionised NH_3, hydrogen sulphide and carbon metabolites (methane) increases when low DO level exists. However, in the presence of optimum level of oxygen, the toxic substances are converted into their oxidised and less harmful forms.

The availability of dissolved oxygen frequently limits the activities and growth of aquatic animals. If DO concentrations are consistently low, aquatic animals will not eat or grow well and will be susceptible to infectious disease. If concentrations fall to very low levels, the animals may die.

Oxygen production in the pond is considerably limited when a plankton die-off occurs or when there are high nutrient loads, large quantities of feed and faecal wastes are found on the pond bottom. Under this type of situation, DO can be maintained at optimum levels by providing additional aerators as well as aerating for additional hours and flushing out decaying plankton. Paddle wheel aerator is capable of elevating the dissolved oxygen level from 0.05 to 4.9 mg/l within 4 h in a 0.5 ha pond. Water exchange is the best solution to prevent low DO problem in the pond where aeration is not practised. The optimum DO concentration for aquatic animal growth is >5 ppm.

Total Alkalinity

Total alkalinity is the sum of titrable bases in water, predominantly bicarbonate and carbonate. Alkalinity of pond water is determined by the quality of water supply and nature of pond bottom soils. Dissolved carbon dioxide combines with water to form carbonic acid (H_2CO_3). A series of reversible equilibrium reactions occur, resulting in the formation of hydrogen ions, bicarbonate ions (HCO_3^-) and carbonate ions (CO_3^{2-}). These ions are the bases providing buffering capacity to water (otherwise called "total alkalinity") against wide swings in pH and enhanced natural fertility of water. Ponds having a total alkalinity of 20–150 ppm have sufficient supply of CO_2 for phytoplankton, and it may improve productivity. It also decreases the potential of metal toxicity. Very high alkalinity (200–250 ppm) coupled with low hardness (less than 20 ppm) results in rise in pH during periods of rapid photosynthesis and causes death of fish. Hence, pH and alkalinity have to be maintained at the optimum level. Dolomite, shell lime and zeolite improve alkalinity and stabilise pond water quality. In tropical regions, lime should be applied at least 1 month before fertiliser application is initiated; otherwise, lime material will precipitate phosphorus.

Total Hardness

Total hardness is the sum of the concentrations of calcium and magnesium in water expressed as mg/l equivalent $CaCO_3$. Total hardness strongly correlates with alkalinity. When the hardness level is equal to the combined carbonate and bicarbonate alkalinity, it is referred to as carbonate hardness. Hardness values greater than the sum of the carbonate and bicarbonate alkalinity are referred to as non-carbonated hardness. The nature of water supply largely determines the hardness of ponds. Total hardness is an indicator of the degree of mineralisation of water, and as total hardness increases, concentrations of most other substances tend to increase. Low-hardness water contains insufficient calcium ions for the protection of fish against acidity and metal toxicity. Agricultural gypsum may be applied to increase the total hardness without affecting the total alkalinity.

Carbon Dioxide

Carbon dioxide is a highly water soluble, biologically active gas. It is produced in respiration and consumed in photosynthesis. It is required for plant growth and its availability may limit the primary productivity of some aquatic ecosystems. In aquaculture ponds, dissolved CO_2 can be a stressor of aquatic animals and influences the pH of water.

Dissolved CO_2 concentrations in aquaculture ponds usually range from 0 mg/l in the afternoon to 5–10 mg/l or more at dawn. Particularly high concentration of carbon dioxide occurs in ponds after phytoplankton die-offs, after destruction of thermal stratification and during cloudy weather. Aeration and application of carbonate buffering material such as $CaCO_3$ and Na_2CO_3 remove all free CO_2 initially and store it in reserve as CO_3 and HCO_3. Experiments have shown that 1.0 mg/l of hydrated lime can remove 1.68 mg/l of free CO_2.

Chlorine

Chlorine is used as disinfectant during preparation for stocking, to destroy disease organisms, control phytoplankton abundance and improve water quality in ponds. Free and combined residual chlorine are extremely toxic to fish. The total chlorine residuals should not exceed 0.002 mg/l as Cl_2 for salmonids and 0.01 mg/l as Cl_2 for other aquatic organisms. Intense aeration, addition of 1 mg/l of sodium thiosulphate for every mg/l of chlorine and exposure to sunlight are some of the management practices.

Plankton

Plankton is comprised of all the microscopic organisms which are suspended in water and includes phytoplankton, zooplankton and bacteria. As plankton is at the base of food chain, there is a close relationship between plankton abundance and fish production. In addition to encouraging fish growth, plankton makes water turbid and prevents the growth of undesirable aquatic weeds through shading. The ability of water to produce plankton depends on many factors, but the most important is the availability of nutrients. Nutrient levels refer to the amount of nitrogen and phosphorus along with carbon and other trace elements, thus accelerating the growth of phytoplankton. Phosphorus regulates phytoplankton production in the presence of nitrogen. Nutrient levels can be increased in the ponds by adding inorganic or organic fertilisers in measured doses. However, increased levels of nutrients may be harmful, causing excessive plankton growth, potential blue-green algae blooms and oxygen depletion (Oberdorff and Porcher 1994). High levels of nutrients can be caused by high stocking densities, overfeeding and dead plant and animal matter. To decrease high nutrient levels, feeding rates should be decreased (or stopped) and the pond may need to be flushed with clean water (Doucha and Lívanský 1995; Zhang et al. 2002).

Toxic Metabolites in Fish Ponds

Ammonia

Ammonia is the principal nitrogenous waste product. It is produced from the decomposition of organic wastes resulting in the breakdown of decaying organic matters such as algae, plants, animals, overfeeding and protein-rich excess feed decays and liberates toxic ammonia gas, resulting in high ammonia levels. In addition to this ammonia, fish's excreted ammonia also accumulates to dangerously high levels. As ammonia in water increases, ammonia excretion by aquatic organism diminishes, and levels of ammonia in blood and other tissue increase. Two forms of ammonia are present in water, one is un-ionised ammonia and the other is ammonium ion (NH_4^+). The gaseous form of ammonia is toxic to aquatic animals and causes gill irritation and respiratory problems due to its ability to diffuse readily across cell membranes. Un-ionised ammonia is determined by total ammonia concentration, pH, water temperature and to a lesser extent salinity. The toxic effect of ammonia may be minimised by maintaining a sufficient level of dissolved oxygen, periodic partial removal of algal blooms and water exchange. The toxic levels for un-ionised ammonia for short-term exposure usually lie between 0.6 and 2 ppm and sublethal effects may occur at 0.1–0.3 ppm. Fortunately, ammonia concentration is seldom high enough in fish ponds to affect fish growth. The greatest concentration of total ammonia nitrogen usually occurs after phytoplankton die-offs.

Nitrite

The source of nitrite is through the addition of feed, fertiliser and manure. It is also an intermediate product in the bacterial nitrification of ammonia and nitrate. Nitrite is highly toxic to fish as it oxidises haemoglobin to form methaemoglobin, which is incapable of transporting oxygen. Nitrite toxicity increases with increasing pH and decreases with increasing calcium and chloride concentrations. Optimum level can be maintained by effective removal of organic waste, adequate aeration and correct application of fertiliser.

Hydrogen Sulphide

Hydrogen sulphide is produced in pond bottom soils under anaerobic conditions and is extremely toxic to aquatic animals. Un-ionised H_2S concentration is dependent on pH, temperature and salinity and is mainly affected by pH. It regulates the sulphur forms (H_2S, HS^- and S^{2-}). Un-ionised H_2S is toxic and it decreases rapidly with increasing pH. H_2S builds up mostly in sediment which is highly reduced (redox potential <150 mv), within a pH range of 6.5–8.5, and low in iron.

Sulphide can be reduced by aeration, water exchange and circulation of water to minimise anaerobic zones in the pond bottom. Application of oxidising agents, periodic pond draining and drying of bottom mud will result in the oxidation of sulphide and enhance the decomposition of organic matter. The safe level of un-ionised H_2S is <1 ppm.

Probiotics

Probiotic bacteria may competitively exclude the pathogenic bacteria or produce substances that inhibit the growth of the pathogenic bacteria. It provides essential nutrients and digestive enzymes to enhance the nutrition and digestion of the cultured animals. It directly uptakes or decomposes the organic matter or toxic material in the water, improving the quality of the water.

When photosynthetic bacteria were added into the water, it could eliminate the NH_3–N, H_2S and organic acids, and other harmful materials rapidly improve the water quality and balance the pH. The heterotrophic probiotic bacteria may have chemical actions such as oxidation, ammonification, nitrification, denitrification, sulphurisation and nitrogen fixation. When these bacteria were added into the water, they could decompose the excreta of fish, remaining food materials, remains of the plankton and other organic materials to CO_2, nitrate and phosphate. These inorganic salts provide the nutrition for the growth of micro-algae, while the bacteria grow rapidly and become the dominant group in the water, inhibiting the growth of the pathogenic microorganisms. The photosynthesis of the micro-algae provides dissolved oxygen for oxidation and decomposition of the organic materials and for the respiration of the microbes and cultured animals. This kind of cycle may improve the nutrient cycle, and it can create a balance between bacteria and micro-algae and maintain a good water quality environment for the cultured animals.

Fish Pond Water Quality Parameters

Parameters	Desirable range	Acceptable range (sp. dependant)
pH	6.5–9.5	5.5–10
Total alkalinity	50–150 ppm as $CaCO_3$	>20 mg/lit and <400 ppm
Total ammonia–N	0–2 ppm	<4 ppm
Un-ionised NH_3	<0.02 ppm	<0.4 ppm
NO_2–N	0–1 ppm	<4 ppm
NO_3–N	<50 ppm as NO_3–N	<90 ppm NO_3–N
Turbidity	50 cm	40–50 cm
Salinity	18–32 ppt	8–32 ppt
Phosphate	0.01–3 ppm	0.01–3 ppm
CO_2	<6 ppm	0–15 ppm
Temperature	25–32 °C	22–35 °C
Dissolved oxygen	>5 ppm	>5 ppm
Total suspended solids	<25 ppm	25–80 ppm
NO_3	0–2.5 ppm	0–2.5 ppm
CO_2	<6 ppm	<6 ppm

Conclusion

To ensure a sustainable fish production, critical water quality parameters should be maintained within optimum levels by using suitable management practices throughout the hatchery and culture period. In recent times, nanotechnology has revolutionised all fields in a big way. Nanotechnology provides opportunity in both aquaculture hatchery and ponds in removing pathogens by its antimicrobial properties. Nanosensors may provide further opportunity in the early detection of metabolites and pathogens to keep the pond environment healthy.

References

Bisson PA, Quinn TP, Reeves GH, Gregory SV (1992) Best management practices, cumulative effects, and long-term trends in fish abundance in Pacific Northwest river systems. In: Watershed management. Springer, New York, pp 189–232

Boyd C, Zimmermann S (2009) Grow-out systems-water quality and soil management. In: New MB, Valenti WC, Tidwell JH, D'Abramo LR, Kutty MN (eds) Freshwater prawns biology and farming. Blackwell, Oxford, pp 239–253

Colt J (2006) Water quality requirements for reuse systems. Aquac Eng 34:143–156

Culver DA, Geddes MC (1993) Limnology of rearing ponds for Australian fish larvae: relationships among water quality, phytoplankton, zooplankton, and the growth of larval fish. Mar Freshw Res 44:537–551

Doucha J, Lívanský K (1995) Novel outdoor thin-layer high density microalgal culture system: productivity and operational parameters. A1gol Stud Stuttg 76:129–147

Hopkins JS, Sandifer PA, Browdy CL (1994) Sludge management in intensive pond culture of shrimp: effect of management regime on water quality, sludge characteristics, nitrogen extinction, and shrimp production. Aquac Eng 13:11–30

Oberdorff T, Porcher JP (1994) An index of biotic integrity to assess biological impacts of salmonid farm effluents on receiving waters. Aquaculture 119:219–235

Qin J, Culver DA, Yu N (1995) Effect of organic fertilizer on heterotrophs and autotrophs: implications for water quality management. Aquac Res 26:911–920

Zhang K, Kurano N, Miyachi S (2002) Optimized aeration by carbon dioxide gas for microalgal production and mass transfer characterization in a vertical flat-plate photobioreactor. Bioprocess Biosyst Eng 25:97–101

An Overview of Aquafeed Formulation and Processing

K.P. Kumaraguru Vasagam, K. Ambasankar, and J. Syama Dayal

Introduction

It has been widely accepted that the growing future demand for aquatic products will have to be met by aquaculture. By year 2020, worldwide aquaculture is projected to supply about 120–130 million tonnes of fish to meet the estimated demands (Rana et al. 2009; FAO 2012). Since feed is the single largest operational expense (between 50 and 60 %) in aquaculture, feed cost determines the profitability and sustainability of this agribusiness activity. Although aquaculture supplied more than 63 million metric tonnes of fish to the world's human food basket in 2011, only two-thirds of this is produced from artificial feeding, and the rest is contributed by non-fed species. However, the percentage contribution by non-fed species is showing a declining trend from more than 50 % in 1980 to the present level of 33.3 %, indicating the significance of the formulated feed in global aquaculture industry and the further increase in demand for formulated feed and ingredients (FAO 2012).

Although global compound aquafeed production represents only 4 % of the total animal feed production (708 million tonnes in 2009), it has increased almost fourfold from 7.6 million tonnes in 1995 to 29.1 million tonnes in 2008, at an average annual rate of 11 %. Compounded aquafeed production is expected to grow to 51.0 million tonnes by 2015 and to 71.0 million tonnes by 2020 (Tacon et al. 2011).

The worthiness of the feed to the farmed animal is to maintain a state of good health and optimal performance by providing all essential nutrients in adequate quantities to prevent deficiency diseases and to support the fast growth in terms of muscle building (Lim and Webster 2001). Expanding list of cultivable species, genetic improvements of existing cultivable species, increasing feed ingredients price and limitations in supply of conventional ingredients are placing constant pressure on fish nutritionists and feed manufacturers to produce quality feeds that meet the requirements.

Aquafeed formulation and manufacturing become complex when we consider the cost, environmental sustainability, nutritional adequacy and constrains in feed manufacturing and storage. Thus, the way in which feed ingredients are selected, processed, mixed and prepared is controlled by many elements, and compromise between the ideal and the practical is a necessary for successful production of compounded aquafeed production (Hardy and Barrows 2002). Though there is no much difference in assortment of nutrients required by terrestrial livestock and aquatic animals for their nutritional wellbeing and health, it is highly challenging to supply those nutrients through a formulated feed for the latter (De Silva and Anderson 1995). In this chapter, we present the recent developments in the aquafeed industry.

K.P. Kumaraguru Vasagam (✉) • K. Ambasankar
J.S. Dayal
Central Institute of Brackishwater Aquaculture, No. 75, Santhome High Road, R.A. Puram, Chennai 600028, Tamil Nadu, India
e-mail: drvasagamguru@gmail.com

S. Perumal et al. (eds.), *Advances in Marine and Brackishwater Aquaculture*,
DOI 10.1007/978-81-322-2271-2_21, © Springer India 2015

For the convenience of the readers, the content of this chapter is divided under three major headings covering ingredients, feed formulation and feed manufacture.

Diversity of Ingredients in Aquafeed Basket

Even though whole feed is a matter of influence, it is a blend of several individual ingredients, and they are the one which determines the cost and quality of the feed. Except for few, most of the aquafeed ingredients are, for the majority, by-products of food processing obtained when high-value food for humans is extracted from a raw material. For decades, fishmeal has been considered as a key protein source for use in aquaculture feeds for both carnivorous and omnivorous species, and many aquaculture formulations still have fishmeal included at a level more than 50 %. Forty years before, poultry and swine diets used 90 % of the fishmeal produced. Now, those industries use less than one-third of fishmeal produced (Barlow 2000), and aquaculture uses the remaining two-thirds (approximately 68 %). Excessive dependence aquaculture on any one particular ingredient could increase the risk associated with supply, price and quality fluctuations. Furthermore, based on the volumes of fishmeal and fish oil used in aquaculture, especially for carnivorous species, production of these species is still viewed as a net fish consumer rather than producer, and this practice has raised concerns about the sustainability of these industrial productions (Naylor et al. 2000). As an approach to reduce risk, the identification, development and use of alternatives to fishmeal in aquafeeds remain a high priority research (Bostock et al. 2010).

Fish In and Fish Out (FIFO) Ratio: A Sizzling Issue in Aquaculture

Increase in demand for the use of fish oil and fishmeal in aquafeed is growing fast on par with the growing aquaculture. On the other side, a marine wild fishery which is a major source of fishmeal is stable for the past decade. Most of the world stocks of pelagic marine fish are learned to be either fully or overexploited (FAO 2009), so further growth in aquaculture production would require reduced dependence upon fishmeal. The demand is further intensified by competitors for fishmeal consumption like poultry. In spite of holding such a distinction in nutritional quality for cultured aquatic species, fishmeal is not only a limited natural resource from ocean but also becoming an expensive resource. In view of this, aquaculture has been highly criticised as net consumer of fish rather than a net producer of fish.

"Fish In Fish Out" (FIFO) ratio is nothing but a numerical ratio which expresses the amount of wild fish used to produce a unit amount of fish through aquaculture. The recent cited figures range from 3:1 to 10:1. FIFO ratio cited by Tacon and Metian (2008) in farmed salmon was 4.9:1, meaning that it needs 4.9 t of wild fish to produce 1 t of farmed salmon. It can be easily understood how they came up with this figure. If we take 1 t (1,000 kg) of wild fish, it is assumed that this would yield 225 kg of fishmeal and 50 kg of fish oil. If we say that on average salmon diets contained 30 % fishmeal and 20 % fish oil, this means that one could produce 250 kg of salmon feed by using up all of the 50 kg of fish oil. Salmon then have a feed conversion ratio (FCR) of 1.25 which therefore will give a harvest volume of 200 kg of salmon. So our starting 1,000 kg of wild fish has been turned into 200 kg of salmon which is a FIFO ratio of 5:1 (1,000:200).

Recently, the International Fishmeal and Fish Oil Organisation (IFFO) indicated the earlier values are misinterpreted and gave an explanation (Jackson 2010) on FIFO ratio clarifying the misunderstanding on the use of fishmeal and fish oil use in aquaculture. In the previous FIFO ratio worked out by Tacon and Metian (2008), while all the fish oil from 1,000 kg wild fish was used to produce the salmon feed, there was 150 kg of fishmeal left over. They assumed it as waste; hence, there is no more fish oil; in their calculation, this is just thrown away and wasted.

Table 1 Comparison of "Fish In Fish Out" (FIFO) ratio worked out by Tacon and Metian (2008) and International Fishmeal and Fish Oil Organisation (IFFO)

		Salomon + shrimp + carp		
	Salmon alone	Salmon	Shrimp	Carp
Weight of pelagic fish used (kg)	1,000	1,000		
Weight of fishmeal (kg)	225	225		
Weight of fish oil (kg)	50	50		
Fish oil in the diet (%)	20	20	2	0
Fishmeal in the diet (%)	30	30	20	5
Requirement of oil (kg)	50	35	15	0
Requirement of fishmeal (kg)	75	52	150	23
Feed that can be produced (kg)	250	175	750	450
FCR	1.25	1.25	1.7	1.8
Fish biomass produced (kg)	200	140	441	250
Fishmeal left as waste (kg)	150	0		
Fish oil left as waste (kg)	0	0		
Total biomass produced	200	831		
FIFO	5	1.2		

Source: Jackson (2010) www.iffo.net/downloads/100.pdf

Realising this as fictional, IFFO rectified this by smartly demonstrating to use the leftover fishmeal in production of shrimp (which needs more fishmeal) and carp. By doing this, they brought down the FIFO ratio of salmon to 2.27 from 4.49, shown by Tacon and Metian (2008). They also indicated the potential for further reduction in considering the use of waste from seafood processing industries for fishmeal production. IFFO specify that FIFO can be brought down by judicious allocation of fishmeal and fish oil among a range of farmed species groups rather than in single. For example, fishmeal and fish oil can be shared for production feed for salmon, shrimp and carps. In this, salmon is a major consumer of fish oil and shrimp is a consumer of more fishmeal. On the other hand, carps need only little of fishmeal. The worked out comparisons of FIFO calculation by Tacon and Metian (2008) and IFFO are presented in Table 1.

From this example, it is clear that there is surplus fishmeal with salmon and surplus fish oil with shrimp production. The excess fishmeal after use in both species can be used for carp which usually needs little. This indicates that mutual sharing among the two species may be more efficient. It also indicated that calculating the FIFO ratio based on just one type of farming does not give the correct picture. Considering this issue in reality, IFFO derived the following new formula to calculate FIFO ratio:

$$\text{FIFO ratio} = \frac{\text{Level of fishmeal in the diet} + \text{level of fish oil in the diet}}{\text{Yield of fishmeal from wild fish} + \text{yield of fish oil from wild fish}} \times \text{FCR}$$

For example, FIFO ratio of salmon will be like this:

$$\text{FIFO ratio} = \frac{30 + 20}{22.5 + 5} \times 1.25 = 2.27$$

Alternates Not Only for Fishmeal But Also for Other Ingredients

Considerable efforts have been made in the past in evaluating a wide range of potential alternatives to fishmeal and fish oil for use in aquaculture feeds (Chitmanat et al. 2009; Glencross 2007). Fluidity of the ingredient price, increasing understanding on the nutrient requirements of the target species, growing competition for food by human and other livestock and increasing awareness on sustainability issues forced the aquafeed industry to seek for alternates not only for fishmeal but also for all other ingredients (NRC 2011; Rana et al. 2009; Glencross 2007). Over the years, poultry nutritionists have been able to develop more than 80 alternative ingredients to fishmeal that can be used in poultry diets on a least-cost basis. But aquaculture has widely experimented only 12–15 commonly used ingredients, and it limits the flexibility to the formulator in maximising the cost benefits in the feed formula.

Potential Feed Ingredients for Use in Aquafeeds

Possible ingredients for use in aquafeeds can be broadly categorised in to three major categories:

1. Vegetable by-products
2. By-products of terrestrial and aquatic animal processing industries
3. Single-cell proteins

1. *Vegetable by-products*

Oil meals

The most important protein supplements of plant origin which could be potential alternates for fishmeal are the oilseed meals, produced from the cake remaining after oil has been extracted. Oils may be mechanically expelled or extracted by using solvents. Most common oil meals are from soybeans, cottonseed, canola, rapeseed, peanuts, sesame seeds, sunflower seeds and coconuts (Table 2).

Vegetable protein concentrates

Vegetable protein concentrates are processed, concentrated form of by-products high in protein content. There are several types of protein concentrates, and their composition highly varies with the processing methods used. Corn gluten meal is the dried concentrated residue from corn after the removal of the larger part of the starch and germ. It may contain fermented corn extractives and/or corn germ meal. They could be potential alternates on par with fishmeal in view of its high protein content. Some common vegetable protein concentrates and their compositions are presented in Table 2.

Cereal by-products

On processing different cereals for human use, the hard outer layer of cereals which consists of combined aleurone and pericarp is a major by-product. As India is one of the major producers of wheat and rice, their respective bran is a major cereal-based ingredient in this country. Cereals are also mainly used as major substrates in brewing industry nowadays to produce alcohol. The remains of the grains after fermentation process are called distillers dried grains. They could be of different origins like barley, sorghum, wheat, rice, corn, etc. Though they are commonly called as dried distillers grains, their variety will be named by adding the grain name as prefix. There are many varieties based on the process involved in its production. They are distillers dried grains, distillers dried solubles (DDS), distillers dried grains with solubles (DDGS) and malt sprouts. Germ meals of corn and wheat are also grouped under this ingredient category. Naturally as by-products of fermentation industry, they are rich in protein content and more bioavailable (NRC 2011).

2. *By-products of terrestrial and aquatic animal processing industries*

Advances in breeding, production and processing technologies have improved the livestock production industry as source of food for humans. On processing these livestock, varieties of rendered animal by-products are produced in large quantities and available in variety of forms for use in aquafeeds. Various by-products of animal processing industries and their compositions are presented in Table 3.

Table 2 Proximate composition (as % DM basis) of vegetable by-products*

	Moisture	Crude protein	Crude fat	Crude fibre	Crude ash
Mustard meal	10.15	32.2	8.9	8.1	9.2
Rapeseed meal	8.30	34.7	7.5	12.3	6.7
Coconut meal	8.7	21.5	3.5	14.8	7.1
Cottonseed meal	10.0	32.9	1.7	21.8	6.0
Groundnut oil meal	8.3	31.8	2.2	27.0	4.7
Safflower meal	8.9	42.7	1.4	13.0	7.1
Sesame meal	7.6	45.0	4.8	6.7	13.0
Soybean meal	10.3	44.7	1.3	6.0	6.7
Sunflower meal	10.0	23.3	1.1	31.6	5.6
Soy protein concentrate	7	59	5.4	1.5	7.9
Canola protein concentrate	10	63	8	4.7	5.9
Corn protein concentrate	10	76.2	4.5	1	1.3
Potato protein concentrate	10	76	1.5	6.3	2
Rice protein concentrate	9	69	10	3	4
Brewers grains, dehydrated	8.4	25.9	6.4	14.3	4.3
Distillers dried grains	8.2	27.1	8	11.2	3.8
Distillers dried solubles (DDS)	6.8	28.7	6	4.3	7.9
Distillers dried grains with solubles (DDGS)	8.2	28.4	8.5	9.4	4.9
Malt sprouts	8.4	25.4	1.7	14.3	4.3
Corn gluten feed	10.1	24.2	2.9	8.2	6.4
Corn gluten meal	8.6	56.1	4	2.9	2.1
Corn germ meal	4.5	47.4	8.5	6.4	0.8
Wheat germ meal	9.7	26.6	7.3	3.3	4.7
Rice bran – full fat	10.0	12.2	11.8	12.3	13.1
Rice bran – de-oiled	0.5	12.3	2.1	14.6	12.6
Rice polishing	10.0	12.1	11.5	4.7	8.8
Wheat bran	12.1	14.7	4.0	9.9	5.8

*Source: Tacon et al. (2009)

(a) *By-products of marine origin*

Condensed fish solubles

Condensed fish solubles are obtained by evaporating excess moisture from the stick water, aqueous liquids, resulting from the wet rendering of fish into fishmeal, with or without removal of part of the oil.

Crab meal

Crab meal (crab process residue meal) is the undecomposed ground dried waste of the crab and contains the shell, viscera and part or all of the flesh.

Shrimp meal

Shrimp meal (shrimp process residue meal) is the undecomposed, ground dried waste of shrimp and contains parts and/or whole of the shrimp.

(b) *By-products of terrestrial animal origin*

Blood meal

Blood meal, flash dried (animal blood meal flash dehydrated), is produced from clean, fresh animal blood, exclusive of all extraneous material such as hair, stomach belching and urine.

Hydrolyzed poultry feather meal

Hydrolyzed poultry feathers meal is a product resulting from the treatment under pressure of clean, undecomposed feathers from slaughtered poultry, free of additives and/or accelerators.

Meat and bone meal

Meat and bone meal is the rendered product from mammal tissues, including bone, exclusive of any added blood, hair, hoof, horn, hide trimmings and stomach and rumen contents.

Table 3 Proximate composition (as % as-fed basis) of various animal meals and by-products as potential alternates of fishmeal*

Ingredient	Moisture	Crude protein	Crude fat	Crude fibre	Crude ash
Animal meal by-products of marine origin					
Fish soluble	7.4	55.9	6.5	3.2	12.6
Shrimp head meal	8.8	46.6	6.4	11.1	26.5
Crab meal	7.1	33.9	2.8	10.7	41.9
Squid viscera meal	10.3	50.3	18.6	1.5	9.8
Animal meals by-products of terrestrial origin					
Blood meal	9.0	85.5	1.4	0.9	5.3
Feather meal	8.4	84.0	4.2	1.0	3.6
Meat and bone meal	7.5	50.1	10.6	2.4	28.8
Poultry by-product meal	7.4	59.0	12.4	2.6	15.3
Terrestrial invertebrate meals					
Silkworm pupae meal	11.1	55.1	23.2	5.5	3.8
Maggot meal	3.0	43.2	23.0	1.0	16.2
Earthworm meal	7.4	56.4	7.8	1.6	8.8
Polychaete worm meal	8.0	55.0	15.0	1.0	12.0

*Source: Tacon et al. (2009)

Poultry by-product meal

Poultry by-product meal consists of the ground, rendered, clean parts of the carcass of slaughtered poultry, such as necks, feet, undeveloped eggs and intestines.

(c) *Terrestrial invertebrate products*

Recently terrestrial invertebrate products have been successfully used in compound aquafeeds. Insect larvae/pupae have been used as traditional supplementary feed items by small-scale farmers in many Asian countries and, together with snails and annelids, offer a potential nonconventional feed source for use by small-scale farmers. Silkworm pupae meal, vermin meal and maggot meal were found to be good source of protein and essential amino acids in diets of aquatic animals (Hertrampf and Piedad-Pascual 2000). In general, invertebrate meals are good dietary sources of animal protein, lipids and energy.

3. *Single-cell proteins*

Single-cell protein (SCP) is a term applied to a wide range of unicellular and filamentous algae, fungi and bacteria which can be produced by controlled fermentation processes. In contrast with conventional plant and animal feed proteins, these microorganisms offer numerous advantages as protein producers, including: (1) their production can be based on raw carbon substrates which are available in large quantities as wastes from other industries which otherwise cause an environmental hazard; (2) the majority of microorganisms cultured are highly proteinaceous; (3) they have a short generation time, and (4) they can be cultivated in a limited land space and produced continuously with good control (Tacon et al. 2009; Hardy and Barrows 2002). Some common single-cell proteins and their compositions are presented in Table 4.

Feed Formulation and Application of Software

For several reasons, formulation and manufacture of compounded aquafeed are commercially unrealisable not only for small and medium farmers but also for large farmers (Tacon and Metian 2008). This is the reason why they mostly hang on commercial feed mills for their feed needs. It is therefore essential that formulations are accurate, to ensure that their cultured animals are not

Table 4 Proximate composition (as % as-fed basis) of various single-cell proteins as potential alternates of fishmeal*

Ingredient	Moisture	Crude protein	Crude fat	Crude fibre	NFE	Ash
Brewer's yeast	7.6	46.1	1.3	2.9	34.0	8.1
Fungal biomass	8.5	44.4	9.4	16.9	16.1	4.7
Spirulina	6.4	62.1	4.8	0.5	17.3	8.9
Chlorella	5.7	47.2	7.4	8.3	20.8	10.6
Methanol substrate	6.4	73.1	5.7	0.4	2.7	11.7

*Source: Tacon et al. (2009)

adversely affected. Though feed manufacturing and process involve several processes, feed formulation is the centre which determines the standing of the manufacturer in the market and the economic returns to the company.

Feed formulation is basically applied nutrition. Feed formulation is the process of enumerating the amounts of feed ingredients that need to be put together, to form a single unvarying mixture further processed as feed that supplies all of the nutrients required by the target animal. A cautiously formulated diet has long been recognised as a necessity in preserving an animal's health and maintaining its ability to resist diseases (Hardy and Barrows 2002; Lall and Olivier 1993). Nutrient deficiencies or excesses may have profound effects on fish survival and disease development (Lim and Webster 2001). Since feed costs account for more than 50 % of the total production costs for most types of aquaculture species, it is important that returns are maximised through use of adequate diets. Since formulation is a central operation in aquafeed production, it should ensure that feed ingredients are economically used for optimum growth of aquatic organisms under culture. It basically requires a good knowledge on the nutrient requirements of the target organism, nutrient composition of a range of feed ingredients and feed manufacturing process involved to assure a maximum nutrient delivery.

The first step in feed formulation is gathering through knowledge about the ingredients in terms of their chemical composition, availability in the market and their consistency in chemical composition, seasonality, etc. This all will directly determine the price of the ingredient and in turn the feed. The other essential component is its nutrient requirement of the candidate species for which the feed is intended. Although requirements generally do not vary greatly among fish or shrimp species, differences in nutrient needs are evident between species (NRC 2011). Therefore, dietary requirements of a variety of aquatic species for energy, protein, lipids, vitamins and minerals have been or are currently being established (NRC 2011; Lim and Webster 2001). The nutrient requirements vary with age, size, level of production, etc. Another significant and most dynamic variable in the feed industry is the ingredient cost. It is important to frequently update and compare the prices of feed ingredients, in order to reduce the overall cost of the diet.

With these above knowledge inputs, mathematical formulae are used to derive the amounts of each ingredient that need to be included in the compounded feed mixture. When using only a few ingredients, the formulae are simple, but a few ingredients are rarely able to supply all the nutrients that will meet the requirements of the aquatic species, so several ingredients are used, requiring complex formulae. Some of these formulae have been built into computer programs, which enable the rapid processing of values that should be included in the formulation. Computer programs also make it easy to check if nutrient requirements are met.

Use of Classical Pearson Square or Box Method and Complex Linear Programming in Aquafeed Formulation

The Pearson square or box method (Box 1) of matching ratios is a simple procedure that has been used for several years (Wagner and Stanton 2012; De Silva and Anderson 1995). A major problem with this method is it capability to

Box 1: Classical Pearson Square Method

- The simplest method for feed formulation, ideally using 2–4 ingredients.
- One nutrient can be balanced at a time, either protein or lipid.
- Sample calculation to prepare a fish feed to have 25 % crude protein using groundnut oil cake (GNOC) and rice bran.
- Protein content of GNOC is 31.8 %, and rice bran is 12.2 %.

  Here it is...

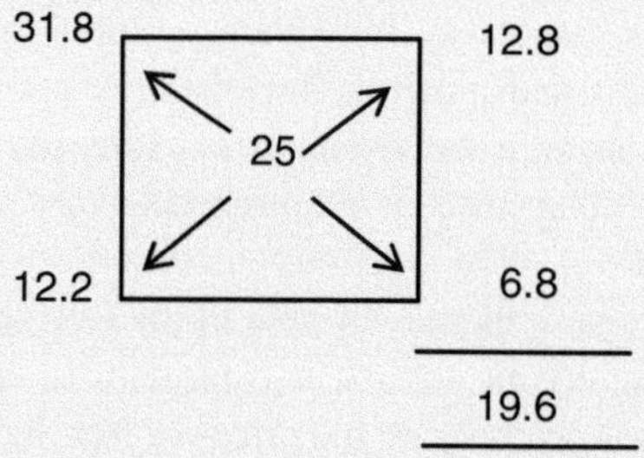

% of GNOC is - (12. 8 × 100)/ 19.6 = 34.7

% of Rice bran is - (6.8 × 100)/ 19.6 = 65.3

- In taking a close look at the square, numbers are in and around the square. Probably one of the more important numbers is the number that appears in the middle of the square. This number represents the nutritional requirement of an animal, here it is protein.
- Subtract the nutrient value from the requirement on the diagonal, and one can arrive at a numerical value entitled parts.
- By summing those parts and dividing by the total, one can determine the % of the ratio that each ingredient should compose in order to provide a specific nutrient level.
- Always double check calculations to make sure that you did not have a mathematical error.
- It also is very important to work on a uniform basis for nutrient composition of ingredients and requirements, preferably dry matter basis, and then convert to an as-fed basis after the formulation.

handle only 2–4 ingredients and single nutrient at a time. Earlier only fishmeal along with some oil meal and cereal by-products was the major ingredient that goes in to the fish diets (mainly farm-made diets); this tool was ideal to use for deriving the ingredient use. Now the entire situations in this industry are changing and getting complicated because of the limitation in the use of fishmeal as a major source of nutrients in fish feed. Several alternate ingredients are used to meet the desired level of nutrients in the feed. In these conditions, Pearson square will be no that much useful to handle expanding list of alternate ingredients and knowledge on nutritional requirements of the target species. Spontaneously the above procedure is almost replaced by a complicated mathematical procedure called linear programming. This is the basic tool of the entire feed industry now.

Linear programming was first introduced to the animal compound feed industry in the mid-1950s. Linear programming is a mathematical procedure by which limited resources are allocated, selected, scheduled or evaluated to achieve an optimal solution to a particular objective. Linear programming has wide application in industrial operations such as blending, mixing and machine tooling and in business activities such as purchasing, planning, bidding, transportation and distribution. Since then, its application in least-cost formulation of feed for livestock and poultry has gained widespread acceptance in most countries with well-developed compound feed industries. On the other hand, least-cost feed formulation based on linear programming is a recent entrance in aquafeed industry.

Least-Cost Formulation

Nowadays, least-cost formulation based on linear programming is achieved through the use of softwares and computers. Use of softwares for day to day feed formulations is recent in aquafeed manufacturing, and it is getting good response.

What Is Least-Cost Formulation?

Least-cost feed formulation is the process of calculations combining many feed ingredients in a certain proportion to provide the target animal a well-balanced nutritional feed at the least possible cost (Rossi 2004). Though least-cost formulation is a mathematical elucidation based on linear programming, it requires the knowledge of animal nutrition to take into consideration the nutrient requirements of the target animal and its competency to digest and assimilate nutrients from various available ingredients (Tacon 1987). Feed formulators also need to be aware of the variations of the nutritional requirements for different species at various stages of their lifespan. Linear programming performs what is designed to do, and it is based on the information put in by the formulator. So the formulation programme is only as good as the nutrient and ingredient parameters entered into it.

Choice of Least-Cost Formulation Software

There are extensive choices of user-friendly softwares been developed and available in the market (Fig. 1). The software ranges from simple spread sheet-based solutions to sophisticated and complex packages designed for large feed manufacturers that require multisite, multiserver and multi-blending capabilities. Some of the most famous softwares of this kind include Brill, Mix-it, Feedsoft, Best Mix, Winfeed, Agrisoft, etc. New and innovative add-on applications are being developed and introduced into the market every year. Some feed formulation software is specifically designed for a certain species, and they may provide tables of nutrient requirements or models of growth for those specific animals. Further, improvements in the look and feel of the software and seamless integration with other functionalities related to formulation also occur routinely. These applications and improvements enable feed formulators to perform their jobs much more efficiently.

Basic Components of Feed Formulation Software

For least-cost feed formulation software to be effective, it should offer the following basic features that are applicable irrespective of species. At this point, it is important to keep in mind that feed formulation runs on data that have been entered into it by the user.

The final feed formula will only be as accurate as the initial information that was input by the user. The various components of formulation software may be explained as follows (Rossi 2004) (Fig. 1):

1. *Available ingredients (ingredient store)*

   All feed formulation software provides a way of entering and managing the ingredients which are available for inclusion in the formulas. Available feed ingredients are listed along with their unit price. Depending on the software being used, optional ingredient properties such as the ingredient types, alternate code names, and applicable species may also be entered.

2. *Nutrient composition*

   Each feed ingredient available for inclusion in the formulae should have corresponding nutrient composition data. The nutrient values are preferably derived from chemical analysis of representative samples of the ingredient. When the nutrient composition is not available, tables of feed composition using average or typical values are used.

3. *Formula specifications*

   Specifications are set for each formula to be solved by the least-cost formulation software. Formula specifications generally define the nutrient levels desired in the formula and the ingredient inclusion levels. Either a lower limit and/or an upper limit for each nutrient and ingredient is set.

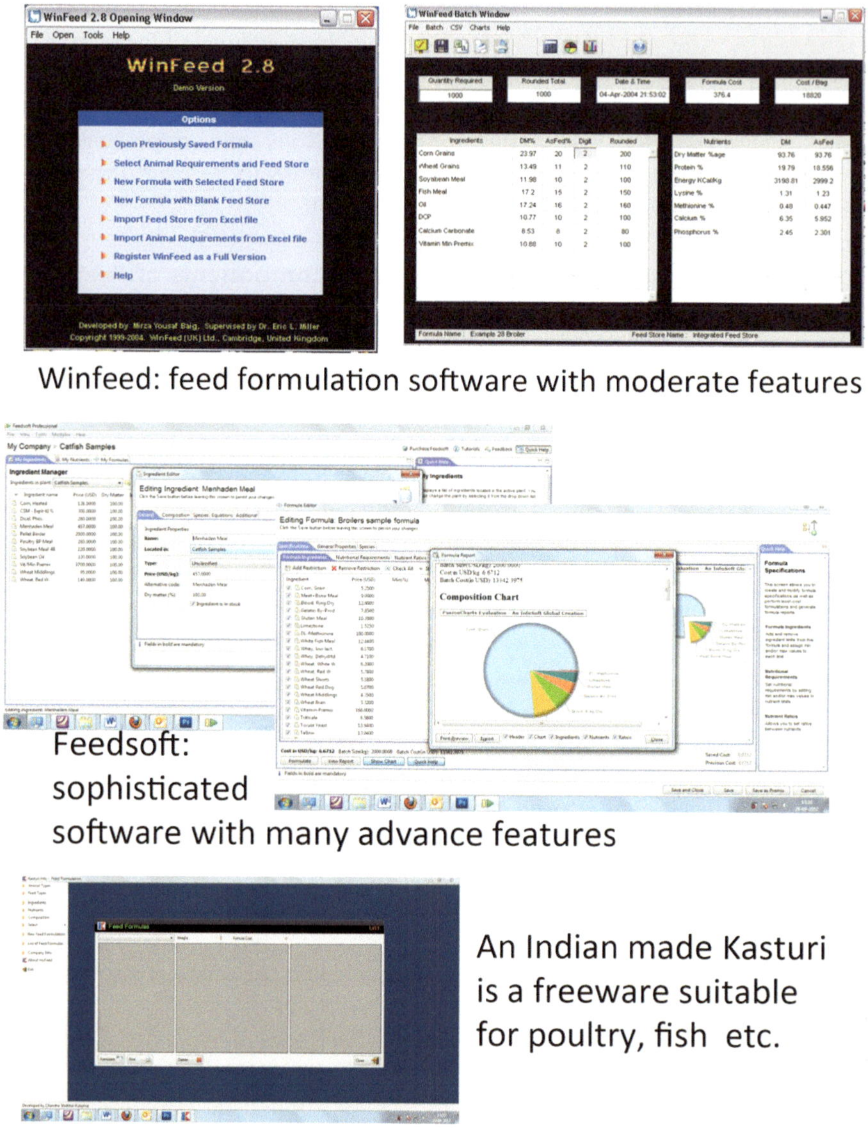

Fig. 1 Screenshots of some feed formulation softwares

Formulation

Once all the above essential information is provided, the formulation software will create formulae that meet the desired specifications at the lowest probable cost. A requirement for proper formulation, however, is that the formula result must be feasible both from a mathematical and a nutritional standpoint. If infeasible results are obtained, the ingredient and nutritional composition should be carefully scrutinised to make sure the solution is nutritionally acceptable for the target species.

One of the most important uses of least-cost feed formulation is in choosing among the available ingredients to be used, based on their nutritional composition and cost. Many times one ingredient can be substituted by another with similar nutritional value. The software helps the user to achieve the highest profit margin when market conditions favour the use of one ingredient

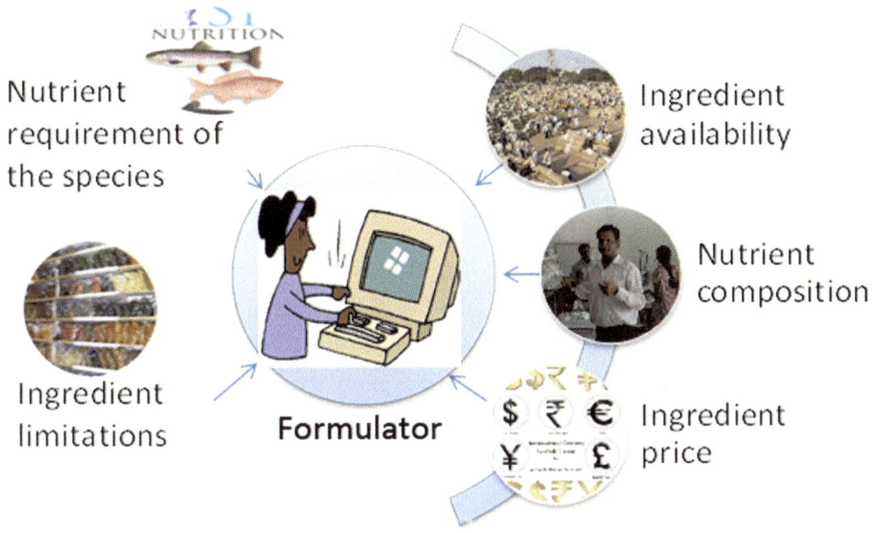

Fig. 2 Dependent components involved in software-assisted feed formulation

over the other. A number of tools are useful in the analysis of formulation results (Fig. 2).

Advantages of Using Software Formulation

- Formula can be adjusted on daily basis.
- Formula can handle any number of ingredients and nutrients.
- Complex restriction on ingredient inclusion and nutrient level is possible.
- Rapid and simple.
- Routine operation may not need profession nutritionist.
- Can be centralised, and formulator can be placed any corner in the world.

Disadvantages in Using Softwares for Feed Formulation

- Formula optimised for nutrient levels alone may not be optimum in terms of production performance.
- Minor error in preliminary database creation may lead serious error in formula.

Different Types of Compounded Aquafeed Manufacturing Practices and Recent Advances

Feeds for farmed aquatic organisms are exposed to water for an extended period of time before being consumed by the target animal under culture. Even with a premium feed formula, the feed manufacturing process determines the magnitude of nutrient delivery to the target animal, and hence the process of feed manufacture must assure a maximum nutrient delivery. Exposing the feed to water not only results in loss of water-soluble nutrients due to leaching but also the breakdown of the feed pellets will degrade the water quality. The challenge here will be to prepare and present the feed to the target animal without much loss of nutrients in to water.

Water-stable dry pelleting is the common compounded feed manufacturing methods to produce grow-out feed for almost all the cultured fin and shell fishes. Since its introduction in the 1930s in poultry and in livestock, pelleting has become an important process to the feed industry. Though loss of some key nutrients on heat cooking is identified as problem in this kind of feed processing, there are several advantages over dry mash feeding and wet feeding practices. Capable of long-term storage under room temperature and maximum nutrient delivery are the competitive advantages over other. Pellet quality issues can be partitioned into several individual components and their contribution of each component. These are formulation (40 %), grind (20 %), conditioning (20 %), die selection (15 %) and cooling/drying (5 %). It should be noted that 60 % of pellet quality is determined before the feed reaches the pellet mill.

The objective in pelleting is to convert the homogeneous blend of ingredients into durable particles having physical characteristics that make them suitable for feeding. There are two major kinds of pelleting in aquafeed manufacturing practices. They are compressed dry pelting and extruded dry pelleting. While compressed dry pelleting is an older method of pelleting, extrusion cooking is a recent technology and has undergone lots of modification. But both technologies still have considerable share in the feed manufacturing business. We present an overview of each processing method, but the reader should be conscious that there are many small modifications in the processes explained below that manufactures employ.

Compressed Dry Pelleting

In compressed pelleting, a feed mixture which has been exposed to steam for about 5–25 s to increase the temperature to about 85 °C is forced through a die. The addition of steam increases the moisture and softens the feed particles, thereby increasing the gelatinization of raw starch, which helps bind the ingredients together in the pellet. On exposure to steam, feed mixture is allowed to attain moisture about 16 %, and the mixture is forced through holes in a metal die by the action of a roller located inside the die. This process is also called as steam pelleting, due to the application of steam to precondition the mix prior to compression. The combination of heat, moisture and pressure forms the mixture into a compressed pellet. As the pellets come out from the outside surface of the die, they are cut off by an adjustable knife to the desired length. Pellet quality is influenced by several factors, including fat level, moisture and humidity. The fat level of the formula should be not lower than 2–3 % to lubricate the holes in the die and to reduce dustiness and no higher than 8–10 % to avoid excessive die lubrication causing insufficient compression of the feed (Picture 1).

The moisture level is critical in that if affects the pellet quality. Either insufficient or excessive moisture in the feed mixture reduces the pellet hardness. Pellets made with insufficient moisture are dry and crumbly, while excessive moisture results in soft pellets due to insufficient compression. The lubricating effects of fat and moisture together lower the electrical power required to operate a pellet mill and extend the operating life of the die. An experienced pellet mill operator becomes familiar with various feed mixtures and manufacturing conditions and makes the adjustments necessary to make high-quality compressed pellets.

Picture 1 Ring die pelletiser displaying the dies (Photo Courtesy: ORB Machinery Company, China)

Pelleting by Extrusion

Cooking extrusion is the most recent development in pelleted fish feed manufacture. It is a process by which a set of mixed ingredients are forced through an opening in a perforated plate or die with a design specific to the food and are then cut to a specified size by blades. The machine which forces the mix through the die is an extruder (Picture 2), and the mix is known as the extrudate.

The basic components of the extrusion cooking look the same as compressed pellets, but the use of different dies and physical conditions result in a very different product. This process is more flexible than compression pelleting, thus allowing for the control of pellet density, which affects buoyancy in water, and for the addition of very high lipid levels. Extrusion technology is used to produce snack foods, such as crackers and breakfast cereals, and pet foods for dogs, cats and caged birds.

The extruder machine functions by increasing the temperature of the feed mixture to 125–150 °C in a pressurised conditioning chamber and increasing the moisture by more than 20 %. The combination of these two steps gelatinizes the starch. The preconditioned mixture is then made into a dough-like consistency in a long barrel by an auger which has tapered flights, which increases the pressure on the mixture as it moves down the barrel towards the die. When sufficient pressure is achieved, the vapour is converted back to liquid water. As the pellet leaves the die, the reduction in pressure causes the liquid water to suddenly expand and convert into vapour. This causes formation of air pockets in the feed pellets, changes the bulk density of the pellet and gets a puffy appearance. By making changes in the formulations and the processing conditions, changes can be achieved in bulk density of the feed so that it floats on, sinks slowly in or sinks quickly in water (Guy 2001). Bonds are formed within the gelatinized starch, which results in a durable, water-stable pellet. After cooling and drying, the pellet density is typically 0.25–0.3 g/cm^3. The extruder consists of a large, rotating screw tightly fitting within a stationary barrel, at the end of which is the removable die. These pellets are formed by extrusion of a moist mixture (20–24 %) followed by drying to reduce the moisture content to 10 % or less.

Cooking extruders are available in two types of screw configurations. They are single and twin screw. The self-wiping design of screws used in the twin-screw configuration allows for greater

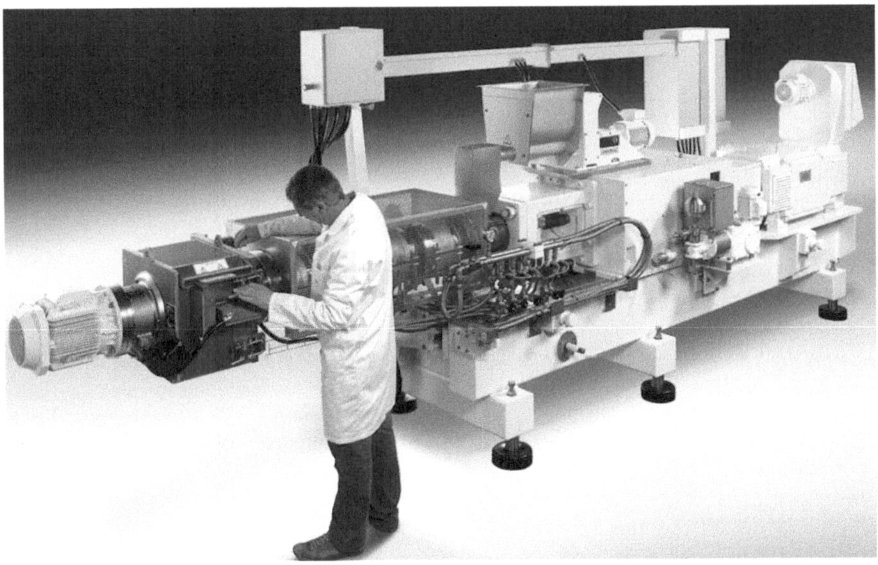

Picture 2 Twin-screw extruder (Photo Courtesy: Clextral Inc.)

flexibility in manufacturing, resulting in pellets that will absorb higher amounts of lipid compared to pellets produced with single-screw extruders (Guy 2001).

Extruded pellets are widely used by the catfish, salmon, trout and shrimp industries and by many other sectors of aquaculture, particularly for fish farmed in sea cages. Extruded pellets can be used in automatic and demand feeders. Extruded pellets are relatively porous and can soak up sprayed oil to reach levels of over 35 %, typical of feed for Atlantic salmon (Hardy and Barrows 2002). The cost of production is slightly higher for extruded pellets than for compressed pellets, but their advantages outweigh the additional cost in many aquaculture applications. For example, extrusion cooking increases the digestibility of carbohydrates in the feed mixture as a result of the exposure of the feed mixture to high temperatures and pressures. Thereby, total energy available to fish is increased by the extrusion process. On extrusion process, low-cost sources like raw and de-oiled rice bran, husks of various legume seeds, etc., can be used as source of carbohydrates. There are also several advantages in feed management. Feed consumption can be easily monitored with floating extruded pellets, which is no way possible with sinking pellets (Goddard 1996).

References

Barlow S (2000) Fishmeal and fish oil: sustainable feed ingredients for aquafeeds. Glob Aquac Advocate 4:85–88

Bostock J, McAndrew B, Richards R, Jauncey K, Telfer T, Lorenzen K, Little D, Ross L, Handisyde N, Gatward I, Corner R (2010) Aquaculture: global status and trends. Philos Trans R Soc B 365:2897–2912

Chitmanat C, Tipin A, Chaibu P, Traichaiyaporn S (2009) Effects of replacing fishmeal with wastes derived from local fisheries on the growth of juvenile tilapia, *Oreochromis niloticus*. Songklanakarin J Sci Technol 331:105–110

De Silva SS, Anderson TA (1995) Fish nutrition in aquaculture. Chapman & Hall, London, 319 pp

FAO (Food and Agriculture Organisation) (2009) The state of world fisheries and aquaculture. FAO, Rome

FAO (Food and Agriculture Organisation) (2012) The state of world fisheries and aquaculture. FAO, Rome

Glencross BD (2007) Harvesting the benefits of grain in aquaculture feeds. In: Proceedings of a workshop, 13 February 2007, Fisheries occasional publications no. 41. Department of Fisheries, North Beach, 102 p

Goddard S (1996) Feed management in intensive aquaculture. Chapman & Hall, New York, 194 pp

Guy RCE (2001) Raw materials for extrusion cooking. In: Guy RCE (ed) Extrusion cooking technologies and applications. Woodhead Publishing Limited, Cambridge, pp 5–28

Hardy RW, Barrows FT (2002) Diet formulation and manufacture. In: Halver JE, Hardy RW (eds) Fish nutrition. Academic, San Diego, pp 506–601

Hertrampf JW, Piedad-Pascual F (2000) Handbook on ingredients for aquaculture feeds. Kluwer Academic Publishers, 624 pp

Jackson A (2010) Fish in-fish out ratios explained. Aquac Eur 34:5–10

Lall SP, Olivier G (1993) Role of micronutrients in immune response and disease resistance in fish. In: Kaushik SJ, Luquet P (eds) Fish nutrition in practice. INRA, Paris, pp 101–118

Lim C, Webster CD (2001) Nutrition and fish health. The Haworth Press, Binghamton

Naylor RL, Goldburg RJ, Primavera J, Kautsky N, Beveridge MCM, Clay J, Folkes C, Lubchence J, Mooney H, Troell M (2000) Effects of aquaculture on world food supplies. Nature 405:1017–1024

NRC (National Research Council) (2011) Nutrient requirements of fish. National Academy Press, Washington, DC

Rana KJ, Siriwardena S, Hasan MR (2009) Impact of rising feed ingredient prices on aquafeeds and aquaculture production, FAO fisheries and aquaculture technical paper 541. FAO, Rome

Rossi R (2004) Least-cost formulation software: an introduction. Aqua Feed Formul Beyond 3:3–5

Tacon AGJ (1987) The nutrition and feeding of farmed fish and shrimp – a training manual. 2. Nutrient sources and composition. FAO field document, project GCP/RLA/075/ITA field document no. 5, Brasilia, Brazil, 129 pp

Tacon AGJ, Metian M (2008) Aquaculture feed and food safety: the role of FAO and Codex Alimentarius. N Y Acad Sci 1140:50–59

Tacon AGJ, Metian M, Hasan MR (2009) Feed ingredients and fertilizers for farmed aquatic animals sources and composition. FAO fisheries and aquaculture technical paper no. 540. FAO, Rome, 209 pp

Tacon AGJ, Hasan MR, Metian M (2011) Demand and supply of feed ingredients for farmed fish and crustaceans: trends and prospects, FAO fisheries and aquaculture technical paper no. 564. FAO, Rome, 87 pp

Wagner J, Stanton TL (2012) Formulating rations with the Pearson square, Extension, fact sheet no. 1.618. Livestock Series, Management, Colorado State University, Fort Collins

Cryopreservation of Fish Gametes and Its Role in Enhancing Aquaculture Production

J. Stephen Sampath Kumar and C. Judith Betsy

Introduction

Cryopreservation of gametes in general has been done to increase the longevity of gametes for several years without any drastic changes in their efficiency or potency. In fish gamete preservation, the technology of cryopreservation was principally adopted for saving endangered species by facilitating the storage of their gametes in the gamete bank. Short-term preservation of fish gametes is the forerunner for the development of cryopreservation technology for fish gametes particularly spermatozoa. The basic concept in this method is the storing of spermatozoa at lower temperature. Long-term preservation specifies the storage of spermatozoa for an indefinite period. This is possible only by 'cryopreservation' of the spermatozoa. Cryopreservation of fish gametes, as a technology, can play an important role in the development of sustainable aquaculture through production of quality seeds. This technology can offer many advantages like breeding possibility during off season, tiding over the effect of monsoon failure on the maturation of males, production of viable and strong offspring by intraspecies hybridization, overcoming the difficulties arising due to the short-time viability of gametes, easy management of brooders through reduction in the number of males to be kept year-round as broodstock, enabling genetic preservation of desired lines and allowing cross-breeding at different times of the year. Cryopreservation could also offer benefits like synchronization of gamete availability of both sexes, sperm economy, simplification of broodstock management, transport of gametes from different fish farms and germplasm storage for genetic selection programmes or conservation of species. Apart from the above advantages, it can also lead to many other avenues such as cryobanking of viable gametes as in the case of animal production and development of gene bank and genetic manipulation in fishes.

Principle of Cryopreservation

The basic principle of cryopreservation is exposure of living cells to sub-zero temperature as low as $-200\ ^\circ$ C through a perfect process thereby arresting its activities without damaging the life of it. It is to be understood that freezing and thawing of biological material involve a series of complex and dynamic processes of heat and water transport between cells and their surrounding medium. The effect of the process depends on the speed at which the cells are frozen or thawed. When cells are subjected to freezing in an aqueous solution, both cells and the solution get super cooled leading to freezing. This is followed by heterogeneous nucleation in the cell solution and

J.S.S. Kumar (✉) • C.J. Betsy
Department of Inland Aquaculture, Fisheries College and Research Institute, Tuticorin 628008, Tamil Nadu, India
e-mail: jstephenkumar@gmail.com

S. Perumal et al. (eds.), *Advances in Marine and Brackishwater Aquaculture*,
DOI 10.1007/978-81-322-2271-2_22, © Springer India 2015

extracellular solution. If such condition occurs intracellularly, the resultant nuclei will be isolated by plasma membranes from the unfrozen cell components. That means there will be separation of ice crystals inside the cell. As water gets frozen, the extracellular solution becomes progressively more concentrated leading to slow dewatering conditions in the cells. This may be due to slow cooling and when cells have sufficient time to lose enough water to remain in osmotic equilibrium with the concentrating extracellular solution leading to water loss inside the cells. This will lead to cell death otherwise called as freeze killing or chill killing. While this may take time in large and multicellular organisms, in small microorganisms and single cells, much of water can be withdrawn during freezing leading to desiccation and the death of the cell instantly.

In contrary to the above situation, if the rate of cooling is faster and rapid, there will be less time for the intracellular water to diffuse out of the cells. A balancing situation will emerge under such fast or rapid cooling. This leads to the survival of the cells by minimizing the time or exposure duration to concentrated solution. The cooling rate also ensures that there is no formation of intracellular ice. This process is called vitrification, and it is the process that is happening inside the cell when cryopreservation is done.

During thawing, the same cellular physiological processes occur in reverse order. The thawing rate should also be rapid and fast enough than that of the corresponding cooling rate. Nevertheless, recrystallization invariably occurs during thawing, forming lethal intracellular ice. A high warming rate is usually employed to minimize the degree of recrystallization when thawing is rapid to provide insufficient time or least possibility for the dehydrated cells to absorb the amount of water lost during freezing.

Cryogens and Cryoprotectants

Cryopreservation can be done only with the help of some chemicals which have the capacity to provide a sub-zero temperature of around −200 ° C. Such agents or media used for the cryopreservation are called as cryogens.

The process of cryopreservation although seemed to be simple, a greater degree of risk to the life of cell is involved as there is a possibility for water loss and formation of intracellular ice crystals to cause cell death if there is any flaw in the protocol.

In order to prevent such losses, during the cryopreservation, some chemicals are used. These chemicals balance between the effects of intracellular ice and concentrated solution, thus protecting the cells from the cryoinjuries or the damages from the freezing thus improving the survival. These chemicals are called as cryoprotectants. They can dissolve in water and lower the melting point of water. During the cryopreservation process, the water present inside the spermatozoa or any somatic cells will tend to form ice crystals. This formation of ice will eventually damage the organelles present inside the cell and will lead to cell death. During exposure to low temperature, the growing ice outside the cells makes smaller pockets of unfrozen liquid. Due to the presence of cryoprotectants in the medium, the pockets thus formed will be larger in size. Larger unfrozen pockets of cells reduce damage from freezing injury, mechanical damage from ice and excessive concentration of salt. The addition of cryoprotectants to minimize cell damage associated with ice formation is termed as cryoprotection.

Scott and Baynes (1980) explained that the cryoprotectants added to the diluents can bind water molecules and reduce pure ice crystal formation in the cell when they are cryopreserved. Cryoprotectants protect the sperm cell from cold and hot shock treatments and prevent cell dehydration. They also provide protection to labile enzymes and stabilize protein in unfrozen and aqueous solutions. But no single substance has been identified as a universal cryoprotectant that can be used for all fish species (Gwo et al. 1991). Several cryoprotectants such as glycerol, DMSO, methanol and DMA have been used as cryoprotectant of fish sperm.

Table 1 Common cryogens and cryoprotectants used for cryopreservation of fish gametes

Cryogens				Cryoprotectants			
Agent	Melting point (°C)	Freezing point (°C)	Efficiency (%)	Agent	Formula	Mol wt.	Density
Ether	102	−171	1.3	Dimethylsulphoxide (DMSO)	CH_3SOCH_3	78.13	1.10
Propane	83	−180	1.0	Glycerol	$C_3H_8O_3$	92.10	1.47
Freon 13	88	−185	0.8	Methanol	CH_3OH	32.04	0.79
Freon 22	118	−155	0.7	1,2-Propanediol	CH_3CHOH CH_2 OH	76.09	1.04
Liquid nitrogen (cooled to FP)	63	−210	0.2				

Suitability of cryoprotectants for fish gametes preservation is largely depending on the species of fishes as there is a great degree of variation in the quality of spermatozoa among different fishes. Therefore, it has become mandatory to standardize the cryopreservation protocol for individual species for different seasons in order to have success after cryopreservation. Some of the cryogens and cryoprotectants along with their qualities have been listed in Table 1. These agents must be non-poisonous and eco-friendly for the handling.

Experiments carried out at Fisheries College and Research Institute of Tamil Nadu Fisheries University of India on the cryopreservation of carp spermatozoa witnessed the usefulness of dimethylsulphoxide (DMSO) as a useful cryoprotectant over glycerol and methanol because it enters and leaves the cells much faster than glycerol. Our experiments with DMSO at concentrations between 5 and 20 % have been found to be successful for carp sperm cryopreservation. However, this does not limit the application of other cryoprotectants as well.

Extenders

Fishes are known to have a sperm density ranging from few million to few thousand million per ml of milt. Raw milt cannot be loaded into the straws and cryopreserved with the same viscosity and density. Therefore, they have to be diluted with a diluent.

The spermatozoa of fishes are basically immotile until they get in contact with water. Motility of the spermatozoa is considered to be the index to ascertain the viability or the ability of the sperm for the fertilization. However, their motile nature will deprive them of the energy and will become inactive after certain period of time depending on the stored up energy within the spermatozoa. Since the surrounding medium affects the spermatozoa greatly, it is necessary to provide a suitable surrounding medium for the spermatozoa to be alive for more time. Some chemicals are known to provide such an effect, which are called as extenders, literally meaning extending the life or stability of the spermatozoa.

An extender is a chemical solution consisting of inorganic chemicals more similar to the body fluid especially blood or seminal plasma in which the viability of spermatozoa can be maintained during in vitro storage. The cryopreservation is greatly supported if an extender is used along with the milt. Composition of some suitable extenders that can be used along with the milt is given in Table 2.

Method for Collection of Milt

Milt from mature brooders is to be collected at room temperature taking care that they are free from urine, faecal matter, blood and water as far as possible. Studies have shown the demerits of milt collection by catheterizing the gonad (Rana et al. 1992) where the motility scores, spermatocrit and the Na, K, and Ca concentrations of the seminal plasma have been found to be significantly lower as compared to the non-catheterized samples.

Table 2 Extender compositions (g/l) used by different researchers for cryopreservation

Extender	NaCl	KCl	$NaHCO_3$	$CaCl_2$	NaH_2PO_4	$MgSO_4.7H_2O$	Glucose	Others	References
Freshwater fish saline	7.5	0.20	0.20	0.20					Chao et al. (1987)
V2 e	7.5	0.38	2.0				1.0	Egg yolk – 20 ml	Bayrle (1982)
V2 f	7.5		2.0				1.0	Egg yolk – 20 ml	Bayrle (1982)
Ringer	6.5	0.14	0.2	0.12	0.2		1.0		Dorier (1951)
Fish Ringer	6.5	0.25	0.2		0.3				Ginsberg (1963)
Modified fish Ringer	6.5	3.0	0.2	0.3					Rana and McAndrew (1989)
Modified Cortland	1.88	7.20	1.0	0.23 (H_2O)	0.41	0.232	1.0		Truscott et al. (1968)
Hfx #10	6.04	1.64		0.14		0.22		Fructose, 0.60; glycine, 6.0	Truscott and Idler (1969)
Hfx #17	5.16	1.64	1.0	0.14		0.22		Fructose, 1.0; BSA, 0.50	Truscott and Idler (1969)
Ext 48	7.3	0.38	1.0	0.23 (H_2O)	0.41	0.23		Fructose, 1.0; lecithin, 5.0	Graybill and Horton (1969)

Before freezing, the milt is diluted with an extender and equilibrated with a cryoprotectant. The common practice is to keep the straws filled with diluted sperm over liquid nitrogen vapours for about 10–15 min before immersing them into liquid nitrogen. Nowadays, programmable or computerized freezer is used for controlled freezing.

Thawing of Cryopreserved Milt

Most of the fish spermatozoa are tolerant to rapid thawing. Frozen spermatozoa of Indian major carps are preferably thawed at 38 ± 2 °C. In different experiments with Indian major carps at this institute, it was found that the thawing temperature may vary between 35 and 38 °C. Nevertheless, there are some instances in which the simple air thawing could also yield good results.

After thawing, the post-thaw motility of the spermatozoa to be checked. Spermatozoa exhibiting motility above 75 % is considered for artificial fertilization (AF). Generally, motility above 75 % is considered and used for artificial fertilization. But it has been observed that the cryopreservation although does not limit the survival of the spermatozoa can cause some damages that will result in the reduction in the motility duration. This can be reflected in the motility duration of the spermatozoa.

Artificial Fertilization with Cryopreserved Spermatozoa

Artificial fertilization is done with egg mass collected from the females either after induction or without induction. Fresh eggs are collected from matured female fish breeders by hand-stripping method. Eggs are collected in dry petri dish or an enamel-coated plate to which the thawed and diluted spermatozoa will be added and mixed well. After thorough mixing, a little water (2–5 ml) is added to activate the spermatozoa for fertilization. Again the eggs and spermatozoa are mixed well for at least 5 min and left for fertilization. The fertilized eggs can be identified by their increase in size and shape. They can be placed in the hapa or incubation pool for hatching.

Normally, hatching occurs after 48 h of fertilization under normal incubation in the incubation pool or hapa. The hatchlings can be collected and stocked in the well-fertilized nurseries for further rearing. There are varied results in these trials, and many attempts are needed to standardize this process.

Points to Remember in the Artificial Fertilization with the Cryopreserved Spermatozoa

Although the artificial fertilization is seemed to be very simple and easy, the success rate in the AF will not be good if not done with care. The following points are to be remembered while doing AF with cryopreserved spermatozoa:

1. The eggs should be freshly collected.
2. The eggs should not get in touch with water.
3. The egg mass must be thoroughly mixed with the milt.
4. Since the number of spermatozoa required for fertilizing unit mass of eggs is not well determined, it is advised to use less quantity of eggs for a given volume of cryopreserved milt with known density of spermatozoa.
5. The thawing time and duration should be well monitored for successful regain of life in the cryopreserved spermatozoa.

Aquaculture Production Enhancement Through Cryopreservation of Fish Gametes

Cryopreservation can help in the collection and preservation of good quality spermatozoa that can be used for artificial fertilization. This can help in getting quality seeds for stocking so that it can lead to higher production in the grow-out tanks. Further cryopreservation can help in attempting production of hybrids with selected species. The artificial fertilization process once standardized can help in the production of more hybrids of desired qualities. Endangered and native species can be protected by the cryopreservation of their gametes and artificial propagation of them. The aquaculture sector needs quality seeds of cultivable species and also other species of fishes and shellfishes so that they can be effectively cultured and production enhanced.

Cryopreservation can provide a year-round supply of seeds from desired species regardless of the spawning season. Cryopreservation of gametes offers immense benefits to selection and cross-breeding programmes and stock improvements. Cryopreservation technique reduces the need for a number of males to be maintained in the hatchery. A very possible application of cryopreservation in the health management of fishes can be immune enhancement through production of quality offsprings through artificial insemination.

References

Bayrle H (1982) Cryopreservation of rainbow trout sperm: effect of equilibration. In: Richter CJJ, Goos HJT (eds) Proceedings of international symposium on reproductive physiology of fish, Wageningen, The Netherlands, Aug 2–6, pp 129–130

Chao NH, Chao WC, Liu KC, Liao IC (1987) The properties of tilapia sperm and its cryopreservation. J Fish Biol 30:107–118

Dorier A (1951) Conservation de la vitalite et du pouvoir fecondant des spermatozoides de triute Arc-en-ciel. Trav Lab Hydrobiol Piscic Univ Grenoble 41:75–85

Ginsberg AS (1963) Sperm-egg association and its relationship to the activation of the egg in salmonid fishes. J Embryol Exp Morphol 11:13–33

Graybill JR, Horton HF (1969) Limited fertilisation of steelhead trout eggs with cryopreserved sperm. J Fish Res Bd Canada 26:1400–1404

Gwo JC, Strawn K, Longnecker MJ, Arnold CR (1991) Cryopreservation of Atlantic croaker spermatozoa. Aquaculture 94:355–375

Rana KJ, McAndrew BJ (1989) The viability of cryopreserved tilapia spermatozoa. Aquaculture 76:335–345

Rana KJ, Gupta SD, McAndrew BJ (1992) The relevance of collection techniques on the quality of manually stripped Atlantic salmon, *Salmo salar* milt. In: Billard R (eds) Workshop on gamete and embryo storage and cryopreservation in aquatic organisms, 3-3-92 to 2-4-92, France, 40 p (Abstracts)

Scott AP, Baynes SM (1980) A review of the biology, handling and storage of salmonid spermatozoa. J Fish Biol 17:707–739

Truscott B, Idler DR (1969) An improved extender for freezing Atlantic salmon spermatozoa. J Fish Res Board Can 26:3254–3258

Truscott B, Idler DR, Hoyle RJ, Freeman HC (1968) Sub-zero preservation of Atlantic salmon sperm. J Fish Res Board Can 25:363–372

Aquaculture Disease Diagnosis and Health Management

R. Ananda Raja and K.P. Jithendran

Introduction

Aquaculture is growing by leaps and bounds and is one of the world's fastest-growing industries in food production. Unlike other terrestrial farm animals and plants, aquatic animals require more attention in order to monitor their health. They live in a complex and dynamic environment and are not readily visible except under tank-holding conditions. Similarly, feed consumption and mortalities are also equally well hidden under water (Bondad-Reantaso et al. 2001). So the problems faced by the aquatic animals are also species and system specific. The complexity of the aquatic ecosystem makes it difficult to understand the difference between health, suboptimal performance, and disease. The range of diseases found in aquaculture is one among the major problems faced by aquaculturists all over the world. Diseases in aquaculture are caused by the outcome of a series of linked events involving the interactions between the host, the environment, and the presence of a pathogen (Snieszko 1974). Environment includes not only the water and its components (such as oxygen, pH, temperature, toxins, and wastes) but also the kind of management practices (e.g., handling, drug treatments, transport procedures, etc.). There are three factors such as stocking density, innate susceptibility, and immunity which are particularly important in affecting host's susceptibility to diseases. The intensive shrimp aquaculture has parallely brought disease problems leading to great economic loss. Diseases may be caused by a single or combinations of multifarious factors. Generally, diseases are broadly classified in to infectious and noninfectious. The former is caused either by virus, bacteria, fungi, parasites, or rickettsia, while the latter is due to environmental stresses, genetic factors, and nutritional deficiencies. The most important steps to reduce or prevent losses due to diseases in aquaculture are monitoring as regularly as possible and appropriate action at the first sign(s) of suspicious behavior, lesions, or mortalities. These fundamental approaches should be followed in many aquatic animal production sectors as in animal husbandry and agricultural production. Some farmers hesitate to reveal the disease problems due to their ignorance that it may result in failure in the competitive market price. It should be made understood that hiding or denying health problems can be as destructive to aquatic animals as it is elsewhere.

Importance of Diagnostics in Aquaculture

Diagnostics play an important role in aquatic animal health management and disease control. Confirmatory diagnosis of a disease is often

R. Ananda Raja (✉) • K.P. Jithendran
Central Institute of Brackishwater Aquaculture, No. 75, Santhome High Road, R.A. Puram, Chennai 600028, Tamil Nadu, India
e-mail: anandarajars@gmail.com

S. Perumal et al. (eds.), *Advances in Marine and Brackishwater Aquaculture*,
DOI 10.1007/978-81-322-2271-2_23, © Springer India 2015

considered as complicated and costly which may be true in some newly emerging diseases, for instance, early mortality syndrome (EMS) outbreak in shrimp aquaculture and its confirmatory etiological diagnosis, but not in all the cases with already standardized and validated diagnostics. Incorrect diagnosis can lead to ineffective or inappropriate control measures which may be even more costly. Disease diagnostics should be made available throughout the entire life cycle of the host till it reaches table for consumption. There are multifarious recent diagnostics available in aquaculture for disease diagnosis at different levels. Some diagnostics are used to screen healthy animals to ensure that they are free from any infection at asymptomatic levels with specific pathogens. This kind of screening is mostly done on aquatic animals which are transferred live or as products from one area or country to another. Such screening reduces the risk of carrying infectious agents including opportunistic pathogens which might proliferate during shipping, handling, or change of environment (Bondad-Reantaso et al. 2001). Further, it reduces the risk of resistant or tolerant animals transferring a significant pathogen to a susceptible population. Diagnostic tests may be applied to diagnose clinically diseased individuals and screen specific disease surveillance and as a confirmatory and calibration tests to validate the other diagnostics and procedures adopted. Valid laboratory results are essential for diagnosis, surveillance, and trade.

Disease Diagnosis in Aquaculture

It is a dynamic field; what found new yesterday becomes dated today, and latest today would become obsolete tomorrow. Disease diagnosis can be basically divided into two types such as *presumptive diagnosis* where a preliminary diagnosis based on gross observations and circumstantial evidence is done and *confirmatory diagnosis* in which the etiological agent is confirmed with a high degree of diagnostic confidence.

Gross and Clinical Signs

Gross observations can be easily made at the farm or pond side. But in per acute and some acute cases, sudden mortality is observed with no gross and clinical signs. Similar the case with asymptomatic carriers (Ananda Raja et al. 2012).Mere gross observations are insufficient for a definite diagnosis. But such information is essential for preliminary understanding of the "case description" or "case history." Accurate and detailed gross observations can also help in effectively reducing the losses or spread of the diseases by means of destruction or isolation of affected stocks and treatments or alterations of husbandry practices. Clinical signs such as behavioral change which includes changes in feeding behavior, weight loss, lethargy, erratic swimming movement or unusual aggregations, parasitism, cuticle softening, discoloration, hemorrhagic lesions, ulcers, predator activity, and unusual mortalities are considered to be the first signs of stress or disease problem in an aquaculture system. Environmental parameters such as temperature, dissolved oxygen, pH, etc., play a significant role in aquaculture both directly (within the ranges of physiological tolerances) and indirectly (enhancing susceptibility to infections or their expression).

Clinical Biochemistry

Clinical chemistry in shrimp/fish pathology is in its infancy state. But routine application of clinical biochemistry will help in arriving at confirmatory diagnosis in future and also identification of any blood-borne parasites. Hematological, immunological, and clinical biochemical values such as bleeding time, coagulation time, total hemocyte count (THC), differential hemocyte count (DHC), bacterial clearance activity, phagocytosis, prophenoloxidase activity, serum acid phosphatase, serum alkaline phosphatase, total serum protein, glucose, cholesterol, total protein, total albumin, alanine transaminase (ALT), aspartate transaminase (AST), triglycerides, and lactate dehydrogenase (LDH) will also give some specific clue in making confirmatory diagnosis.

Environmental Parameters

Often environmental parameters are not included in routine diagnostic procedures done in aquaculture. But it is essential to assess the water and soil quality parameters such as salinity, temperature, pH, dissolved oxygen (DO), ammonia nitrogen (NH_3–N), nitrite nitrogen (NO_2–N), nitrate nitrogen (NO_3–N), phosphate phosphorus (PO_4–P), and microbial load since they play vital role in deciding any disease outbreak in aquaculture system. Sometimes any one of these environmental factors alone can lead to high mortality, but mere presence of certain pathogenic organism in the host and pond ecosystem can mislead our confirmatory diagnosis.

Necropsy Examination

Necropsy examination is performed to inform farmer, clinical staff, researcher, academicians, or legal authorities about the cause of death. It is essential for getting new information and guidance for future. Post mortem examinations can provide information about illness and health that would not be discovered in any other way and help to understand why the animal died. The rare pathological conditions can be preserved, and retention of whole animal/organ/tissue would benefit to future needs. Much of what we know about illness today came from such examinations. They help to:

- Identify the cause of death.
- Confirm the nature of the illness and/or the extent of the disease.
- Identify other conditions that may not have been diagnosed.
- Identify complications or side effects of treatments and drugs.

It is also possible that the information gained may benefit future generations in the family, or other animals suffer similar problems. Before proceeding to post mortem examination, one should ascertain when the fish first showed signs and the treatment given (Noga 2010; Roberts 2012).

Isolation and Identification of Pathogen

The organ of choice for isolating systemic bacterial pathogens in fish is kidney which can be approached either dorsally or ventrally, and in shrimp it is hepatopancreas being a vital organ. Fish/shrimp pathogens should be cultured at room temperature (22–25 °C), not at 37 °C, as is routinely done in many microbiology laboratories since some of the fish pathogens grow poorly or not at all at 37 °C (Bondad-Reantaso et al. 2001). For example, *Vibrio salmonicida* grows at 17 °C. Samples from marine and brackishwater source to be cultured on a medium that has high salt content at least 1.5 % (Bruno 1996). Special media like thiosulfate–citrate–bile salt–sucrose agar (TCBS) can also be used. Live specimens should be used for culture whenever possible. Identification of an obligate pathogen (*Aeromonas salmonicida*) (Drinan 1985) in a dead fish is a stronger diagnosis than the isolation of an opportunist (*A. hydrophila*). The other pathogens like virus, fungi, parasite, etc. should be isolated as per the standard protocol for each species of the organisms. It is very important to understand that mere isolation and identification of pathogen from any host do not warranty that the disease and mortality are due to its presence in the system. The specific cause of death should only be ascertained when the Koch's postulate is proven.

Bioassay

It is a quantitative procedure that uses susceptible organisms to detect toxic substances or pathogens. Bioassay is done with samples collected from suspected or asymptomatic carriers and tested using a highly susceptible host (life stage or species) as the indicator for the presence of the pathogen. In this assay Koch's postulate is well proven.

Microscopy

Bright-field microscopy is the simplest of all the light microscopy techniques where the sample is illuminated with white light from below and observed from above. The technique is very easy and simple to do with minimal sample preparation, but it requires expertise in reading the slides. Low contrast of most biological samples and low apparent resolution are the limitations. Dark-field microscopy is yet another technique commonly used for improving the contrast of unstained, transparent specimens. But this technique suffers from low light intensity in the final image of many biological samples and continues to be affected by low apparent resolution. Many times for on-farm diagnosis, the presence of virus can be detected by tissue squash preparation and staining. This can then be observed under a microscope for a particular viral infection like Monodon baculovirus (MBV) by hepatopancreas or fecal squash preparation stained with 0.05 % aqueous malachite green for detection of large, single, or multiple roughly spherical, eosinophilic, polyhedral, intranuclear occlusion bodies (OBs). Moreover, microscopy plays a crucial role in the identification of bacterial pathogens by using the special stains like Gram's staining and acid fast staining.

Histopathology

Histopathology holds its importance from the day of its invention in the field of diagnostics. Proper sampling and fixation are the most important steps for correct disease diagnosis. The moribund or very recently dead animals are suitable for histopathology, while putrefied or frozen animals are unsuitable. Fish/shrimps are usually fixed in 10 % neutral buffered formalin (NBF) fixative in a wide-mouth plastic bottle. The fixative volume should be at least 10 times more than the volume of sample to get the tissues properly fixed. The samples collected should be as small as possible not more than 0.5 cm^2 thickness. For shrimps, Davidson's fixative is commonly used, and the composition of the common fixatives used is listed below (Bell and Lightner 1988; Lightner 1996).

Common Fixative Used for Histopathology

Davidson's fixative		
95 % ethanol	–	330 ml
37 % formaldehyde	–	220 ml
Glacial acetic acid	–	115 ml
Distilled water	–	335 ml
4 % Formal saline (for parasites)		
37 % formaldehyde	–	40 ml
Distilled water	–	960 ml
Sodium chloride	–	8.5 g
10 % Formal saline (for tissues)		
37 % formaldehyde	–	100 ml
Distilled water	–	900 ml
Sodium chloride	–	8.5 g

10 % neutral buffered formalin		
37 % formaldehyde	–	100 ml
Distilled water	–	900 ml
Sodium dihydrogen phosphate	–	4 g
Disodium hydrogen phosphate	–	6 g

The presence of virus in different tissues can be detected by histopathology. However, proper histopathological techniques and expertise in reading slides are necessary to interpret the results. If properly detected, this will be the most accurate diagnostic method. But it will be difficult to detect any low levels of infections by this method. The most well-defined common viral diseases affecting shrimp and fish are listed below with the details of the inclusion bodies with respect to the specific diseases seen in the histopathology.

In addition, immunohistochemical staining methods have also been developed with paraffin-embedded tissue sections for the detection of viruses such as infectious pancreatic necrosis virus (IPNV), infectious salmon anemia virus (ISAV) and nodavirus (Bondad-Reantaso et al. 2001). Viral antigen is localized by an antibody raised against the virus, and subsequent addition of colored substrate results in a colored product that can be visualized by light microscopy.

Sl. No.	Disease	Etiology	Inclusions
1.	Monodon baculovirus (MBV) disease	Family: *Baculoviridae*, dsDNA type A monodon baculovirus (MBV)	Large, single, or multiple roughly spherical, eosinophilic, polyhedral, *intranuclear* occlusion bodies (OBs) in the epithelial cells of the hepatopancreas tubules and the anterior midgut (Lester et al. 1987; Lightner 1988; Vogt 1992; Bondad-Reantaso et al. 2001)
2.	White spot disease (WSD)	Family: *Nimaviridae*, dsDNA *Whispovirus*, white spot syndrome virus (WSSV)	Ectodermal (epidermis, gills, fore and hind gut, antennal gland, and neurons) and mesodermal (hematopoietic tissue, hemocytes, striated muscle, heart, lymphoid organ, and connective tissues) tissues with eosinophilic to basophilic *intranuclear* inclusions (Momoyama et al. 1994; Wongteerasupaya et al 1995)
3.	Infectious hypodermal and hematopoietic necrosis (IHHN)	Family: *Parvoviridae*, ssDNA infectious hypodermal and hematopoietic necrosis virus (IHHNV)	Cowdry type A *intranuclear* inclusion bodies (IBs) in cells of ectodermal and mesodermal origin (Morales-Covarrubias and Chavez-Sanchez 1999)
4.	Hepatopancreatic disease	Family: *Parvoviridae* ssDNA hepatopancreatic parvovirus (HPV)	Single, prominent, basophilic, *intranuclear* inclusion bodies in the hypertrophied hepatopancreatic epithelial cells (Promjai et al. 2002)
5.	Yellowhead disease	Family: *Roniviridae*, ssRNA yellowhead/gill-associated virus/lymphoid organ virus (YHV/GAV/LOV)	Basophilic, *intracytoplasmic*, Feulgen-positive inclusions in the lymphoid organs, interstitial tissues of the hepatopancreas, connective tissues underlying the midgut, cardiac tissues, hematopoietic tissues, hemocytes, and gill tissues (Chantanachookin et al. 1993)
6.	Taura syndrome	Family: *Dicistroviridae*, ssRNA Taura syndrome virus (TSV)	Eosinophilic then changes to basophilic, *intracytoplasmic*, Feulgen-negative inclusion bodies in the cells in areas of necrosis (Lightner et al. 1995; Lightner 1996; Hasson et al. 1999).
7.	Infectious myonecrosis	Family: *Totiviridae*, dsRNA infectious myonecrosis virus (IMNV)	*Perinuclear*, pale, basophilic to dark basophilic inclusion bodies are evident in muscle cells, connective tissue cells, hemocytes, and cells that comprise lymphoid organ spheroids (Lightner et al. 2004; Poulos et al. 2006)
8.	Monodon slow growth syndrome	Family: *Luteoviridae*(?), ssRNA Laem–Singh virus (LSNV)	LSNV is detected in the fasciculated zone and in onion bodies of the organ of Bellonci (Sritunyalucksana et al. 2006)
9.	Muscle necrosis disease	Family: *Nodaviridae*, ssRNA *Penaeus vannamei* nodavirus (PvNV)	*Perinuclear*, pale, basophilic inclusion bodies are evident in muscle cells, connective tissue cells, hemocytes, and cells that comprise lymphoid organ spheroids (Melena et al. 2012)
10.	White tail disease (WTD) or white muscle disease (WMD)	Family: *Nodaviridae*, RNA *Macrobrachium rosenbergii* nodavirus (MrNV) and its associate extra small virus (XSV)	Pathognomonic oval or irregular basophilic *intracytoplasmic* inclusion bodies are demonstrated in the target tissues by histology (Arcier et al 1999; Hsieh et al. 2006)

(continued)

Sl. No.	Disease	Etiology	Inclusions
11.	Koi herpesvirus disease (KHVD)	Family: *Alloherpesviridae*, DNA herpesvirus	Eosinophilic *intranuclear* inclusions in branchial epithelial cells, leucocytes, kidney, spleen, pancreas, liver, brain, gut, and oral epithelium (Bergmann et al. 2006)
12.	Viral encephalopathy and retinopathy (VER) or Viral nervous necrosis (VNN)	Family: *Nodaviridae*, ssRNA piscine nodavirus of the genus *Betanodavirus*	*Intracytoplasmic* inclusion in nervous cells (Munday et al. 2002).
13.	Iridovirus infection	Family: *Iridoviridae*, dsDNA virus of genera *Lymphocystivirus* and *Ranavirus*	Basophilic *intracytoplasmic* inclusion bodies seen in liver, kidney, heart, pancreas, gastrointestinal tract, gill, and pseudobranch and positive indirect fluorescent antibody test – IFAT in spleen, heart, kidney, intestine, and gill (Jung et al. 1997)
14.	Epizootic hematopoietic necrosis	Family: *Iridoviridae*, dsDNA epizootic hematopoietic necrosis virus of genus *Ranavirus*	Basophilic *intracytoplasmic* inclusion bodies seen in liver, kidney, heart, pancreas, gastrointestinal tract, gill, and pseudobranch (Reddacliff and Whittington 1996)
15.	Infectious hematopoietic necrosis (IHN)	Family: *Rhabdoviridae*, ss RNA infectious hematopoietic necrosis virus	*Intracytoplasmic* inclusion bodies seen in hematopoietic tissues, kidney, spleen, liver, pancreas, and digestive tract (Wolf 1988; Bootland and Leong 1999)
16.	Spring viraemia of carp (SVC)	Family: *Rhabdoviridae*, spring viraemia of carp virus (SVCV), a species in the genus *Vesiculovirus*	*Intracytoplasmic* inclusion bodies seen in hematopoietic tissues, kidney, spleen, liver, pancreas, and digestive tract (Haghighi Khiabanian Asl et al. 2008)
17.	Viral hemorrhagic septicaemia (VHS)	Family: *Rhabdoviridae*, viral hemorrhagic septicaemia virus (VHSV) belonging to the genus *Novirhabdovirus*	*Intracytoplasmic* inclusion bodies seen in hematopoietic tissues, kidney, spleen, liver, pancreas, and digestive tract (Evensen et al. 1994)

Transmission or Scanning Electron Microscopy

It requires special methodology to be followed in fixing and processing of tissues for electron microscopy. Transmission electron microscopy (TEM) is very much useful and a great boon to diagnostic pathology to identify and determine the structure of an unknown virus that is characterized for the first time. This can also be used as a confirmatory test for the detection of already known virus or any intracellular parasites. Moreover, it is used in studying the ultrastructural changes during the progress of diseases. Scanning electron microscopy (SEM) is useful in identifying the surface level changes on the cell, and moreover it gives the structure of the cell as a whole in 3D view. The latest technology made scanning transmission electron microscope (STEM) as a dual-mode instrument by combination of both TEM and SEM principles. All of the images seen up to now provide information about the structure of a specimen, but it is also possible to analyze chemical composition of the particles by analytical electron microscopy (AEM) (Egerton 2005).

Antibody-Based Assays

Antibody-based tests for pathogen detection using immune sera polyclonal antibodies

(PAb's) or monoclonal antibodies (MAb's) can be used in fish disease diagnosis. Since crustaceans do not produce antibodies, antibody-based diagnostic tests are limited in their application to pathogen detection in shrimp diseases. Moreover, since crustacean viruses cannot be routinely produced in tissue culture, purified virus from infected hosts must be used to produce antibody. This has severely limited the development and availability of this diagnostic tool in shrimp disease diagnosis. Antibody-based diagnostic methods have been developed with mouse or rabbit antibodies generated to viruses purified from infected hosts. The recent application of MAb technologies to this problem has begun to provide a few antibody-based tests. MAb's are available for three of the OIE listed crustacean viruses such as TSV, IHHNV, and WSSV (Bondad-Reantaso et al. 2001).

Molecular Methods

Accurate, easy, and convenient availability of rapid and reliable diagnostic methods plays an important role in any disease control and health management programs in aquaculture. Treatment regime is well developed in human, animal husbandry, and agriculture for each and every specific disease, but it is still in growing phase in aquaculture. Proper early diagnosis is as good considered as treatment in aquaculture. So the molecular diagnostics based on polymerase chain reaction (PCR) principles have been extensively used to control the spread of major shrimp and fish pathogens (Ananda Raja et al. 2012), but they have the disadvantage of requiring sophisticated equipment and highly trained personnel. There are so many molecular diagnostics in aquaculture. It is appropriate to use well-proven, validated, and frequently used techniques. Recently, lateral flow chromatographic immunodiagnostic strips similar to common drugstore pregnancy tests have begun to appear for some shrimp diseases (Flegel et al. 2008). Using this kind of strips, unskilled farm personnel can easily diagnose shrimp or fish disease outbreaks at the pond side. The strips are relatively cheap and give an answer within 10 min.

Health Management in Aquaculture

The proverb "prevention is better than cure" is well suited to the health management in aquaculture. The disease prevention and control strategy is the best practice for successful hatchery and grow-out culture practices. Quarantine measures should strictly be adopted to import broodstock to avoid entry of existing or emerging pathogen. The following salient points are considered very important to get successful grow-out culture:

- Seasonal factors and crop planning based on the disease incidence.
- Ponds should be dried before starting the culture.
- Strict biosecurity measures should be adopted.
- Sieve should be used at water inlet, and the water should be bleached before stocking to weed out wild shrimp, fishes, and intermediate hosts.
- Good water quality should be maintained throughout the culture.
- Zero-water exchange or minimal-water exchange from reservoir ponds in case of shrimp culture.
- Disease-free stock should be used from good genetic strain of broodstock.
- Development and use of disease-resistant stocks will help in prevention of catastrophic disease outbreak and loss.
- Coastal Aquaculture Authority (CAA) guidelines should be followed for optimum shrimp stocking density in grow-out culture system.
- Quarantine measures should strictly be adopted to import broodstock to avoid entry of existing or emerging pathogen.
- Adequate balanced good nutrition to be made available to avoid problems associated with cannibalism and horizontal spread of diseases.
- Proper destruction and disposal of infected as well as dead animals to be regularly monitored.

- Animals should be handled with good care to avoid unwanted stress.
- Proper chemical prophylaxis and vaccine development are needed for immunological protection.
- Regulations are required to prevent transfer of pathogens from one host population to another, nationally or internationally.
- Sanitation and disinfection of hatchery and equipments are to be strictly followed.
- Despite all the precautions, disease outbreak may occur. Handling a disease outbreak with least economic loss is an art of farm management. Prompt action is essential in such circumstances to rectify the problems, reduce the losses, and minimize the impacts on neighboring farms.
- Record keeping is necessary to identify problems in the pond environment and animal health and to rectify those problems at the earliest during the production cycle. It also helps the farmer to learn from the past.

References

Ananda Raja R, Panigrahi A, Kumar S (2012) Epidemiological investigation of brackish water culture systems in West Bengal, India. J Appl Aquac 24:49–59

Ananda Raja R, Kumar S, Sundaray JK, De D, Biswas G, Ghoshal TK (2012) Hematological parameters in relation to sex, morphometric characters and incidence of white spot syndrome virus in tiger shrimp Penaeus monodon Fabricius, 1798 from Sunderban, West Bengal. Indian J Fish 59:169–174

Arcier JM, Herman F, Lightner DV, Redman RM, Mari J, Bonami JR (1999) A viral disease associated with mortalities in hatchery-reared post larvae of the giant freshwater prawn *Macrobrachium rosenbergii*. Dis Aquat Org 38:177–181

Bell TA, Lightner DV (1988) A handbook of normal shrimp histology, Special publication no. 1. World Aquaculture Society, Baton Rouge, 114 pp

Bergmann SM, Kempter J, Sadowski J, Fichtner D (2006) First detection, confirmation and isolation of koi herpesvirus (KHV) in cultured common carp (*Cyprinus carpio* L.) in Poland. Bull Eur Assoc Fish Pathol 26:97–104

Bondad-Reantaso MG, Mcgladdery SE, East I, Subasinghe RP (2001) Asia diagnostic guide to aquatic animal diseases, FAO fisheries technical paper 402, supplement 2. FAO, Rome, 240 pp

Bootland LM, Leong JC (1999) Infectious hematopoietic necrosis virus. In: Woo PTK, Bruno DW (eds) Fish diseases and disorders, vol 3: viral, bacterial and fungal infections. CAB International, Oxon, pp 57–121

Bruno DW (1996) Cold water vibriosis caused by *Vibrio salmonicida*, Aquaculture information series no. 15. The Scottish Office Agriculture, Environment and Fisheries Department. Marine Laboratory, Aberdeen

Chantanachookin C, Boonyaratpalin S, Kasornchandra J, Direkbusarakom S, Aekpanithanpong U, Supamattaya K, Sriuraitana S, Flegel TW (1993) Histology and ultrastructure reveal a new granulosis-like virus in *Penaeus monodon* affected by yellow-head disease. Dis Aquat Org 17:145–157

Drinan EM (1985) Studies on the pathogenesis of furunculosis in salmonids. Ph.D. thesis, National University of Ireland, Dublin

Egerton RF (2005) Physical principles of electron microscopy. An introduction to TEM, SEM, and AEM. Springer, New York, 202 pp

Evensen Ø, Meier W, Wahli T, Olesen NJ, Jørgensen PEV, Håstein T (1994) Comparison of immunohistochemistry and virus cultivation for detection of viral haemorrhagic septicaemia virus in experimentally infected rainbow trout *Oncorhynchus mykiss*. Dis Aquat Org 20:101–109

Flegel TW, Lightner DV, Lo CF, Owens L (2008) Shrimp disease control: past, present and future. In: Bondad-Reantaso MG, Mohan CV, Crumlish M, Subasinghe RP (eds) Diseases in Asian aquaculture VI. Fish Health Section, Asian Fisheries Society, Manila, pp 355–378, 505 pp

Haghighi Khiabanian Asl A, Azizzadeh M, Bandehpour M, Sharifnia Z, Kazemi B (2008) The first report of SVC from Indian carp species by PCR and histopathologic methods in Iran. Pak J Biol Sci 11:2675–2678

Hasson KW, Lightner DV, Mohney LL, Redman RM, Poulos BT, White BL (1999) Taura syndrome virus (TSV) lesion development and the disease cycle in the Pacific white shrimp *Penaeus vannamei*. Dis Aquat Org 36:81–93

Hsieh CY, Wu ZB, Tung MC, Tu C, Lo SP, Chang TC, Chang CD, Chen SC, Hsieh YC, Tsai SS (2006) In situ hybridization and RT-PCR detection of *Macrobrachium rosenbergii* nodavirus in giant freshwater prawn, *Macrobrachium rosenbergii* (de Man), in Taiwan. J Fish Dis 29:665–671

Jung S, Miyazaki T, Miyata M, Danayadol Y, Tanaka S (1997) Pathogenicity of iridovirus from Japan and Thailand for the red sea bream *Pagrus major* in Japan, and histopathology of experimentally infected fish. Fish Sci 63:735–740

Lester RJG, Doubrovsky A, Paynter JL, Sambhi SK, Atherton JG (1987) Light and electron microscope evidence of baculovirus infection in the prawn *Penaeus plebejus*. Dis Aquat Org 3:217–219

Lightner DV (1988) Diseases of cultured penaeid shrimp and prawns. In: Sindermann CJ, Lightner DV (eds)

Disease diagnosis and control in North American marine aquaculture. Elsevier, Amsterdam, pp 8–127

Lightner DV (1996) A handbook of shrimp pathology and diagnostic procedures for diseases of cultured Penaeid shrimp. World Aquaculture Society, Baton Rouge, 304 pp

Lightner DV, Redman RM, Hasson KW, Pantoja CR (1995) Taura syndrome in *Penaeus vannamei* (Crustacea: Decapoda): gross signs, histopathology and ultrastructure. Dis Aquat Org 21:53–59

Lightner DV, Pantoja CR, Poulos BT, Tang KFJ, Redman RM, Pasos De Andrade T, Bonami JR (2004) Infectious myonecrosis: new disease in Pacific white shrimp. Glob Aquac Advocate 7:85

Melena J, Tomala J, Panchana F, Betancourt I, Gonzabay C, Sonnenholzner S, Amano Y, Bonami J-R (2012) Infectious muscle necrosis etiology in the Pacific white shrimp (*Penaeus vannamei*) cultured in Ecuador. Braz J Vet Pathol 5:31–36

Momoyama K, Hiraoka M, Nakano H, Koube H, Inouye K, Oseko N (1994) Mass mortalities of cultured kuruma shrimp, *Penaeus japonicus*, in Japan in 1993: histopathological study. Fish Pathol 29:141–148

Morales-Covarrubias MS, Chavez-Sanchez MC (1999) Histopathological studies on wild broodstock of white shrimp *Penaeus vannamei* in the Platanitos area, adjacent to San Blas, Nayarit, Mexico. J World Aquac Soc 30:192–200

Munday BL, Kwang J, Moody N (2002) Betanodavirus infections of teleost fish: a review. J Fish Dis 25:127–142

Noga EJ (2010) Fish disease diagnosis and treatment, II edn. Wiley-Blackwell, Ames, 519 pp

Poulos BT, Tang KFJ, Pantoja CR, Bonami JR, Lightner DV (2006) Purification and characterization of infectious myonecrosis virus of penaeid shrimp. J Gen Virol 87:987–996

Promjai J, Boonsaeng V, Withyachumnarnkul B, Flegel TW (2002) Detection of hepatopancreatic parvovirus in Thai shrimp *Penaeus monodon* by in situ hybridization, dot blot hybridization and PCR amplification. Dis Aquat Org 51:227–232

Reddacliff LA, Whittington RJ (1996) Pathology of epizootic haematopoeitic necrosis virus (EHNV) infection in rainbow trout (*Oncorhynchus mykiss* Walbaum) and redfin perch (*Perca fluviatilis* L.). J Comp Pathol 115:103–115

Roberts RJ (2012) Fish pathology, IV edn. Wiley-Blackwell, Hoboken, 590 pp

Snieszko SF (1974) The effects of environmental stress on outbreaks of infectious diseases of fishes. J Fish Biol 6:197–208

Sritunyalucksana K, Apisawetakan S, Boon-nat A, Withyachumnarnkul B, Flegel TW (2006) A new RNA virus found in black tiger shrimp *Penaeus monodon* from Thailand. Virus Res 118:31–38

Vogt G (1992) Transformation of anterior midgut and hepatopancreas by monodon baculovirus (MBV) in *Penaeus monodon* postlarvae. Aquaculture 107:239–248

Wolf K (1988) Infectious hematopoietic necrosis. In: Fish viruses and fish viral diseases. Cornell University Press, Ithaca, pp 83–114

Wongteerasupaya C, Vickers JE, Sriurairatana S, Nash GL, Akarajamorn A, Boonsaeng V, Panyim S, Tassanakajon A, Withyachumnarnkul B, Flegel TW (1995) A non-occluded, systemic baculovirus that occurs in cells of ectodermal and mesodermal origin and causes high mortality in the black tiger prawn *Penaeus monodon*. Dis Aquat Org 21:69–77

Index

S. Perumal et al. (eds.), *Advances in Marine and Brackishwater Aquaculture*,
DOI 10.1007/978-81-322-2271-2, © Springer India 2015

D

E

F